Models for Capitalizing on Web Engineering Advancements:

Trends and Discoveries

Ghazi I. Alkhatib
Princess Sumaya University for Technology, Jordan

Managing Director:	Lindsay Johnston
Senior Editorial Director:	Heather Probst
Book Production Manager:	Sean Woznicki
Development Manager:	Joel Gamon
Development Editor:	Hannah Abelbeck
Acquisitions Editor:	Erika Gallagher
Typesetters:	Adrienne Freeland
Print Coordinator:	Jamie Snavely
Cover Design:	Nick Newcomer, Greg Snader

Published in the United States of America by
Information Science Reference (an imprint of IGI Global)
701 E. Chocolate Avenue
Hershey PA 17033
Tel: 717-533-8845
Fax: 717-533-8661
E-mail: cust@igi-global.com
Web site: http://www.igi-global.com

Library of Congress Cataloging-in-Publication Data

Models for capitalizing on web engineering advancements: trends and discoveries / Ghazi I. Alkhatib, editor.
    p. cm. -- (Advances in information technology and web engineering series)
  Includes bibliographical references and index.
  Summary: "This book contains research on new developments and existing applications made possible by the principles of Web engineering, focusing on a broad range of applications - from telemedicine to geographic information retrieval"--Provided by publisher.
  ISBN 978-1-4666-0023-2 (hardcover) -- ISBN 978-1-4666-0024-9 (ebook) -- ISBN 978-1-4666-0025-6 (print & perpetual access) 1. Web site development. 2. Information technology--Technological innovations. I. Alkhatib, Ghazi, 1947-
  TK5105.888.M577 2012
  006.7--dc23
                          2011045667

British Cataloguing in Publication Data
A Cataloguing in Publication record for this book is available from the British Library.

Table of Contents

Section 1
Web Engineering Trends: Clouds, Location-Aware, and Agents

Liliana Ardissono, Università di Torino, Italy
Gianni Bosio, Università di Torino, Italy
Anna Goy, Università di Torino, Italy
Giovanna Petrone, Università di Torino, Italy
Marino Segnan, Università di Torino, Italy

Sorin A. Matei, Purdue University, USA
Lance W. Madsen, Innervision Advanced Medical Imaging, USA
Robert Bruno, Purdue University, USA

Jairo Francisco de Souza, Federal University of Juiz de Fora, Brazil
Sean Wolfgand Matsui Siqueira, Federal University of the State of Rio de Janeiro, Brazil
Rubens Nascimento Melo, Pontifical Catholic University of Rio de Janeiro, Brazil

Section 2
Web Engineering Discoveries

Mehdi Adda, University of Quebec at Rimouski, Canada

Section 4
Web-Based Technologies for Improving QoS

Detailed Table of Contents

Section 1
Web Engineering Trends: Clouds, Location-Aware, and Agents

Chapter 1

Liliana Ardissono, Università di Torino, Italy
Gianni Bosio, Università di Torino, Italy
Anna Goy, Università di Torino, Italy
Giovanna Petrone, Università di Torino, Italy
Marino Segnan, Università di Torino, Italy

The ubiquitous and pervasive availability of the Internet, and of Web-based services, is dramatically changing people's lives. Meanwhile, man's presence on the Web is growing fast, enabled by the offerings of low-cost devices and wireless broad-band Internet connections. On the other hand, users are adopting on-line applications and services more and more, both for personal and business purposes, in order to get in touch with other people and to manage documents, pictures, and many other types of artifacts in a ubiquitous environment that does not impose the usage of a specific end-user device. The naturally arising question is thus the following: if these projects, activities, and collaboration teams are developed all around, why do we need to deal with large numbers of separate services, each one operating in a separate representation of reality, instead of performing tasks in a unified environment integrating such different services automatically? This chapter discusses this aspect of current user collaboration and presents an infrastructure for the development of user-centered clouds of services that enable the user to interact with her favorite applications from a unified environment managing her workspaces and collaborations. The management of such collaborative service clouds is enabled by an open architecture, which provides services and data synchronization coupled with a set of core applications supporting essential aspects of the functionality of the environment, such as workspace awareness, user groups definition, and user cooperation to the management of shared tasks involving the usage of heterogeneous services.

Chapter 2

Sorin A. Matei, Purdue University, USA
Lance W. Madsen, Innervision Advanced Medical Imaging, USA
Robert Bruno, Purdue University, USA

This paper examines the potential cognitive impact of location aware information systems compared to that of search engines using a dual coding and conjoint retention theoretical framework. Supported by virtual reality or mobile devices, location aware systems deliver information that is relevant for a specific location. Research questions and hypotheses formulated under the assumption that location aware systems are better prepared to contextualize and make information memorable are explored using a planned comparison repeated measures 3 (2 treatment; 1 control) x 3 (pre-test, post-test, one week post-test) design. The results indicate that information acquisition in location-aware systems is just as powerful as that facilitated by search engines and that information recall (after 1 week) of facts is superior when using location-aware systems. The findings reinforce and extend dual coding theory suggesting that spatial and three-dimensional indexing can be one of the channels used in indexing and recalling information. The results also indicate that location-aware applications are a promising technology for distributing information in general and for learning in particular.

Chapter 3

Jairo Francisco de Souza, Federal University of Juiz de Fora, Brazil
Sean Wolfgand Matsui Siqueira, Federal University of the State of Rio de Janeiro, Brazil
Rubens Nascimento Melo, Pontifical Catholic University of Rio de Janeiro, Brazil

In order to perform its tasks on the Semantic Web, software agents must be able to communicate with other agents using domain ontologies, even when considering different ontologies. Thus, it's necessary to address the semantic interoperability issue to enable agents to recognize common concepts and misunderstandings. This work proposes the use of GNoSIS, a tool for composing ontology similarity functions, and specific modules in Goddard agent architecture in order for software agents to negotiate meanings of terms not defined in its ontology.

Section 2
Web Engineering Discoveries

Chapter 4

Mehdi Adda, University of Quebec at Rimouski, Canada

Ontologies are used to represent data and share knowledge of a specific domain, and in recent years they tend to be used in many applications such as database integration, peer-to-peer systems, e-commerce, semantic web services, bioinformatics, or social networks. Feeding ontological domain knowledge into those applications has proven to increase flexibility and inter-operability and interpretability of data and knowledge. As more data is gathered/generated by those applications, it becomes important to analyze and transform it to meaningful information. One possibility is to use data mining techniques to extract patterns from those large amounts of data. One challenging general problem in mining ontological data is taking into account not only domain concepts, properties and instances, but also hierarchical structures of those concepts and properties. In this paper, the authors research the specific problem of extracting ontology-based sequential patterns.

Web crawlers specialize in downloading web content and analyzing and indexing from surface web, consisting of interlinked HTML pages. Web crawlers have limitations if the data is behind the query interface. Response depends on the querying party's context in order to engage in dialogue and negotiate for the information. In this paper, the authors discuss deep web searching techniques. A survey of technical literature on deep web searching contributes to the development of a general framework. Existing frameworks and mechanisms of present web crawlers are taxonomically classified into four steps and analyzed to find limitations in searching the deep web.

The number of internet web applications is rapidly increasing in a variety of fields and not much work has been done for ensuring their quality, especially after modification. Modifying any part of a web application may affect other parts. If the stability of a web application is poor, then the impact of modification will be costly in terms of maintenance and testing. Ripple effect is a measure of the structural stability of source code upon changing a part of the code, which provides an assessment of how much a local modification in the web application may affect other parts. Limited work has been published on computing the ripple effect for web application. In this paper, the authors propose, a technique for computing ripple effect in web applications. This technique is based on direct-change impact analysis and dependence analysis for web applications developed in the .Net environment. Also, a complexity metric is proposed to be included in computing the ripple effect in web applications.

As open systems persist, garbage collection (GC) can be a vital aspect in managing system resources. Although garbage collection has been proposed for the standard LINDA, it was a rather course-grained mechanism. This finer-grained method is offered in LINDACAP, a capability-based coordination system for open distributed systems. Multicapabilities in LINDACAP enable tuples to be uniquely referenced, thus providing sufficient information on the usability of tuples (data) within the tuple-space. This paper describes the garbage collection mechanism deployed in LINDACAP, which involves selectively garbage collecting tuples within tuple-spaces. The authors present the approach using reference counting, followed by the tracing (mark-and-sweep) algorithm to garbage collect cyclic structures. A time-to-idle (TTI) technique is also proposed, which allows for garbage collection of multicapability regions that are being referred to by agents but are not used in a specified length of time. The performance results indicate that the incorporation of garbage collection techniques adds little overhead to the overall performance of the system. The difference between the average overhead caused by the mark-and-sweep and reference counting is small, and can be considered insignificant if the benefits brought by the mark-and-sweep is taken into account.

Chapter 8

Ali A. Alwan, Universiti Putra Malaysia, Malaysia
Hamidah Ibrahim, Universiti Putra Malaysia, Malaysia
Nur Izura Udzir, Universiti Putra Malaysia, Malaysia

Checking the consistency of a database state generally involves the execution of integrity tests on the database, which verify whether the database is satisfying its constraints or not. This paper presents the various types of integrity tests as reported in previous works and discusses how these tests can significantly improve the performance of the constraint checking mechanisms without limiting to a certain type of test. Having these test alternatives and selecting the most suitable test is an issue that needs to be tackled. In this regard, the authors propose a model to rank and select the suitable test to be evaluated given several alternative tests. The model uses the amount of data transferred across the network, the number of sites involved, and the amount of data accessed as the parameters in deciding the suitable test. Several analyses have been performed to evaluate the proposed model, and results show that the model achieves a higher percentage of local processing as compared to the previous selected strategies.

Section 3
Web-Engineered Applications

Chapter 9

Xiaoling Dai, Charles Sturt University, Australia
Kaylash Chaudhary, The University of the South Pacific – Laucala, Fiji
John Grundy, Swinburne University of Technology, Australia

Micro-payment systems are becoming an important part of peer-to-peer (P2P) networks. The main reason for this is to address the "free-rider" problem in most existing content sharing systems. The authors of this chapter have developed a new micro-payment system for content sharing in P2P networks called P2P-Netpay. This is an offline, debit based protocol that provides a secure, flexible, usable, and reliable credit service in peer-to-peer networks ensuring equitable participation by all parties. The authors have carried out an assessment of micro-payment against non-micro-payment credit systems for file sharing applications. The chapter reports on the design of our experiment and results of an end user evaluation. The chapter then discusses the performance of the credit model, comparing it to a non-micro-payment credit model. Through evaluation of the proposed system and comparison with other existing systems, the authors find that the new approach eliminates the "free-rider" problem. The chapter analyses a heuristic evaluation performed by a set of evaluators and presents directions for research aiming to improve the overall satisfaction and efficiency of this model for peers.

Chapter 10

Shazia Kareem, The Islamia University of Bahawalpur, Pakistan
Imran Sarwar Bajwa, University of Birmingham, UK

Telemedicine is modern technology that is employed to provide low cost, high standard medical facilities to the people of remote areas. Store-and-Forward method of telemedicine suits more to the progressive countries like Pakistan as not only is it easy to set up but it also has a very cheap operating cost. However, the high response time taken by store & forward telemedicine becomes a critical factor in emergency cases, where each minute has a price. The response time factor can be overcome by using virtual telemedicine approach. In virtual telemedicine, a Clinical Decision Support System (CDSS) is deployed at rural station. The CDSS is intelligent enough to diagnose a patient's disease and prescribe proper medication. In case the CDSS cannot answer a query, the CDSS immediately sends an e-mail to a medical expert (doctor), and when the response is received, the CDSS knowledge-base is updated for future queries. In this chapter, the authors not only report a NL-based CDSS that can answer NL queries, but also present a complete architecture of a virtual telemedicine setup.

Chapter 11

Current Workflow Management Systems (WfMS) are capable of managing simultaneous workflows designed to support different business processes of an organization. These departmental workflows are considered to be interrelated since they are often executed concurrently and are required to share a limited number of resources. However, unexpected events from the business environment and lack of proper resources can cause delays in activities. Deadline violations caused by such delays are called temporal exceptions. Predicting temporal exceptions in concurrent workflows is a complex problem since any delay in a task can cause a ripple effect on the remaining tasks from the parent workflow as well as from the other interrelated workflows. In addition, different types of loops are often embedded in the workflows for representing iterative activities, and presence of such control flow patterns in workflows can further increase the difficulty in estimation of task completion time. In this chapter, the authors describe a critical path based approach for predicting temporal exceptions in concurrent workflows that are required to share limited resources. This approach allows predicting temporal exceptions in multiple attempts while workflows are being executed. The accuracy of the proposed prediction algorithm is analyzed based on a number of simulation scenarios. The result shows that the proposed algorithm is effective in predicting exceptions for instances where long duration tasks are scheduled (or executed) at the early phase of the workflow.

Chapter 12

The World Wide Web (WWW) offers an enormous wealth of information and data, and assembles a tremendous amount of knowledge. Much of this knowledge, however, comprises either non-structured data or semi-structured data. To make use of these unexploited or underexploited resources more efficiently, the management of information and data gathering has become an essential task for research and development. In this paper, the author examines the task of researching a hostel or homestay using the Google search web service as a base search engine. From the search results, mining, retrieving and sorting out location and semantic data were carried out by combining the Chinese Word Segmentation System with text mining technology to find geographic information gleaned from web pages. The results obtained from this particular searching method allowed users to get closer to the answers they sought and achieve greater accuracy, as the results included graphics and textual geographic information. In the future, this method may be suitable for and applicable to various types of queries, analyses, geographic data collection, and in managing spatial knowledge related to different keywords within a document.

Section 4
Web-Based Technologies for Improving QoS

Chapter 13

F. Albalas, Jadara University, Jordan
B. Abu-Alhaija, Middle East University, Jordan
A. W. Awajan, Al-Balqa' Applied University, Jordan
A. A. Awajan, Princess Sumaya University for Technology, Jordan
Khalid Al-Begain, University of Glamorgan, UK

New web technologies have encouraged the deployment of various network applications that are rich with multimedia and real-time services. These services demand stringent requirements are defined through Quality of Service (QoS) parameters such as delay, jitter, loss, etc. To guarantee the delivery of these services QoS routing algorithms that deal with multiple metrics are needed. Unfortunately, QoS routing with multiple metrics is considered an NP-complete problem that cannot be solved by a simple algorithm. This paper proposes three source based QoS routing algorithms that find the optimal path from the service provider to the user that best satisfies the QoS requirements for a particular service. The three algorithms use the same filtering technique to prune all the paths that do not meet the requirements which solves the complexity of NP-complete problem. Next, each of the three algorithms integrates a different Multiple Criteria Decision Making method to select one of the paths that have resulted from the route filtering technique. The three decision making methods used are the Analytic Hierarchy Process (AHP), Multi-Attribute Utility Theory (MAUT), and Kepner-Tregoe KT. Results show that the algorithms find a path using multiple constraints with a high ability to handle multimedia and real-time applications.

Chapter 14

Asmaa Alsumait, Kuwait University, Kuwait
Sami Habib, Kuwait University, Kuwait

This paper presents a software tool for integrating a child-friendly computer system based on commercial off-the-shelf (COTS) components. The effective selection of COTS components, which meet a child's requirements and expectations, is a non-trivial and challenging optimization problem. However, many published papers consider the functional requirements while ignoring usability requirements. The functional requirements are concerned with what the computer should be able to do, whereas the usability requirements are concerned with the extent to which the child is able to learn effectively and efficiently throughout the COTS based computer. In this paper, the authors propose an iterative five-task selection and integration of COTS process, including both hardware devices and software modules, to be automated. The core of the automated tool is employing Simulated Annealing (SA) to search the design space to match, select, and integrate COTS components with a maximal satisfaction while neither exceeding a given budget nor violating child and performance constraints. A Monte Carlo simulator was utilized to evaluate the goodness of the COTS based computer design. Computational results based on building a computer for a child handwriting e-learning application show feasibility of SPACots in finding a solution satisfying all constraints while reducing the cost by 58%.

Ivaylo Atanasov, Technical University of Sofia, Bulgaria
Evelina Pencheva, Technical University of Sofia, Bulgaria

The chapter investigates the capabilities for open access to quality of service management in the Evolved Packet System. Based on the analysis of requirements for policy and charging control in the Evolved Packet Core, functions for quality of service (QoS) management and charging, available for third party applications, are identified. The functionality of Open Service Access (OSA) and Parlay X interfaces is evaluated for support of dynamic QoS control on user sessions. An approach to development of OSA-compliant application programming interfaces for QoS management in the Evolved Packet System is presented. The interface's methods are mapped onto the messages of network control protocols. Aspects of interface implementation are discussed, including interface to protocol conversion.

Sorin Adam Matei, Purdue University, USA
Anthony Faiola, Indiana University, USA
David J. Wheatley, Motorola Applied Research, USA
Tim Altom, Indiana University, USA

As designers of mobile/media-rich devices continue to incorporate more features/functionality, the evolution of interfaces will become more complex. Meanwhile, users cognitive models must be aligned with new device capabilities and corresponding physical affordances. In this paper, the authors argue that based on HCI design theory, users approach objects by building mental models starting with physical appearance. Findings suggest that users who embrace a device's multifunctionality are prevented from taking full advantage of an array of features due to an apparent cognitive constraint caused by a lack of physical controls. The authors submit that this problem stems from established mental models and past associated behaviors of both mobile and non-mobile interactive devices. In conclusion, users expressed a preference for immediate access and use of certain physical device controls within a multi-tasking environment, suggesting that as mobile computing becomes more prevalent, physical affordances in multifunctional devices may remain or increase in importance.

Preface

INTRODUCTION

This is the fifth book of the series Advances in Information Technology and Web Engineering series containing updated articles published in volume 5 (2010) of the *International Journal of Information Technology and Web Engineering*, with the title of "Models for Capitalizing on Web Engineering Advancements: Trends and Discoveries." This preface reports on current and future trends in Information Technology and Web engineering, such as the evolution of Web usage, tools and methods for linking data over the Web, cloud computing models and approaches, and finally, Information Technology infrastructure to support these new developments. Several suggestions on application scenarios, challenges, and research directions are interspersed based on these new trends.

Web 2.0 is a term used to describe the plethora of websites that exists currently addressing the needs of Internet users in the cyberspace where they can network and participate in a more interactive way. Examples of Web 2.0 based technologies are Flickr, YouTube, Twitter, Facebook, where users can share photos, videos, and interact with each other, and Wikipedia, a place where users can help to contribute to an article's content either by editing or adding to it. Blogging is also included in the Web 2.0 family. Compared to the conventional fashion of publishing, it allows readers to share their views by commenting on it. Recently there's a discussion of the third wave, the Web 3.0 to be followed by Web 4.0.

WEB 3.0

Web 3.0 based applications are expected to be a virtual reality location where users can try several applications interactively such as mapping and gaming. An example would be the Second Life, where more than 1 million players, including offline merchants, participate. Another application where map application, geographic information system (GIS), and global positioning system (GPS) are linked together to draw maps for a particular location for the purpose of urban planning and development, areas relocations, and utilities infrastructures. Figure 1 depicts the evolution of Webx.x from the first Web1.0, to Web 2.0, to current and future Web 3.0, and the future Web 4.0. The left side of the line shows tools, while the bottom of the line gives concepts and method.

The social graph just connects people; the semantic graph connects people, companies, places, interests, activities, projects, events, groups, multimedia, documents, Web pages, services, products, and emails. Figure 2 presents example of semantic graphs of linked data.

Figure 1. The road of the Webx.x (Source: www.novaspivack.com, used by permission)

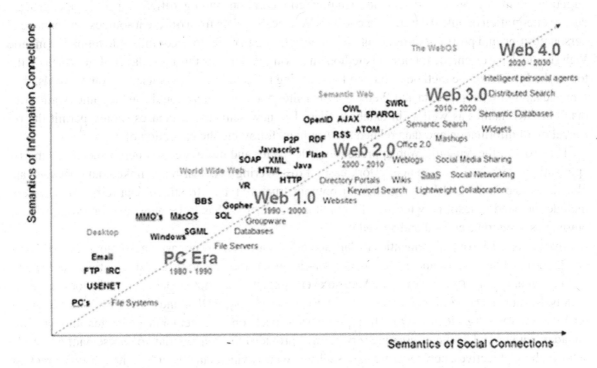

Figure 2. Sample Semantic Web graph (Source: www.novaspivack.com, used by permission)

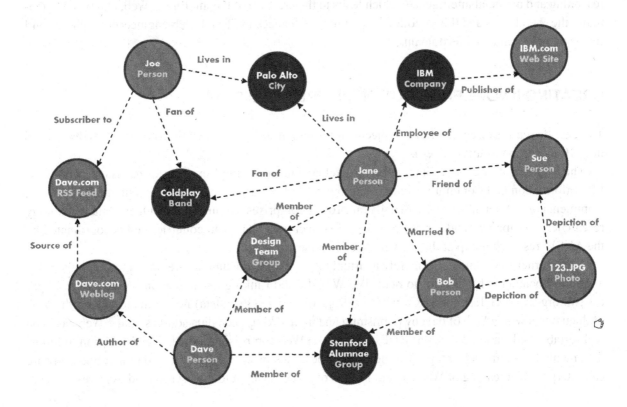

This new trend in data links uses richer semantics to enable better search mechanisms, effective targeting of marketing advertisements, smarter collaboration among different people and groups, deeper intergration of linked data, richer contents by accessing data from different sources, and enhanced personalization and profiling through intelligent interfaces. In order to accomplish such links, semantic Web should depart from the traditional method on linking Web pages through a layer of metadata on the top of the Internet to the method of having the meaning reflected in the data itself. In other words, data = metadata. For example, data and Web pages and sites about Amman should use the same word "Amman" in URL names as well as in database files. The new semantic databases should permit the retrieval of all information and data related to "Amman" following the execution of a query.

The pros for the semantic Web approach include: better and more precise query execution, smarter applications with less development effort, ease of discovering and sharing of linked data between applications, and facilitates the integration of both unstructured and structured data retrieval. The cons include the need to learn new tools and technologies, metadata has to be generated or extracted, and ontologies need to be created and agreed upon.

Two paths exist to attach semantics to data and information (www.novaspivack.com, accessed October 22, 2011): bottom-up standard-based, and top-down services and tools based. The first one requires adding semantic databases all over the Web so every website becomes semantic, and the tools used are mainly Resource Description Framework (RDF) (www.w3.org/RDF/) and Ontology Web Language (OWL) (www.w3.org/TR/*owl-ref/*). The top-down approach, on the other hand, generates automatically metadata for vertical domains, creating services that provide this as an overlay to non-semantic Web. To achieve these objectives, new tools are used such as Twine (twine.com – later bought by Evri), Freebase (Freebase.com), and Evri (Evri.com).

The last future phase in the Web 4.0 of 2020+ is intelligence and personalization of the Web using reasoning and artificial intelligence, which leads to the creation of the intelligent Web, and where eventually the Web learns and thinks collectively. Future advances in IT and Web engineering volumes will report on new trends in this domain.

CREATING KNOWLEDGE OUT OF INTERLINKED DATA

This section reports on collaborative project in information and communication technologies that started in 2010 in Europe (http://lod2.eu).

The LOD2 is a large-scale integrating project co-funded by the European Commission within the FP7 Information and Communication Technologies Work Programme (Grant Agreement No. 257943). Commencing in September 2010, this 4-year project comprises leading Linked Open Data technology researchers, companies, and service providers from across 7 European countries and is coordinated by the AKSW research group at the University of Leipzig.

The Internet as we know it is merely a virtual library where documents are linked to each other; it is a Web of documents for humans to read. The W3C-hosted Linking Open Data initiative (LOD, http:// esw.w3.org/SweoIG/TaskForces/CommunityProjects/LinkingOpenData) now aims to extend the Web of documents with a Web of data by publishing and interlinking open data sources on the Web based on well-established standards such as RDF and URIs. A Web comprising linked data in addition to linked documents brings the advantage that the content becomes machine-readable. Computer and software can interpret the meaning of Web content instead of just offering it to the user to read. As a result, more intelligent search engines can combine information from different sources, mash-ups integrating heterogeneous information can be more easily built, and many more, currently unforeseen creative uses of information on the Web become possible.

With more than 20 billion facts already published as Linked Open Data (LOD), the Data Web is not just a vision, but becoming reality right now. For example, all BBC programming, Wikipedia as a structured knowledge base (DBpedia), and statistical information from Eurostat and the US census are, in addition to hundreds of other datasets, readily available on the Web of Data.

Over the past 3 years, the semantic Web activity has gained momentum with the widespread publishing of structured data as RDF. The Linked Data paradigm has therefore evolved from a practical research idea into a very promising candidate for addressing one of the biggest challenges in the area of intelligent information management: the exploitation of the Web as a platform for data and information integration in addition to document search. To translate this initial success into a world-scale disruptive reality, encompassing the Web 2.0 world and enterprise data alike, the following research challenges need to be addressed: improve coherence and quality of data published on the Web, close the performance gap between relational and RDF data management, establish trust on the Linked Data Web, and generally lower the entrance barrier for data publishers and users. With partners among those who initiated and strongly supported the Linked Open Data initiative, the LOD2 project aims at tackling these challenges by developing:

1. Enterprise-ready tools and methodologies for exposing and managing very large amounts of structured information on the Data Web.
2. A test-bed and bootstrap network of high-quality multi-domain, multi-lingual ontologies from sources such as Wikipedia and OpenStreetMap.
3. Algorithms based on machine learning for automatically interlinking and fusing data from the Web.
4. Standards and methods for reliably tracking provenance, ensuring privacy and data security, as well as for assessing the quality of information.
5. Adaptive tools for searching, browsing, and authoring of Linked Data.

LOD2 will integrate and syndicate linked data with large-scale, existing applications and showcase the benefits in the three application scenarios of media and publishing, corporate data intranets, and e-Government. The resulting tools, methods, and data sets have the potential to change the current Web 2.0 to Web 3.0.

LOD2 Partners

The partners in LOD2 project include:

Universität Leipzig: Project coordination; develop knowledge structuring and enrichment algorithms as well as browsing, visualization and authoring interfaces; collaborate with OKFN in order to employ LOD2 results in the PublicData.eu use case.

Centrum Wiskunde & Informatica: CWI will be primarily involved in WP2 and work together with OpenLink on improving RDF data management with state-of-the-art database research approaches. CWI will be involved with a minor stake in WP5 in order to evaluate and adapt browsing and navigation in large-scale knowledge bases.

Exalead: Exalead will contribute and advance components of its search engine infrastructure, with emphasis on semantic linked data and search on linked data. Exalead search technology will be adapted and integrated as a component in the LOD2 Stack. An open search API will be developed to browse and access to the semantic linked data. As an (application) service provider in corporate environments, Exalead will lead the specification, setup, and implementation of the enterprise use case (WP8). In addition to this, Exalead will provide a prominent channel for exploiting this use case and the outcomes of the LOD2 project as a whole.

Freie Universität Berlin (FUB): Freie University Berlin will bring in expertise, tools and outreach capabilities to LOD2: (1) FUB has developed and maintains the Silk – Link Discovery Framework and Link Quality Assurance Workbench, which will be significantly extended and integrated into the LOD2 Stack. (2) FUB has developed D2R Server, the most widely used tool for publishing relational databases as Linked Data on the Web. D2R Server will be used for the domain complementation task in WP3 and will be included together with Pubby and Silk into the LOD2 Stack (WP6). (3) Within WP10 Training, Dissemination, community building, FUB will use its existing community building (initiator of W3C LOD) and outreach capabilities (Linked Data on the Web (LDOW) workshop series, Semantic Web Challenge competition series) to maximize the impact of LOD2.

Digital Enterprise Research Institute: NUIG will guarantee technical excellence in reliable large-scale data processing with the same practices which have been daily driving the works behind the Sindice and Sig.ma projects. NUIG will provide the relevance, feasibility, and consensus of the initiative thanks to the continuous interaction between the Linked Data community and the Linked Data Research Centre, a cross institute initiative.

Open Knowledge Foundation: Adaptation of the LOD2 Stack for PublicData.eu Support with legal mechanisms for knowledge sharing and with open standards; research and consultation with end-user communities; and dissemination and communication with relevant knowledge users and providers.

OpenLink Software: OpenLink contributes in particular to developing the scalable LOD2 knowledge store (WP2); to track and infer about data provenance and reliability; to support personalized views on knowledge and spatial data; alert on data; and to contribute to standardization activities regarding the integration of semantic and spatial technologies.

Semantic Web Company: SWC adopted the tool Poolparty, a self-developed modeling tool for corporate thesauri, as a component for the LOD2 stack. SWC will provide its expertise in technology assessment and business development when it comes to evaluate the economic rationale, organizational effects, and the commercial potential of semantic Web technologies. SWC will investigate governance and regulatory issues such as competition, IPR issues, and privacy, thereby contributing to the use cases in WP 7, 8 & 9.

TenForce: TenForce brings in LOD2: (1) thorough expertise in industrial implementation of taxonomies and metadata for automatic categorization and content management, (2) hands-on experience in conducting large-scale projects in this matter, such as portals for Wolters Kluwer Europe and the European Commission, (3) the capacity to deliver product quality.

Wolters Kluwer: WKD will primarily work on adapting and evaluating the LOD2 Stack for media and publishing, as well as contribute to the PublicData.eu use case, due to the experience as a publisher with governmental information.

LOD2 Technology Stack Projects

The LOD2 technology stack projects include:

Comprehensive Knowledge Archive Network (CKAN): CKAN is a registry or catalogue system for datasets or other "knowledge" resources. CKAN aims to make it easy to find, share, and reuse open content and data, especially in ways that are machine automatable.

D2R Server: D2R Server is a tool for publishing relational databases on the Semantic Web. It enables RDF and HTML browsers to navigate the content of the database, and allows applications to query the database using the SPARQL query language.

DBpedia Extraction: DBpedia is a community effort to extract structured information from Wikipedia and to make this information available on the Web. It currently already contains a tremendous amount of valuable knowledge extracted from Wikipedia. The DBpedia knowledge base will be used for evaluation LOD2's interlinking, fusing, aggregation, and visualization components. The DBpedia multi-domain ontology will be used as background-knowledge for the LOD2 applications (WP7, WP8 and WP9), and as an alignment and annotation ontology for LOD in general.

DL-Learner: DL-Learner is a tool for supervised Machine Learning in OWL and Description Logics. It can learn concepts in Description Logics (DLs) from user-provided examples. Equivalently, it can be used to learn classes in OWL ontologies from selected objects. It extends Inductive Logic Programming to Descriptions Logics and the Semantic Web. The goal of DL-Learner is to provide a DL/OWL-based machine learning tool to solve supervised learning tasks and support knowledge engineers in constructing knowledge and learning about the data they created.

MonetDB: MonetDB is an open-source high-performance database system that allows to store relational, XML, and RDF data, downloadable from monetdb.cwi.nl. While being well-known for its columnar architecture and CPU-cache optimizing algorithms, the crucial aspect leveraged in the scope of this project is its unique run-time query optimization framework, which provides a unique environment to crack the recursive-correlated-self-join queries caused by semantic Web queries to triple stores.

OntoWiki: OntoWiki is a tool providing support for agile, distributed knowledge engineering scenarios. It facilitates the visual presentation of a knowledge base as an information map with different views on instance data. It enables intuitive authoring of semantic content, with an inline editing mode for editing RDF content, similar to WYSIWIG for text documents.

PoolParty: PoolParty is a thesaurus management system and a SKOS editor for the Semantic Web including text mining and linked data capabilities. The system helps to build and maintain multilingual thesauri providing an easy-to-use interface. PoolParty server provides semantic services to integrate semantic search or recommender systems into systems like CMS, DMS, CRM, or Wikis.

SemMF: SemMF is a flexible framework for calculating semantic similarity between objects that are represented as arbitrary RDF graphs. The framework allows taxonomic and non-taxonomic concept matching techniques to be applied to selected object properties. Moreover, new concept matchers are easily integrated into SemMF by implementing a simple interface, thus making it applicable in a wide range of different use case scenarios

Sig.ma: Sig.ma is a tool to explore and leverage the Web of Data. At any time, information in Sigma is likely to come from multiple, unrelated websites – potentially any website that embeds information in RDF, RDFa or Microformats (standards for the Web of Data). Sig.ma is a semantic Web browser as well as an embeddable widget and also provides a Semantic Web API.

Silk Framework: The Silk Linking Framework supports data publishers in setting explicit RDF links between data items within different data sources. Using the declarative Silk - Link Specification Language (Silk-LSL), developers can specify which types of RDF links should be discovered between data sources as well as which conditions data items must fulfill in order to be interlinked. These link conditions may combine various similarity metrics and can take the graph around a data item into account, which is addressed using an RDF path language.

Sindice: Sindice is a state of the art infrastructure to process, consolidate and query the Web of Data. Sindice collates these billions of pieces of metadata into a coherent umbrella of functionalities and services.

Sparallax: Sparallax is a faceted browsing interface for SPARQL endpoints, based on Freebase Parallax. This demonstrator showcases the benefits of intelligent browsing of Semantic Web data and represents a good starting point for LOD2 interfaces developed in WP 5.

Triplify: Triplify provides a building block for the "semantification" of Web applications. As a plugin for Web applications, it reveals the semantic structures encoded in relational databases by making database content available as RDF, JSON, or Linked Data. Triplify makes Web applications more easily mashable and lays the foundation for next-generation, semantics-based Web searches.

OpenLink Virtuoso: Virtuoso is a knowledge store and virtualization platform that transparently integrates data, services, and business processes across the enterprise. Its product architecture enables it to deliver traditionally distinct server functionality within a single system offering along the following lines: Data Management & Integration (SQL, XML and EII), Application Integration (Web Services & SOA), Process Management & Integration (BPEL), and Distributed Collaborative Applications. The open-source data integration server and the highly efficient and scalable RDF triple store implementation in Virtuoso will be the basis for the knowledge store component in the LOD2 Stack.

WIQA: The Web Information Quality Assessment Framework is a set of software components that empowers information consumers to employ a wide range of different information quality assessment policies to filter information from the Web. Information providers on the Web have different levels of knowledge, different views of the world, and different intentions. Thus, provided information may be wrong, biased, inconsistent, or outdated. Before information from the Web is used to accomplish a specific task, its quality should be assessed according to task-specific criteria.

Examples of Projects

The following is a list of projects LOD2 selected from 2010:

1. Umweltbundesamt GmbH (Environment Agency Austria), Austria - Team contact: Bastiaan Deblieck, Tenforce, Belgium

Abstract: The Federal Environmental Agency of Austria (UBA) is the leading expert organization for all environmental issues and media. It works for the conservation of nature and the environment and thus contributes to the sustainable development of society. Its core tasks include the monitoring, management and evaluation of environmental data. The UBA intends to learn how Linked Open Data can help them to aggregate, share and publish data. As the UBA primarily deals with measurements, statistics, research results and geo-location, they face the following specific challenges:

- How to align meta-data
- How to connect data in general
- How to geo-locate the data

2.   Greater London Authority, U.K. - Team contact: Hugh Williams, OGL, U.K.

Abstract: The Greater London Authority (GLA) is home to the Mayor of London and the London Assembly. As part of its commitment to openness and transparency, the GLA has published a number of datasets in their currently available formats on its data store site (http://data.london.gov.uk/). They would now like to consolidate this effort around a concrete data model enabling a general deployment as Linked Open Data. Initially, the Greater London Assembly members' data is being considered for the Proof of concept dataset, with some of the challenges to be addressed being:

- Is this a suitable dataset for initial publication?
- How do we model the data?
- What do we do with the temporal aspect of the data?
- What is the best format to present this data to a community of developers?
- What is the technology for publishing linked open data?

3.   Deutsche Bibliographie, HistorischeKommission, Germany - Team contact: Thomas Riechert, ULEI, Germany

Abstract: The German Biography is an online project of the Historical Commission at the Bavarian Academy of Science. The original print version of two biographical lexica contains information about 47.000 biographies, including 45.000 additional persons and over 12.000 places. Funded by the German Research Foundation (Deutsche Forschungsgemeinschaft), some 55 volumes in print have already been digitalized and tagged according to TEI-P5, while all persons have been aligned to the open data authority file PND. This information is publicly available at http://www.deutsche-biographie.de/.

The aim of the current project with PubLink is to provide metadata about the individual biographies to enable the visualization of interpersonal relations, for instance. The publication of the metadata in RDF will make the retrieval of such information and inference of new statements not only possible, but also very flexible. Likewise, the integration of the biographical metadata into the Linked Data Cloud will also enhance the use of the biography for researchers. In this way, it is also an example of how European cultural heritage is merged into the digital world. The project is supported by PubLink in the transformation and publication of the data as RDF and helps to foster the establishment of a knowledge engineering methodology.

You can see the result of the support activities by LOD2/PUBLINK here: http://ndb.publink.lod2.eu.

4.   InstitutoCanario des Estadística - ISTA, Canary Islands - Team contact: Michael Hausenblas, NUIG/ DERI, Ireland

Abstract: The InstitutoCanario de Estadística (ISTAC) http://www.gobiernodecanarias.org/istac is the central organ of the regional statistical system and official research center of the Canary Islands. ISTAC is extending their Dissemination Environment JAXI-2 with Linked Data capabilities. JAXI-2, based on a combination of Tomcat, Alfresco and an Oracle DB, allows the publication of meta-report statistical resources based on PC-Axis. The main open questions are around conversions of statistical data from PC-Axis and in future from SDMX as well as integration aspects into the JAXI-2 workflow.

5.   Digital Agenda Scoreboard, Belgium - Team contact: BastiaanDeblieck, Tenforce, Belgium

Abstract: The EC Directorate General Information Society and Media (DG Infso) is one of the larger DGs in the European Commission and aims at supporting the development and use of information and communication technologies (ICTs) for cultural, societal and economic benefits. The PubLink project with the DG Infso has mainly focused on the publication of the Digital Agenda Scoreboard as Open Data with a flexible but pragmatic visualization through the support from the LOD2 consortium. This statistical information had previously been published as a report in PDF format. Yet, this format used to have extremely limited capabilities for reuse and browsing.

In May 2010, the European Commission adopted the Digital Agenda for Europe (DAE) - a strategy to take advantage of the potential offered by the rapid progress of digital technologies. The DAE is part of the overall Europe2020 strategy for smart, sustainable and inclusive growth. The Digital Agenda contains commitments to undertake 101 specific policy actions (78 actions to be taken by the Commission, including 31 legal proposals, and 23 actions proposed to the Member States) that are intended to stimulate a virtuous circle of investment in and usage of digital technologies. It identifies thirteen key performance targets to show whether Europe is making progress in this area. The present Scoreboard only addresses policy actions planned for the last twelve months in the Digital Agenda.

Readers can see the result of the support activities by LOD2/PUBLINK at: http://ec.europa.eu/information_society/digital-agenda/scoreboard/graphs/index_en.htm

First Release of the LOD2 Stack

The LOD2 consortium announced the first release of the LOD2 stack available at: http://stack.lod2.eu. The LOD2 stack is an integrated distribution of aligned tools that support the life-cycle of Linked Data from extraction, authoring/creation over enrichment, interlinking, and fusing, to visualization and maintenance. The stack comprises new and substantially extended existing tools from the LOD2 partners and third parties.

The LOD2 stack is organized as a Debian package repository making the tool stack easy to install on any Debian-based system (e.g. Ubuntu). A quick look at the stack and its components is available via the online demo at: http://demo.lod2.eu/lod2demo. For more thorough experimentation a virtual machine image (VMware or VirtualBox) with pre-installed LOD2 Stack can be downloaded from: http://stack.lod2.eu/VirtualMachines/. More details and the instructions on installing the LOD2 Stack from scratch are available in the HOWTO Start document. The first release of the LOD2 stack contains the following components (available as Debian packages):

- LOD2 demonstrator, the root package (LOD2)
- Virtuoso, RDF storage and data management platform (Openlink)
- OntoWiki, semantic data wiki authoring tool (ULEI)
- SigmaEE, multi-source exploration tool (DERI)

- D2R, RDF wrapper for SQL databases (FUB)
- Silk, interlinking engine (FUB)
- ORE, ontology repair and enrichment toolkit (ULEI)

Online services were integrated into the LOD2 Stack: PoolParty (taxonomy manager by SWCG) and Spotlight (annotating texts w.r.t. DBpedia by FUB). The LOD2 Stack also makes use of dataset metadata repositories, such as thedatahub.org and http://publicdata.eu. Selections of datasets have been packaged and are available in the LOD2 stack repository.

The LOD2 stack is an open platform for Linked Data components. LOD2 welcomes new components. Detailed instructions how to integrate your component into the LOD2 Stack as Debian package are available in the HOWTO Contribute. For assistance or any questions related to the LOD2-stack contact support-stack@lod2.eu. From now on we will regularly release improved and extended versions of the LOD2 Stack. Major releases are expected for Fall 2012 and 2013. Just to notice that leading Web 3.0 technologies are combined in the LOD2 projects into the coherent LOD2 stack (e.g. DBpedia, Virtuoso, Sindice, Silk).

LOD2 in a Nutshell

Several challenges still face successful implementation of LOD-based projects. These include:

1. **Coherence:** Relatively few, expensively maintained links
2. **Quality:** partly low quality data and inconsistencies
3. **Performance:** Still substantial penalties compared to relational
4. **Data consumption:** large-scale processing, schema mapping and data fusion still in its infancy
5. **Usability:** Missing direct end-user tools

While these challenges will continue to be addressed by new development in LOD2 tools (figure 3) and applications, the last one will lead the way to the next generation of Internet use: the Web 4.0. By adding intelligence to the front end and have knowledge and intelligence built into the Web, a new Semantic Intelligent Web will emerge. Intelligent software agents will be faster to develop than currently done, leading to lean and light software that will be easier to surf the Internet and interact with other agents.

This preface identifies several challenges facing the realization of Web 4.0 semantic Web of knowledge and intelligence (K&I):

- Indentifying relevant K&I
- Capturing such K&I
- Validation and verification of K&I to build trust in using Web 4.0
- Human factors of experts inclination not to share K&I
- Tools to codify K&I for the creation of semantic Web of K&I
- Dealing with K&I decay and maintenance

LOD2 Major use Cases

The following is a brief description of the three major application scenario of LOD2:

Figure 3. LOD2 in a nutshell (Source: http://lod2.eu, PowerPoint presentation by Sören Auer, used with permission)

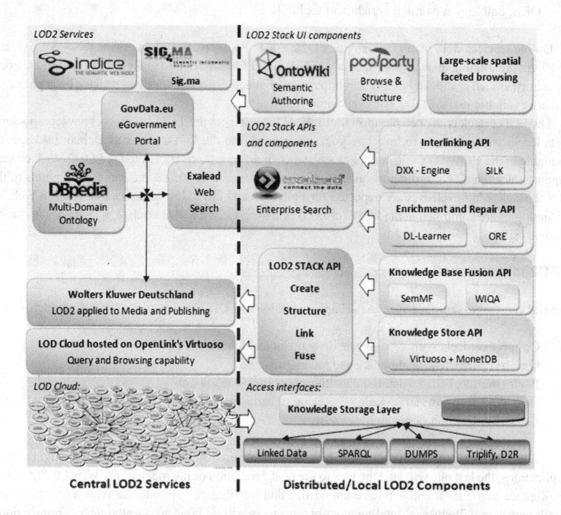

Use Case I – Media & Publishing: Large amounts of data resources from the legal domain are used to test and explore the commercial value of linked data in media and publishing. This data will be interlinked and merged automatically. Data from external sources will be used to semantically enrich the existing datasets. Adequate licensing and business models are also investigated with respect to the management of interoperable metadata.

Use Case II – Enterprise Data Web: Linked Data is a natural addition to the existing document and Web service intranets and extranets. Corporate data intranets based on Linked Data technologies can help to substantially reduce data integration costs. Using the LOD2 Stack for linking internal corporate data with external references from the LOD cloud will allow a corporation to significantly increase the value of its corporate knowledge with relatively low effort.

Use Case III – Linked Governmental Data: The project will showcase the wide applicability of the LOD2 Stack through the design, specification, implementation, testing, and user evaluation of a case study targeting ordinary citizens of the European Union. LOD2 will establish a network of European governmental data registries in order to increase public access to high-value, machine readable data sets generated by European, national as well as regional governments and public administrations. The semi-automatic classification, interlinking, enrichment and repair methods developed in LOD2 will create a significant benefit, since they allow governmental data to be more easily explored, analyzed and mashed together.

This preface recommends scenarios for implementing Use Case II and III. The Enterprise Use Case may comprise supply chain, integration of different enterprise systems, and virtual enterprises. The government Use Case may include:

- Linking entry/exit border points of a particular country or group of related countries, such as airports and land border entry points.
- A group of government units belonging to a major ministry, such as defense and foreign affairs may link its data for faster access and security reasons.
- Custom border points may be linked to identify potential risk of material and product movement between states or countries in the same region.

CURRENT TECHNIQUES FOR LINKING DATA

In pre-LOD era, Web services (WS) technologies were used to link loosely coupled data in order to facilitate access over the Internet. Web services linking strategies include native language-based, such as XML and JEE2, and standard-based.

The core standards are related to the three basic functions of defining, publishing, and accessing WS:

- **Accessing:** SOAP (Simple Object Access Protocol)
- **Registering:** UDDI (Universal Description, Discovery and Integration) OASIS latest release 3.0
- **Describing:** WSDL (Web Services Description Language)

Extended standards supporting composite WS include:

- Messaging
- Business processes and workflow
- Database update of transactions

Other standards for supporting WS functionally include:

- Security
- Management
- Reliability
- Addressing
- Ontology, Semantics, and Metadata, and finally

Companion standards to support platform functionally:

- **Portal technology:** Coverage of portal technology may be found atnet.educause.edu/ir/library/pdf/pub5006k.pdf
- **Grid computing and enterprise grid:** Coverage of grid computing may be found at www.oracle.com/us/technologies/grid/index.html
- **Enterprise Service Bus (ESB):** Complete coverage of ESB technology may be found at http://go.techtarget.com/r/15236108/10927021/6
- And all of the above could be stored and accessed through cloud computing platform

A WS may deal with one application, such as car reservation, or composite WS that deals with more than one application, such as making car, airline, restaurant, and hotel reservations in one WS application. This preface is concerned with custom developed WS applications, such as E-business, supply chain, virtual enterprises, and enterprise computing, rather than the general purpose WS applications, such as weather, currency conversion, and the like.

Figure 4 contains a proposed framework for describing the relationships among these standards:

CLOUD COMPUTING (CC)

The discussion on data linking approaches will not be complete without exposing new trends in cloud computing.

Cloud computing comprises three service models:

- **Software as a service (SaaS)**, which delivers device-independent Web apps with Web services extensions;
- **Platform as a service (PaaS)**, which hosts a development environment for mashing up composite processes and apps; and
- **Infrastructure as a service (IaaS)**, which deploys high-end apps configured for elastic computing resources.

Implementing cloud computing may follow one the following approaches:

- **Public clouds** where all services provided by CC service provider
- **Private clouds** where a particular company or enterprise uses CC as a host while custom development or deployment of applications by company IT personnel.
- **Community clouds** where a limited geographical area is serviced by a CC service provider using one or more of the above stated services or as a private cloud service.
- **Hybrid clouds** where required services are deployed in more than one of the above preceding models. For example, common data center may be deployed on public clouds with some non-critical applications, critical applications deployed on private cloud, and selected services deployed and provided to other parties as a community cloud.

Figure 4. A framework for linking WS standards

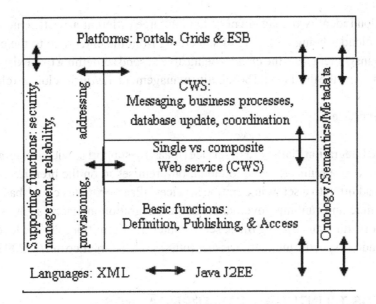

In such environments, cloud computing architecture may expand or shrink as needs increase or diminish. This dynamic elasticity dictates different design approaches for developing new applications and scaling and re-engineering existing applications to fit the cloud platform. A new research paradigm has emerged to deal with the selecting and evaluation of different alternatives for mapping CC services to models in light of current and future applications of a particular company. Method of evaluation may use cost/benefit analysis and multi-attribute utility models such Analytical Hierarchical Process (AHP). AHP may use software program such as ExpertChoice (www.expertchoice.com)

Current Trends in Cloud Computing

• Merging the clouds with ESB messaging

For example, the Microsoft Azure cloud supports relational as well as non-relational data. One new Azure aspect is represented in the Windows Azure Service Bus September Release. This software is intended to help developers build distributed and loosely-coupled applications in the cloud, as well as hybrid applications across on-premises and the cloud. Enhancements enable asynchronous cloud eventing, event-driven SOA, and advanced intra-app messaging.

In this environment, many future cloud computing applications will be completely new. However, others will attempt at relocating current systems on the clouds. For example, many enterprises want to deploy the data transformations and message brokering being done in today's land data centers, and place that on a cloud platform. Full ESBs on cloud have been discussed. While it is not alone, Microsoft's effort to meet architects' messaging needs places it in the forefront of cloud vendors (Vaughan, 2011).

- Security

Gartner views cloud identity management as having three different aspects. One would be identity management *to* the cloud – being able to send something from the enterprise to the cloud. The second would be identity management *from* the cloud – being able to send something that exists somewhere else, to your organizations. And the third would be identity management *within* the cloud to cloud (Earls, 2011).

- iPlatform as a service

Gartner proposed Integration Platform as a service (iPaaS) to support services integration on the cloud. These include cloud to on-premises, cloud to cloud, on-premises to on the premises, and E-commerce B2B integration. In addition to a set of integration services, iPaaS provides cloud-based services aimed at enabling design time and runtime governance of the integration artifacts, such as process models, composition, transformation and routing rules, service interface definition, service level agreement, and policies utilized to address specific integration issues (http://go.techtarget.com/r/15209212/10927021/8).

LINKING CLOUDS TO INTEGRATION APPROACHES

In this section, the preface proposes a framework for linking integration approaches with the cloud computing platform, as in the figure 5.

IT INFRASTRUCTURE FOR THE CLOUD AND DATA INTEGRATION

This section contains current development in IT infrastructure that support cloud computing environments and data integration strategies. Both hardware made by IBM and Oracle employs blades to achieve scalability, virtualization, and elasticity. However, Oracle used Sum blades, while IBM uses its own.

IBM zEnterprise Hardware and System z Software (www.ibm.com)

Highlights of zEnterprise include:

- A "System of Systems," integrating leading technologies from IBM to dramatically improve productivity of today's multi-architecture data centers and tomorrow's private clouds
- First-of-a-kind design that embraces multiple technology platforms—mainframe, UNIX® and x86, integrated within a centrally managed unified system
- Unique hybrid computing capabilities powered by the industry's premier enterprise server, offering breakthrough innovation, virtualization and unrivalled scalability, reliability, and security
- Rapidly deploy services using prepackaged solutions and preintegrated technologies designed to meet the needs of specific workloads

Figure 5. A proposed framework for linking cloud computing to integration strategies

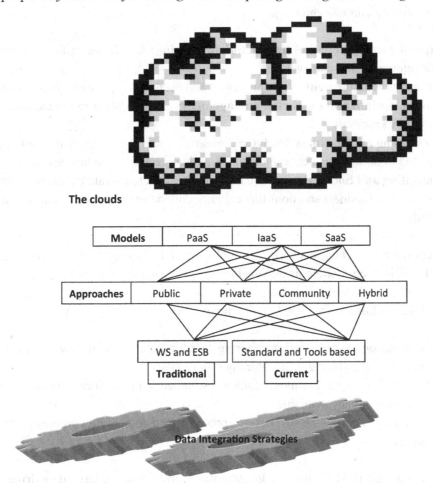

The demands of customers, partners and employees to smarter computing systems—systems that raise the bar on efficiency, performance and cost savings while lowering management complexity. Furthermore, it extends the strengths and capabilities of the mainframe—such as security, fault tolerance, efficiency, virtualization and dynamic resource allocation—to other systems and workloads running on AIX® on POWER7, and Microsoft Windows or Linux® on System x®. The zEnterprise System includes a central processing complex (CPC)—either the zEnterprise 196 (z196) or the zEnterprise 114 (z114), the IBM zEnterpriseBladeCenter Extension (zBX) with its integrated optimizers and/or select IBM blades, and the zEnterprise Unified Resource Manager (http://www-03.ibm.com/systems/z/hardware/zenterprise/).

New IBM Software for System z enables hybrid and cloud environments for smarter computing, enabling clients to:

- Handle all variety of workloads
- Address the full application lifecycle
- Exploit the zEnterprise hardware and OS capabilities
- Modernize any environment

System z software announcements address critical new and modern workloads (http://www-01.ibm. com/software/os/systemz/announcements/):

- **Multiplatform development and transaction processing:** Accelerate agility with intelligent application development and management
- **Business Process Management:** Agile processes and decisions to optimize business performance
- **Virtualization, optimization and risk management:** Consolidation to reduce cost, complexity and help align IT resources
- **Data Warehousing and Business Analytics:** Integrating and transforming data into trusted information, helping organizations better understand, anticipate and shape business outcomes
- **Cloud computing and Social Business:** Cloud applications to elevate business performance and productivity; Social Business solutions that enhance collaboration and community. both internally and externally

zEnterprise configured an entry level offer for deploying an Infrastructure as a Services (IaaS) Cloud delivery model for zLinux environments that enables the provisioning of zLinux images (under z/VM) through a self-service portal quickly.

zEnterprise™ Starter Edition for Cloud deployments offers:

- Advanced automation and optional monitoring to dramatically speed new service provisioning (measured in minutes) reducing datacenter operations costs
- The industry's highest RAS and most efficient virtualization to ensure multi-tenancy cloud deployments are continuously available
- The industry's most secure platform (EAL 5) protecting customer and corporate data in a shared cloud infrastructure

Cloud Computing with IBM System z addresses the need to manage large data driven workloads is achieving broad adoption of cloud computing. By infusing clouds with security and manageability, companies are provided with the agility to move quickly in highly competitive environments; to activate and retire resources as needed; to manage infrastructure elements in a dynamic way; and to move workloads for more efficiency—while seamlessly integrating with their traditional computing environment. IBM Cloud Computing on System z is transforming the business, delivering proven qualities of service, advanced workload optimization, and efficient resource consolidation.

Oracle Cloud Suites (www.Oracle.com)

Oracle's strategy is to offer a broad portfolio of software and hardware products and services to enable public, private and hybrid clouds, enabling customers to choose the right approach for them. Unlike competitors with narrow views of the cloud, Oracle provides the broadest, most complete, and integrated cloud offerings in the industry. Oracle offers the following cloud models:

Public clouds (PC): in PC the application services area offers Fusion customer relationship management (CRM), Fusion human capital management (HCM), and social network for enterprise collaboration. It offers the following as platform services: Java and its database.

IaaS: For Infrastructure as a Service (IaaS), Oracle offers a complete selection of computing servers, storage, networking fabric, virtualization software, operating systems, and management software. Unlike other vendors with partial solutions, Oracle provides all the infrastructure hardware and software components needed to support diverse application requirements. Oracle's robust, flexible cloud infrastructure supports resource pooling, elastic scalability, rapid application deployment, and high availability. The unique ability to deliver application aware virtualization and management integrated with compute, storage, and network technologies enables the rapid deployment and efficient management of public and private IaaS. The Sun ZFS Storage Appliance seamlessly integrates with Oracle VM and VMware features providing a powerful breakthrough solution for deploying storage in a cloud computing infrastructure.

PaaS: The Oracle Platform as a Service (PaaS) provides a shared and elastically scalable platform for consolidation of existing applications and new application development and deployment. The Oracle PaaS platform delivers cost savings through standardization and higher utilization of the shared platform across multiple applications. The Oracle PaaS also delivers greater agility through faster application development leveraging standards-based shared services, and elastic scalability on demand. The Oracle PaaS includes database services based on Oracle Database and Oracle Exadata Database Machine, as well as middleware service based on Oracle Fusion Middleware and Oracle Exalogic Elastic Cloud. With engineered systems such as Exadata and Exalogic providing extreme performance and efficiency for mixed workloads, Oracle provides the best foundation for PaaS.

Cloud Management: Oracle Enterprise Manager is Oracle's complete cloud lifecycle management solution. It is the industry's first complete solution including self-service provisioning balanced against centralized, policy-based resource management, integrated chargeback and capacity planning and complete visibility of the physical and virtual environment from applications to disk.

Cloud Integration: As more organizations use a mix of private and public cloud services they need a better way to integrate those services into an effective, secure hybrid cloud environment. Oracle takes a unified approach to loading and replicating data, as well as integrating transactions and business processes to ensure that organizations retain optimal control, fast time-to-market, and flexibility within their infrastructure. At Oracle, this solution for cloud integration includes Oracle SOA and Oracle Data Integration products. Oracle's cloud integration is the only solution with the following design principles:

- **Unified:** Comprehensive and unified set of integration components seamlessly integrate on-premise and public cloud applications and services
- **Proven:** Deployed by thousands of leading organizations to ensure high reliability, real-time performance, and trusted integration
- **Open:** Leverages existing investments in Oracle database, middleware, applications and hardware systems, while working with third party cloud applications

Cloud security: Oracle can uniquely safeguard information and allow organization to benefit from the reduced costs and complexity of consolidation and cloud computing. Areas include Data Security, Identity Management, and Governance, Risk, and Compliance – offering solutions you can rely on to deploy private clouds, public clouds, or outsource, with the following benefits:

- **Complete:** Provides a comprehensive set of solutions to mitigate threats across your databases and applications

- **Proven:** Deployed by thousands of leading organizations to address compliance for multiple government and industry regulations
- **Cost-Effective:** Leverages existing investments in Oracle database, middleware, applications and hardware systems

GRID COMPUTING AND CLOUD COMPUTING

Redhat announced a PaaS offering called open-shift, followed by the release of Jboss Enterprise Data Grid 6, cloud integration is greatly enhanced. The enterprise data grid (EDG) is a solution that (Sharma, 2011):

- Is cloud ready
- Comes with highly scalable distributed data cache
- Will reduce response times in applications
- Will provide additional failure resilience

Given the fact that EDG is based on infispan and pieces borrowed from NoSQLetc, it supports mutitenacy, scalability, elasticity and distributed code execution. With these features, the EDG manages to achieve cloud readiness within its core, the architecture.

This preface presented current and future trends of information technology and Web engineering, such as Web 3.0 and Web 4.0, linking data strategies with coverage of a major project in Europe, cloud computing. Throughout the preface, the discussions included a proposed framework for linking clouds to integration strategies, challenges facing the realization of knowledge and intelligent semantic Web, and several suggested applications and scenarios to the implementation of clouds and data integration strategies.

This volume contains 16 chapters classified into four sections, each containing multiple articles as follows:

Section 1: Web Engineering Trends: Clouds, Location-aware, and Agents

This section contains three articles on using cloud computing for Web collaboration, improving location aware and search engine retrieval systems, and using ontologies to improve software agent communication.

Section 2: Web Engineering Discoveries

With five articles, coverage in this section include proposing a language for knowledge discovery in a semantic Web, searching beyond Web pages into deep Web retrieval system, discovering the rippling effects in Web application projects, developing finer garbage collection in object oriented systems, and modeling of ranking and selection of integrity tests in distributed databases.

Section 3: Web-engineered Applications

These applications include micro payment of peer to peer systems, virtual telemedicine and virtual telehealth, handling exceptions in concurrent workflows, and geographic information retrieval and text mining on Chinese tourism Web pages.

Section 4: Web-based Technologies for Improving QoS

This last section contains article dealing mainly with Quality of Service for Multimedia and real-Time services and in packet system as managed by third party, discussion on tool to assist in selecting COTS components to improve quality of software performance, and considerations of physical affordance to improve the design of multifunctional mobile devices.

Ghazi I. Alkhatib
Princess Sumaya University for Technology, Jordan

REFERENCES

Earls, A. (2011, Tuesday, October 25). *Gartner takes on cloud identity management*. Retrieved from SearchSOA.com

IBM. (n.d.). *Website*. Retrieved October 23, 2011, from www.ibm.com

LOD2. (n.d.). *PowerPoint presentation*. Retrieved October 24, 2011, from http://lod2.eu

Nova Spivack. (n.d.). *Website*. Retrieved October 22, 2011, from www.novaspivack.com

Oracle Technologies. (n.d.). *Home page*. Retrieved October 23, 2011, from www.Oracle.com

Sharma, R. (2011, May 5). *JBoss enterprise data grid takes cloud computing to the next level*. Retrieved from http://cloudtimes.org/jboss-enterprise-data-grid-takes-cloud-computing-to-the-next-level/

TechTarget. (n.d.). *Website*. Retrieved from http://go.techtarget.com/r/15209212/10927021/8

Vaughan, J. (2011). *Microsoft Azure cloud get messaging bus*. Retrieved from SearchSOA.com

W3 Consortium. (n.d.). OWL-ref. Retrieved October 24, 2011, from www.w3.org/TR/**owl**-ref

W3 Consortium. (n.d.). RDF Framework. Retrieved October 24, 2011, from www.w3.org/RDF/

Section 1
Web Engineering Trends:
Clouds, Location–Aware, and Agents

Chapter 1
Integration of Cloud Services for Web Collaboration:
A User–Centered Perspective

Liliana Ardissono
Università di Torino, Italy

Gianni Bosio
Università di Torino, Italy

Anna Goy
Università di Torino, Italy

Giovanna Petrone
Università di Torino, Italy

Marino Segnan
Università di Torino, Italy

ABSTRACT

The ubiquitous and pervasive availability of the Internet, and of Web-based services, is dramatically changing people's lives. Meanwhile, man's presence on the Web is growing fast, enabled by the offerings of low-cost devices and wireless broad-band Internet connections. On the other hand, users are adopting on-line applications and services more and more, both for personal and business purposes, in order to get in touch with other people and to manage documents, pictures, and many other types of artifacts in a ubiquitous environment that does not impose the usage of a specific end-user device. The naturally arising question is thus the following: if these projects, activities, and collaboration teams are developed all around, why do we need to deal with large numbers of separate services, each one operating in a separate representation of reality, instead of performing tasks in a unified environment integrating such different services automatically? This chapter discusses this aspect of current user collaboration and presents an infrastructure for the development of user-centered clouds of services that enable the user to interact with her favorite applications from a unified environment managing her workspaces and collaborations. The management of such collaborative service clouds is enabled

DOI: 10.4018/978-1-4666-0023-2.ch001

by an open architecture, which provides services and data synchronization coupled with a set of core applications supporting essential aspects of the functionality of the environment, such as workspace awareness, user groups definition, and user cooperation to the management of shared tasks involving the usage of heterogeneous services.

INTRODUCTION

The availability of Internet connections almost everywhere, not only at office and at home, but also in the streets, coffee-shops, airports and on the beach, has changed our life. Reliable and fast wireless connections, and a wide range of mobile devices such as small laptops, PDAs and smartphones, enable us to navigate the Web and ubiquitously access on-line services.

Thanks to the availability of such infrastructures, many services and tools, supporting business and everyday activities at different levels, are offered on the Web. Cloud computing (Creeger, 2009; Dikaiakos et al., 2009) is emerging as a lead paradigm for distributing infrastructures, platforms and resources over the Net, avoiding the need of owning expensive and "heavy" hardware or software systems. Moreover, the concept of *service cloud* is emerging, as a paradigm for offering Web-based services based on an open, distributed architecture: many computer-supported activities are provided on the so-called Web 2.0 (O'Reilly, 2007), which enhances the possibility of sharing information and resources, and thus the support to on-line collaboration. Office automation and business management software are offered in a Software-as-a-Service (SaaS) modality (Turner et al., 2003); collaboration among colleagues and partners takes place on the Web (e.g., ActiveCollab, 2011); communication is mediated by the Net (e.g., Skype: http://www.skype.com); e-learning and e-commerce are growing fast; social interaction takes place mostly on-line; videos, music and pictures are published (and shared) on the Web (e.g., Facebook: http://www.facebook.com); and so on. In other words, the Web, with its *clouds*, is becoming the place where most human interaction and collaboration take place.

Services provided on the Web make our work and life much easier than before: transferring money electronically, chatting with far away friends, sharing the pictures and videos of your holidays, participating in a meeting with your colleagues in Berlin without moving from your office are activities that have become easy and fast when performed on-line. However, in order to manage these activities, people usually have to store lots of different bookmarks, pointing to the various services; they have to remember different login/ password couples and to define many times the lists of contacts to be involved in activities (in an email thread, in a chat, in a document sharing tool, etc.). People usually receive lots of notification messages from the services they are subscribed to and they have to install different clients (on their laptops or smartphones), in order to access such services.

This heterogeneity impose a heavy overload on the end-user. To quote Jacob Nielsen: "a bunch of stand-alone tools will provide a disconnected user experience, causing employees to waste inordinate amounts of time moving between environments" (Nielsen, 2009). Although Nielsen is talking about Intranet users, this claim also holds for generic Internet users, especially because it is important "to avoid burdening users with double work. Don't, for example, force users to update their profile or photo in both the traditional employee directory and a Facebook-like social connection tool" (Nielsen, 2009).

We think that a major step towards this goal is to offer a single place on the Net where all the services needed to coordinate people life are available and share the required information. Moreover, the groups of friends and collaborators are defined once and are available in all the services in use; notifications messages are filtered in order to be delivered at the appropriate time and to be easy to browse through; and finally, people will be able to connect to their personal on-line environments not only from their laptop or smart-phone, but also from any Internet kiosk they can find in airports, stations, or from the PC at a friend's place, using a Web browser. Such an on-line environment can be seen as a virtual shared desktop where the available services (offered by different providers) run remotely, and represents a *user-centered cloud of services*, in which services are available "in the cloud", are selected and configured on the basis of the user's needs, and primarily support information sharing and user collaboration.

In this chapter we propose a framework for the management of *Collaborative Service Clouds* that support user collaboration and provide a user-centered unified view of business and private activities across applications. Such Collaborative Service Clouds are open, Web-based environments that ensure the possibility of integrating heterogeneous services. The management of such environments is enabled by an open architecture, that provides services and data synchronization by handling a shared context, coupled with a set of core applications supporting an integrated management of user collaborations and communications (by handling workspace awareness, user groups definition support, advanced calendar and task sharing features).

In order to show how a Collaborative Service Cloud like the one sketched so far could support people in organizing and scheduling their business and private activities, as well as in collaborating with colleagues and friends, we provide a brief usage scenario, taken from (Ardissono et al., 2010).

Alex works for a Telecom company, where he is the coordinator of the team that participates in the national project *Wow*. He and his wife Lucy have two children, Tom and Jenny. Alex has a keen interest in photography: he is a member of the local chapter of a national photography association and he is organizing a picture exhibition about the fall of the Berlin wall. Today he is going to Milano to attend a *Wow* project meeting; while he is on the train, he connects to the service cloud using his laptop. He uploads the draft of the *Call for Pictures* on a document sharing tool and shares it with the *Berlin_exhibition_group* (i.e., the members of the association collaborating with him in the organization of the event) in order to get feedback from them. Then, he sends an invitation to the same group for a meeting to discuss the organization of the exhibition setting; for this purpose, he exploits a smart calendar application that supports him by automatically accessing the other members' e-calendars and proposing dates in which everyone is available.

While Alex is on the train, he is notified that his wife is on-line, so he contacts her through an instant messaging application in order to decide who is going to take Tom to the football training session and Jenny to the dance class. They involve grandma' in the chat, to see if she is available: grandma' will take Jenny and Alex will contact his wife later on in order to decide who takes Tom. Alex, by means of a click, transforms the chat in a task, *Contact_Lucy*, with deadline at 17.00, automatically copied as a calendar event. He also assigns a new task, *Take_Jenny*, to grandma'. In the meantime, Alex receives an alert from Facebook saying that his friend Marc cannot participate to the trip next Sunday, so he enters Facebook and proposes to move the trip to the following weekend: the deadlines of the tasks related to the organization of the trip are automatically moved accordingly.

Alex then switches to the task management application and notices that there is a pending task, i.e. *Write_HCI_survey*, assigned to him, and he decides to further detail it. Since Lora and Mark (two colleagues involved in the project) are on-line, he involves them in the editing. By discussing in a parallel chat, they define a sub-thread (*HCI_survey*) composed by three tasks assigned respectively to Daisy (who is immediately notified about it), Lora, and Mark. They also set the task assigned to Daisy as preceding the other two, which can be handled in parallel.

Then Alex notices that there is another pending task, assigned to him, with deadline in a few days (*Write_introduction*): by means of a click he enters the on-line document editing application: the document *Introduction* is automatically opened and labeled "task: write introduction". While he is writing, he is notified about an email from the father of a schoolmate of Tom's saying that the school portal published news about a parent meeting on Friday afternoon. Alex realizes that it would be easier to be automatically notified about news concerning school activities. Since the school portal is registered to a public notification server, Alex enters the configuration page of his service cloud and subscribes to the news published by the portal on that server.

The rest of this chapter presents our proposal for supporting the kind of collaboration described in the above scenario. Specifically, we will present a survey of the related work, in various research fields, that are relevant to the management of Collaborative Service Clouds as described above (Section *Background and Related Work*). Then, we will describe our proposal (Section *A Framework for Managing Collaborative Service Clouds*), starting with a clarification of the main principles underlying its design; we will then describe the framework architecture, introduce the applications that implement its core functionality, and describe the steps required for the integration of external services, presenting some exploitation scenarios. Finally, in order to validate the framework, we will briefly present a proof-of-concept implementation of it. We will conclude the chapter (Section *Conclusions*) by discussing future work directions.

BACKGROUND AND RELATED WORK

The need for open environments supporting the integration of heterogeneous services and a flexible management of user collaboration has been advocated for in very different scenarios, involving both private and corporate users. For instance, as far as private users are concerned, (Grimes & Brush, 2008) reports the need of several working parents to integrate different Web calendars in order to holistically manage their home and work schedules. Moreover, Yahoo Pipes (Yahoo, 2011), and other similar mashup environments, demonstrate the interest of non-technical users towards a simple integration of external services in order to create their own apps. From the industrial point of view, the situation is somehow similar: as discussed in (Prinz et al., 2006), the chain production model is evolving towards dynamic collaborations among spontaneously assembled groups of people working together. Moreover, "collaborative tasks are often ill-structured at the outset, emerge in the course of the collaborative process, and need to respond flexibly to changing goals or situations" (Haake et al., 2010). Thus, besides traditional project management tools, which are suitable for the execution of stable, well-structured projects, there is a need for tools supporting lightweight user cooperation and flexible team management. A related approach is that of *open enterprise* (Corso & Mainetti, 2008), i.e. an enterprise model in which content and information exchange with external actors is constant and fruitful. This model, coupled with job globalization, and supported by the enhanced technological opportunities, raises the need for an increased flexibility, that requires workers to collaborate and communicate also "beyond home and office", in a mobility context (e.g., while traveling), and across the enterprise boundaries (Hislop & Axtell, 2007).

In the following, we will present the related work concerning such critical aspects of cloud service integration and Web-based user collaboration.

Integration of Heterogeneous Remote Services

As far as the provision of a unified environment offering different services is concerned, many "big actors", like Google (http://www.google.com), Amazon (http://www.amazon.com), Mozilla (http://www.mozilla.com), and Microsoft (http://home.live.com) provide solutions that claim to support such an integration. Google, for instance, provides a unified environment that, with a single login, enables users to access many communication and collaboration services (http://www.google.com/apps). Moreover, Google recently presented Google+ (Google, 2011d), a social platform fully integrated with many Google services (GoogleMail, GoogleDocuments (Google, 2011a), GoogleTalk (http://www.google.com/talk), Picasa (http://picasaweb.google.com), and so on), and supporting virtual video-rooms where people can meet on-line to chat or to see a YouTube (http://www.youtube.com) video all together. Although Google+ represents a key step in the direction of a tighter integration among different services, it suffers a lack of integration with heterogeneous applications. Specifically, the applications you can include in an environment like the one proposed by Google are those provided by the same provider; if you are used to use Skype for chatting, MySpace (http://www.myspace.com) to meet friends, organize your dinners or trips, and you have an email account on Hotmail (http://www.hotmail.com), you cannot really integrate such tools, and instead you have to abandon them in favor of GoogleMail, GoogleTalk, and Google+.

Human Collaboration Support: Task Management

In a unified environment supporting Collaborative Service Clouds, a major role should be played by a service that supports the user in handling tasks. As David Allen shows in his best-seller (Allen, 2003), the need to keep things-to-be-done clearly organized is very common, not only for work tasks, but also in our private life. Moreover, the need to have an integrated view of all the things-to-be-done, including all the interest spheres of your life, is a major one, since it enables you to handle them in a more efficient way.

Probably, the most common e-tools people use to try to achieve this goal are e-calendars. However, not all things-to-be-done correspond to a calendar event: there are tasks without a precise deadline (e.g., "as soon as you find some time, repair the roller shutter"), or things that have a deadline, but require quite a lot of work before (e.g., project reports with many contributions). This means that e-calendars are not enough to keep a structured view of your things-to-be-done. What is needed is a user-centered, easy to use way to organize and handle things-to-be-done, that should: (a) be closely related to an e-calendar; (b) be more structured than a "to-do list", but without the complexity of a workflow management tool; (c) offer the possibility of sharing tasks, for instance when they require a collaborative action; (d) support the user in the execution of the task itself; e.g., it should be integrated with the tools supporting the execution of activities related to the task.

Some recent applications seem to go in this direction. For example, Google Tasks (Google, 2011b) is a Web-based service, integrated with GoogleCalendar and GoogleMail, that enables users to quickly convert an email message into a task. However, the task resulting from it is a simple text note, with no structure and no integration with other applications. Another example is GTDInbox (InBox Foundry, 2010), a GoogleMail add-on for Firefox that enables you to directly transform your email messages into structured lists of things-to-be-done, linked to your GoogleCalendar. However, GTDInbox has some strong limitations: it works only for GoogleMail/GoogleCalendar; it is an add-on of Firefox, and thus imposes the usage of a specific mail client; it does not take into account other possible sources of tasks, like instant messages, messages coming from social networks, and so on.

There are some other interesting task managers, most of them based on the Allen (2003) methodology. The most popular are Things (Cultured Code, 2011), for Mac and iPhone, and DoIt (DoIt. im, 2011). Both applications offer little imposed structure and enable you to assign tasks to other people, with or without deadlines. These tools have nice functionality for personal tasks management, but their collaborative features are very poor. Moreover, they require a specific client to be installed on the user's device.

Moreover, task management is evolving towards the coordination of multi-application activities. For example, Teambox (http://teambox. com/) is an open source tool, offered in a SaaS modality, which enables the user to coordinate team members by assigning tasks, managing projects and collaborating by sharing ideas and documents. Furthermore, there are many suites dedicated to the enterprise, which support email, messaging services, chat, and video-conferencing; collaborative word processing; file sharing, shared calendars, and often also Web site development and maintenance. Microsoft Office 365 (www. microsoft.com/office365) and Google Apps for Business (http://www.google.com/apps/intl/en/ business/index.html) are two examples, just to mention the major players. These solutions are directed to an enterprise target and thus fail to support an integrated management of private and business activities. Moreover, they are closed platforms, which do not allow the integration of external, heterogeneous services.

Human Collaboration Support: Team and Project Management

Another interesting family of applications that should be taken into account are groupware and project management tools (see the AppAppeal Web site, http://www.appappeal.com, for a thor-

ough review of the hundreds of existing on-line project management and collaboration tools). These applications typically support workflow management and user collaboration and are often integrated with communication tools, such as email and chat. Most Web-based applications for workflow management, which also support user collaboration and document sharing (e.g., Alfresco Software, 2011 and BSCW (OrbiTeam Software GmbH & Co. KG, 2011)), are oriented to workflow and enterprise content management, other than supporting users in an integrated management of their business and everyday activities.

As far as groupware/project management tools are concerned, a lot of them require a client application to be installed on the user device (e.g. Collanos, 2011). Other groupware tools are completely Web-based: for instance, EGroupware (Stylite, 2011), Feng Office (2011), and Active-Collab (2011). However, their main drawbacks are the lack of an actual support to collaboration, their exclusive business/enterprise orientation, and their closed architectures: these tools offer many services (e.g. document sharing, email, etc.), but the user has to adopt the whole system, abandoning her favorite applications; moreover, no new service can be plugged in to enhance the system functionality in response to a user need.

A few recent environments, offered as cloud services, partially address the integration and collaboration support goals we pursue. For instance, Cohuman (http://www.cohuman.com) helps users to coordinate and plan their daily tasks to complete projects on time and it has a strong integration with Google Apps, which makes it interesting for a large amount of users, given their popularity. The application uses simple concepts and combines them to create a smart approach to task prioritization. The application works with the understanding that many people work on multiple projects each day. The system uses its task score system to prioritize each user's tasks from across all of their projects. The system automatically updates the order of tasks as changes are made to the task lists of others. Because task dependencies can track across different projects, prioritization can be based on information that is not visible to the person intended to perform a task. This prioritization is an emergent property of the network of tasks and their relations to each other based on blocking, as well as direct actions taken by users. However, Cohuman features are limited to project/task management instead of addressing a comprehensive management of users different aspects of life. Kohive (http://kohive.com), on the other hand, is an on-line desktop where users can collaborate with others. It is suitable for small businesses, students and groups with similar interests. It shows interesting features but at the moment it is a closed environment. Differently, OpenACircle (http://www.openacircle.com) is a Web2.0 collaboration tool, based on the notion of *circle*. A circle is a private, persistent team-space, in which members can video conference, chat, share files and virtually collaborate. The circle owner has the ability to delete the circle, any of its members or any shared content. Invited members can edit or delete content that they added to the circle. When the user "opens a circle" it provides an environment for the team to collaborate. Moreover, it supports a *library*, which is a collection of all the files that are a part of a circle. The user can upload and view pictures, documents, videos, articles and meeting notes in one central location. OpenACircle is focused again on business needs, but its main drawback is that it works only with some operating systems (i.e., Windows and Mac OS) and within some browsers (e.g., it does not work with Google Chrome).

A similar group management feature is offered by Google+, which enables the user to create private, user-defined and possibly overlapping groups of contacts, incidentally called "circles", as well: users can publish content available only to some circles, thus increasing their control on content visibility, and they can decide to see only the contents published by a limited number of

contacts. Both solutions enable users to quickly define their personal, thematic contact lists but they fail to support user collaboration in a general sense. In fact, each group (circle) exists only from the viewpoint of its creator (it is a reference to a list of accounts); thus, it cannot be used by the group members for managing the execution of new shared activities, but only for reacting to the threads of activity started from the circle creator.

A FRAMEWORK FOR MANAGING COLLABORATIVE SERVICE CLOUDS

We propose a framework supporting the management of an open, on-line environment based on the integration of heterogeneous services, that supports user collaboration and provides a user-centered management of business and private activities. Such a Web-based environment is what we called, in Section *Introduction*, a user-centered, Collaborative Service Cloud (CSC). The management of a CSC requires:

- An open architecture that provides a mechanism for handling a *shared context*, i.e. a "place" where applications can share and synchronize data and a mechanism to manage it.
- A set of *core applications*, belonging to the CSC infrastructure and implementing essential aspects of the functionality of the cloud, such as a smart support to the integrated organization of user tasks, and an integrated view and management of user collaborations and communications (thanks to workspace awareness, user groups definition support, and advanced calendar sharing features).

Before describing our work, we summarize the main principles that drove the design of our proposal:

- Our goal is *not* to propose yet another application replacing the communications and collaboration tools the user is used to (e.g., Mozilla Labs, 2010). In contrast, we present an open architecture, equipped with a set of core applications, and a methodology to integrate new services. Our architecture provides users with an integrated management of activities using their preferred tools.

- Our framework enabling Collaborative Service Clouds aims at improving the support to on-line collaboration among users, not only at the level of single applications (document sharing, on-line meetings, and so on), but also in the management of the cloud itself, i.e. the (shared) context in which these applications are used.

- Following the principles of SaaS and cloud computing (Creeger, 2009; Dikaiakos et al., 2009), our framework is "almost client-less" because it enables the user to access her service cloud from any device equipped with a standard Web browser, without the need of installing a specific client. To support this claim, imagine that Alex is at the airport, had his smart phone stolen and needs to connect to his Collaborative Service Cloud in order to modify the schedule of his activities for today. His laptop is at the hotel but in the airport there are some Internet kiosks available. If the system Alex has to connect to is completely Web-based, i.e. the services he needs are really "in the cloud", he will be able to easily reschedule his activities, from the airport.

- Our goal is to enable the user to access a unified and structured view of all her activities, that takes into account both job commitments and personal life plans. Obviously, the model does not force the user to adopt such an integration, but enables it. This unified management of business and private activities supports the future vision of a work not constrained in physical spaces (e.g. the office buildings), but that can be performed remotely, maybe without moving from home. Moreover, it supports the vision of an employee that sees her/his company as one of the possible sources of social relations, which can support work collaborations, but span over private communications and activities. This idea is supported by studies that demonstrate how the daily usage of social networks (and in particular of collaborative web-based tools, forums, and resource sharing tools) within the work environment could increase the efficiency of workers (Dynamic Markets, 2008; see also http://www.att.com/gen/press-room?pid=2815).

Infrastructure: Synchronization of Heterogeneous Applications in the Collaborative Service Cloud

The open architecture supporting the management of the shared context is based on SyncFr, a platform for the synchronization of applications based on the publish-and-subscribe model, and described in (Ardissono et al., 2009c). A *Coordination Middleware* (CM) manages the *Shared Context*, i.e. a shared dataspace (Papadopoulos & Arbab, 1998), and mediates all data exchanges between core applications in the CSC. The core applications interact with the CM through a publish-and-subscribe mechanism: when the application (autonomously or as a consequence of the user interaction) performs some significant action, it publishes an event on the Shared Context; moreover, the application is notified when an event is published that the application itself is subscribed to.

External applications (i.e. remote services, provided by different providers) can be integrated within the CSC by wrapping them with *adapters* that mediate the interaction between application and CM, or by exploiting a public publish-and-subscribe service. The details about the interoperability with external applications will be described in Section *Heterogeneous Service Integration* and in Section *Proof-of-Concept Implementation* we will provide some implementation details.

Cloud Functionality: The Role of Core Applications

While the synchronization of applications by means of the Shared Context and of the publish-and-subscribe interaction supports the basic synchronization of applications, i.e. their capability to receive information about the user, her working context, etc., the provision of a user-centered environment supporting user collaboration requires the management of the user's workspaces and collaboration groups. In the Collaborative Service Cloud, such functions are provided by four core applications, described in the following. In particular, the Collaborative Task Manager provides users with an enhanced tool to manage their activities and collaborations, and it plays an important role within the framework.

Notifications, Surveys, Groups, and Calendars

The *Notification&Survey Manager* supports the workspace awareness (Dourish & Bellotti, 1992; Gutwin et al., 1996). In particular, it filters notifications by providing a contextual organization of the awareness information, in order to reduce the disruptive effect of alerts on the user's attention; see (Ardissono et al., 2009a, 2009b, Ardissono & Bosio, 2011) for further details. Moreover, it manages an asynchronous awareness space which the user can browse to view the notifications, organized according to her activity contexts (Ardissono et al., 2011). The Notification&Survey Manager also provides the basic support to the *simple, subscription-based integration* of external applications, as we will describe in

Section *Heterogeneous Service Integration*. The Notification&Survey Manager is subscribed to all the event types the user agrees to be notified. It can exploit different mechanisms to send the notifications to the user, like email, instant messages, or a specific Web-based user interface (e.g. a customized GoogleGadget: http://code.google.com/intl/it/apis/gadgets). When the received event requires to perform a survey (e.g. a meeting to be scheduled) the Notification&Survey Manager proposes the user a standard survey and manages users answers: it collects them and publishes an event with the results of the survey (e.g., a confirmation event if everybody has agreed).

The *User&Group Manager* supports individual user registration and preference setting, as well as the definition of user groups at CSC level. By interacting with this application, a CSC user can define collaboration groups, e.g. people working with Alex in the *Wow* project, people from the photography association involved in the organization of the exhibition, his family environment for children care (including, for instance, his wife, grandparents, and the baby sitter), and so on (see Section *Introduction*). Collaboration groups represent interest *spheres*, i.e. "general topics" the user aims at sharing with friends and collaborators and are "inherited" in the integrated applications in order to support a unified management of groups within the CSC. Specifically, when the user performs an operation within the User&Group Manager (i.e., she modifies the composition of a group), an event (e.g. *GroupModified*) is published on the Shared Context and the CSC applications subscribed to that event type will be notified, thus automatically "inheriting" the performed change in the group definition. In this way, there is no need to manually propagate the group definitions in other applications.

The *Smart Shared Calendar Manager* accesses user e-calendars and finds the slots in which a set of users are available. For instance, if Alex aims at planning a meeting with his friends from the photography association, he can specify the involved people, possibly selecting a CSC group (like the *Berlin_exhibition_group* of our scenario), a range of dates and a description of the activity. The Smart Shared Calendar Manager will automatically intersect all their e-calendars and suggest Alex a set of possible dates. When Alex selects a date, a survey is automatically sent to the invited people (exploiting the publish-and-subscribe mechanism and the Notification&Survey Manager functionality). When a date with a sufficient number of positive answers is found, the meeting is automatically inserted in the involved users e-calendars. The Smart Shared Calendar Manager avoids the user to: (a) manually define (possibly long) lists of users to be invited by sharing user groups definitions; (b) manually check all involved user calendars to find a slot when (almost) everybody is available; (c) manually update the involved users calendars when the meeting has been fixed.

Collaborative Task Management

The CSC framework offers a *Collaborative Task Manager* that manages structured lists of tasks, which can be set up by the user, or originated from emails, instant messages, chats, and possibly other sources. The Collaborative Task Manager enables the user to share tasks, to partially order them in simplified workflow-like structures, and to link each task to the CSC applications needed to perform it. The ultimate goal of the Collaborative Task Manager is to support a unified management of tasks across applications and to support user collaboration.

In particular, the Collaborative Task Manager enables users to manage things-to-be-done as a simple list of notes, but offers the interested users the possibility of handling their activities in a more structured way. In fact, users can choose the level of complexity they prefer since the Collaborative Task Manager offers the possibility of handling both professional (possibly more complex) and personal (usually less structured) activities in an integrated, and thus more efficient, way.

The most important aspects of our proposal concerning task management and its collaborative features could be summarized in three features, described in the following, which characterize our approach with respect to other task management tools.

Structure. The main concept that enables us to provide a structure to tasks is the *activity frame*, which represents a partially ordered set of tasks, related to a single topic; it can be shared by a set of users and it offers a "container" for supporting collaboration on a set of related tasks. When a user organizes tasks within a frame, she can give them a temporal ordering (e.g. *"write_section1* must be performed before *write_section2"*), and she can state whether two (or more) tasks, A (e.g., *write_intro*) and B (e.g., *write_TOC*), that precede another task C (e.g., *write_section1*), must be performed both (i.e., they represent an AND split in workflow terms) or whether performing only one of them is enough for enabling C (i.e., they represent and OR split in workflow terms); see Figure 1. Moreover a hierarchical relation can hold between a task and a sub-frame: if Alex defined a frame, shared with the *International_Wow_group* (that includes people involved in the *Wow* project from all the partner organizations), to handle the preparation of a project report, he can delegate part of the work concerning a task within the frame (e.g., *write_section1*, assigned to Alex) by creating a new sub-frame ("child" of the *write_section1* task) containing the details of the management of *write_section1*.

Integration. An important aspect of the Collaborative Task Manager is the integration between tasks and CSC applications: each task can be linked to the applications needed to perform it and such applications are activated "within the task context"; e.g. their activation is labeled with the task name, they "inherit" the group of people the user wants to share the activity with. Moreover, the task context supports a structured management of workspace awareness: when the user performs some significant activity by interacting with a

Figure 1. Example of an activity frame for structuring tasks; source: (Ardissono et al., 2010)

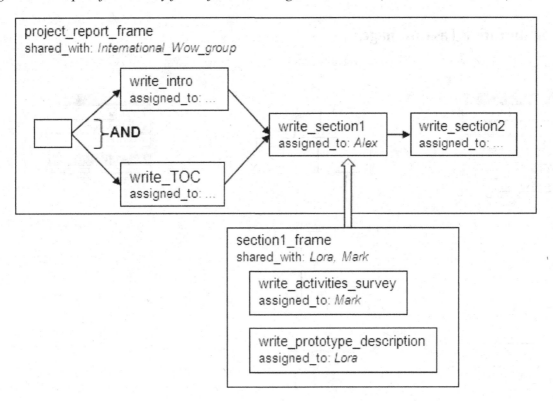

CSC application (e.g. updating a document) the event published in the Shared Context can be labeled with task information; in this way, the Notification&Survey Manager can organize the user's notifications (e.g., by grouping them) by task. Finally, the Collaborative Task Manager exploits the publish-and-subscribe mechanism to provide a run-time support to user activities. When a task A has been marked as completed by a user, an event (*taskDone[name=A,...]*) is published on the Shared Context. The Collaborative Task Manager instances of the involved users (which are subscribed to this type of events) receive the event and, as a consequence, they "enable" all those tasks having A as precondition (e.g., task B). Moreover, all the users sharing task A are notified that it has been completed, and all the assignees of task B are notified that B has been "enabled". The user interface of our current proof-of-concept prototype (see Section *Proof-of-Concept Imple-*

mentation), shown in Figure 2, takes this information into account by providing different visualization clues for open enabled tasks (names in green, e.g. "write TOC"), open but not enabled tasks (names in red, e.g. "write section 1/2"), and closed tasks (names in grey, e.g. "write INTRO").

Collaboration. The last aspect playing a major role within our Collaborative Task Manager is task sharing and collaboration. In contrast to most project management tools, that reflect a rigid hierarchical organization of enterprise work, we take into account the theories, based on SCUM and Extreme Programming, emerging in modern software development approaches (e.g., Beck, 2000), which favor a collaborative approach to project management suggesting shared team responsibility in project plan building. Thus, we introduced the concept of frame sharing. Frames, being "containers" of a small set of closely related tasks, represent the minimal meaningful unit

Figure 2. User interface of the Collaborative Task Manager of the proof-of-concept prototype

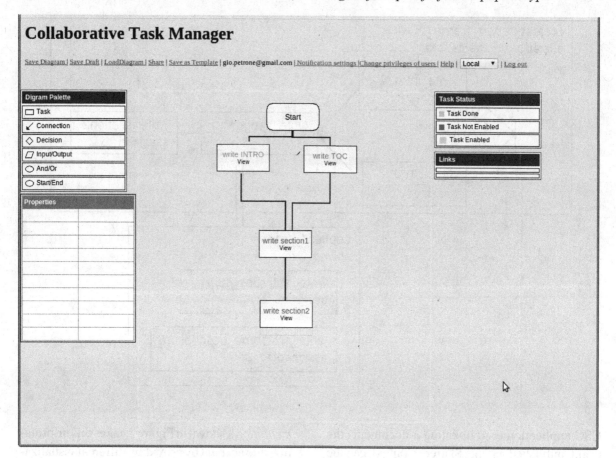

that can be shared and enable users to collaborate to the definition of tasks and task structure within a frame, in a "democratic" way, even though they are not assignees of those tasks. Referring to the scenario described before, Alex can open an interactive working session with the colleagues he wants to involve in the new sub-frame ("child" of *write_section1* task) in order to define the details of the task assignment and management (e.g., *write_activities_survey* and *write_prototype_description*; see Figure 1).

Heterogeneous Service Integration

As described in the Section *A Framework for Managing Collaborative Service Clouds*, the applications belonging to the CSC communicate by interacting with the Coordination Middleware (CM) through a publish-and-subscribe mechanism. The interoperability with remote services can be achieved by exploiting the CM used by core applications, or a public publish-and-subscribe (P&S) service, as described below. All applications that exploit the CM to communicate will be called "internal applications"; the term "external applications" will be used to refer to those remote services whose interoperability is guaranteed by a public P&S service.

Figure 3 depicts an example of a CSC that includes a *Coordination Middleware* (CM) and two Web-based *public P&S services*, on which external applications (*Remote Service 1/2/3*) publish information. In the figure, grey boxes represent internal applications (i.e., applications

Figure 3. Event-based interaction and open API invocation in a collaborative service cloud

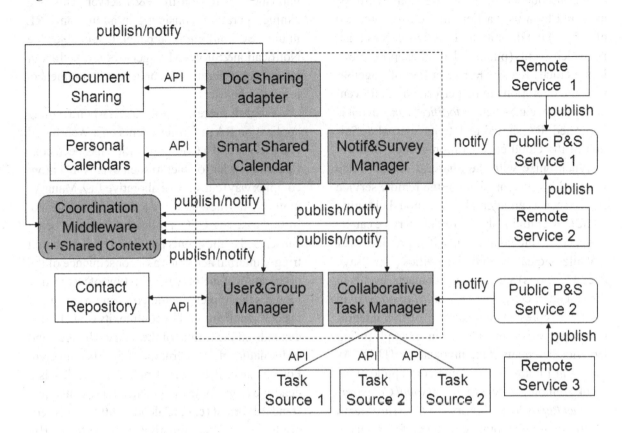

exploiting the CM) and white boxes represent external ones (i.e., remote services exploiting a public P&S server); arrows labeled as "publish" represent the publication of events on public P&S servers; arrows labeled as "notify" represent the event notification from public P&S servers to cloud applications (see below). Moreover, arrows labeled as "publish/notify" represent event flow between internal applications and the CM. Finally, arrows labeled as "API" that internal applications use the open API offered by remote external services.

Our framework enables two different possibilities to support interoperability with remote services: the *complete, API-based integration* transforms a remote service into an internal application and is represented in Figure 3 by the connection between the *Coordination Middleware* and the *Document Sharing* service (the role of the

Doc Sharing adapter is described below); the *simple, subscription-based integration* is supported by *Public P&S Services* and is represented by the connections between *Public P&S Service 1* and the *Notification&Survey Manager* (case (a)) and between *Public P&S Service 2* and the *Collaborative Task Manager* (case (b)).

Complete, API-Based Integration

This possibility represents the integration of a remote service within the cloud functionality. This type of integration requires the design and implementation of an adapter (e.g., the *Doc Sharing adapter* in Figure 3) for the remote application (RA), which includes: (a) the definition of the events to be published on the Shared Context, together with their relation with RA internal activities; e.g., a *ModifiedDocument* event is published as a consequence of a "save & exit" activity, triggered by a user action on the user interface of RA; (b) the definition of the subscription list, i.e. which events (published by other applications in the cloud) RA has to be notified of, together with the operations to be performed upon event notification; e.g., when a *ModifiedGroup* event is published, RA should update its internal definition of the cloud collaboration groups, in order to synchronize with the Shared Context. Once wrapped by such an adapter, the remote service can be seen as an internal application, belonging to the CSC. Obviously, the remote service can be wrapped by the adapter only if it provides API that allow to catch internal activities (e.g., "save & exit" actions) and data (e.g. contacts, events in the calendar, etc.). The *Doc Sharing adapter* in Figure 3 represents an example of this integration type: we implemented an example using GoogleDocuments API, invoked on HTTP. As shown in Figure 3, also some core applications invoke remote services (e.g., *Personal Calendar*, *Contact Repository*, *Task Source 1/2/3*) in order to exploit their collaborative features; for instance, GoogleCalendar, via open API, provides a direct access to personal calendar instances; Google-Contacts provides the User&Group Manager with a repository accessible through HTTP, that enables the User&Group Manager to share user and group information among applications within the cloud; GoogleMail, Twitter, or Facebook represent sources for new tasks handled by the Collaborative Task Manager.

Simple, Subscription-Based Integration

This possibility represents a light integration of a remote service within the cloud functionality.

In the simplest case, if a remote application RA is registered to a public Web-based publish-and-subscribe (P&S) server (e.g., Google, 2011c) on which it publishes information in the form of events, the user can subscribe her Notification&Survey Manager instance to events

published by RA on the P&S server. This is a simple operation, consisting in adding an URL in the CSC configuration panel, that enables the Notification&Survey Manager to be notified about events published by RA, and, as a consequence, to notify the user.

A slightly more complex case is when a remote application RA publishes events on a Web-based P&S server and a CSC application has been configured in order to react to such events in a more complex way (e.g. the Collaborative Task Manager might create a new task in reaction to some events). In this case the CSC application, besides being subscribed to the events published by RA, should trigger internal activities as a consequence of the notification of such events. This kind of integration requires the implementation of some code within the CSC application. The complexity of such code depends on the format of the data exchanged and on the nature of the internal activity to be triggered. For instance, if the event published by RA is a deadline (e.g., for tax payment), represented in a standard format (e.g., iCalendar: http://tools.ietf.org/html/rfc5545), the activity to be performed by the CSC application is to write the deadline on the user's personal e-calendar.

Discussion

Both the *complete, API-based integration* and case (b) of the *simple, subscription-based integration* include the definition of some reaction to events published on the CM or on a public Web-based P&S service; this implies that they have to handle the data these events are about, and face possible format mismatch issues. As we mentioned above, in some cases the use of a standard format, such as iCalendar, can solve the problem, but in other cases this is not possible or not enough. In order to face this issue, the current version of the CSC framework relies on ad-hoc conversion rules, an approach adopted also in some commercial middleware, such as WebSphere (Budinsky, 2004). In our framework, the conversion rules are stored in a declarative knowledge base, so that all the adapters that have to face the same conversion problem can exploit the same rules.

Before concluding this section, it is worth discussing the applicability of our approach. The *complete, API-based integration* (and partially also the most complex cases of the *simple, subscription-based integration*) requires the design, implementation and configuration of adapters. Thus, except for the simplest case (in which the interoperability consists in a rough notification to the user), our framework does not support a "plug and play" mechanism to integrate remote services in the CSC. This solution is thus appropriate for companies or organizations that act as "integration providers", by adopting the CSC framework and implementing the adapters for remote services; examples of such integration providers can be universities that aim at offering collaborative services to their students and staff members, or software houses that sell the customization of the CSC to small and medium-sized enterprises or to communities of interest (amateur photographers, sport fans, etc.). In particular, our framework could be exploited to support a *private enterprise cloud* within an Intranet (or Extranet), or a *public cloud*, open to the integration of external services (e.g., a municipality, aiming at offering an cloud services to its citizens). In the former case, the CM could be based on an enterprise middleware, including a publish-and-subscribe functionality, like GigaSpaces (Cohen, 2004); in the latter case, the cloud interoperability should be supported by a public Web-based publish-and-subscribe server, like Pubsubhubbub (Google, 2011c).

Proof-of-Concept Implementation

In this section we describe a proof-of-concept implementation of the most important aspects of the CSC we developed in order to demonstrate the feasibility of our approach.

We implemented a Coordination Middleware based on GigaSpaces (Cohen, 2004), a robust and scalable platform offering direct support for many types of data transfer, coordination primitives, server side event filtering. Moreover, we exploited Pubsubhubbub (Google, 2011c), for implementing a public Web-based publish-and-subscribe service that supports the communication with external remote services (see the *simple, subscription-based integration* described in Section *Heterogeneous Service Integration*). In particular, the CSC applications subscribe to Pubsubhubbub for event notifications: when an application publishes an event on the server, it notifies the subscribers by means of an HTTP POST request, through a callback mechanism called Web Hook (http://wiki.Webhooks.org).

In order to demonstrate the case of the *complete, API-based integration* (see Section *Heterogeneous Service Integration*), we developed a sample adapter for GoogleDocuments (Google, 2011a), that exploits GoogleDocuments open API (http://code.google.com/apis/documents/overview.html) to publish events on the Shared Context (e.g. a *ModifiedDocument* event is published when the user performs a "save & exit" action) and is subscribed, and thus reacts, to events published on the CM by other CSC applications (e.g., when a *ModifiedGroup* event is published by the Group Manager, the GoogleDocuments adapter updates its corresponding list of collaborators).

Concerning the single sign-on issue, we tested a solution based on OpenID (Tsyrklevich & Tsyrklevich, 2007), enabling users to access the CSC by a single login on their Google account.

Finally, we implemented prototypes of the CSC core applications as Web applications, by exploiting the Google Web Toolkit (http://code.google.com/webtoolkit). In particular:

- The Notification&Survey Manager prototype can exploit different instant messaging applications to notify users, thanks to the open API offered by the following services: GoogleTalk and Facebook Chat, both through the Smack library (http://

www.igniterealtime.org/projects/smack); Msn Messenger, through the JML library (http://sourceforge.net/apps/trac/java-jml); Yahoo, through the free Java Openymsg (http://sourceforge.net/projects/ope-nymsg); Twitter, through the Jtwitter library (http://www.winterwell.com/soft-ware/jtwitter.php); Skype, through its Java API.

- The User&Group Manager prototype exploits GoogleContacts API (http://code. google.com/apis/contacts/) to manage shared collaboration groups.

- The Smart Shared Calendar Manager can interact with different calendar formats; in particular we tested a solution that exploits GoogleCalendar API (http://code.google. com/apis/calendar).

- The first Collaborative Task Manager prototype was integrated with GoogleWave, thanks to the Wave APIs (http://code. google.com/apis/wave), that allows to manage events of creation, modification and deletion of data generated by the users participating to the wave. This solutions provides the Collaborative Task Manager with a mechanism for handling: (a) collab-orative editing (definition, structures, as-signments, properties, etc.) of tasks within a frame; (b) presence awareness; (c) syn-chronous chat-like tool available; (d) no-tification of changes to people sharing the modified activity frame. Since Google dis-continued development of GoogleWave, we are working on a different implemen-tation for the mentioned features of our Collaborative Task Manager, that experi-ments different solutions. As far as the ex-ternal sources of tasks are concerned, we exploited the same API mentioned above to retrieve messages from instant messaging applications. Moreover, Facebook events and email messages are retrieved thanks to Facebook API (http://developers.face-

book.com) and JavaMail API (http://java. sun.com/products/javamail) respectively.

CONCLUSION

This chapter presented a service integration model supporting the development of user-centered service clouds that enhance user collaboration in the Web, using Web 2.0 tools and services. The core element of our model is a Shared Context, exploited to synchronize heterogeneous services, and the management of the service synchroniza-tion by means of loosely-coupled interactions, based on the publish-and-subscribe protocol. In this chapter we presented a framework support-ing Collaborative Service Clouds, which offer, on top of the integration of heterogeneous services, a few core applications helping users to establish collaborations with their partners and to operate in shared activities using distributed workspaces and flexible task management functions. Collaborative Service Clouds support alternative models for inte-grating remote services, such as those based on the development of adapters enabling the applications to interact with the Shared Context of the cloud, and other, looser types of integration, based on the exploitation of public publish-and-subscribe servers. These integration models are suitable for different exploitation scenarios, discussed in the paper. Moreover, concerning the functionality offered to the end-user, Collaborative Service Clouds offer a Web-based user interface which does not require the installation on the user's device of any particular client, so that it can be accessed from any computer, not only from the personal one of the user, and thus from any place offering an Internet access.

We conclude by mentioning two main is-sues not addressed in this chapter that represent interesting possible enhancement directions of our work. The first issue concerns the *semantic* interoperability of data exchanged in the Shared Context. The current approach, relying on ad-hoc conversion rules, should be replaced by some support for automatic format conversion and for the interpretation of the meaning of data. The second aspect concerns privacy and security issues: safe mechanisms supporting data exchange and sharing have to be applied in order to preserve the privacy of user data. The implementation of such mechanisms represents a mandatory enhancement of our framework.

REFERENCES

ActiveCollab. (2011). *Project collaboration software*. Retrieved September 8, 2011, from http://www.activecollab.com

Alfresco Software. (2011). *Alfresco*. Retrieved September 8, 2011, from http://www.alfresco.com

Allen, D. (2003). *Getting things done: The art of stress-free productivity*. Harlow, UK: Penguin.

Ardissono, L., & Bosio, G. (2011). Context-dependent awareness support in open collaboration environments. *User Modeling and User-Adapted Interaction - The Journal of Personalization Research*, in press.

Ardissono, L., Bosio, G., Goy, A., & Petrone, G. (2009a). Context-aware notification management in an integrated collaborative environment. In A. Dattolo, C. Tasso, R. Farzan, S. Kleanthous, D. Bueno Vallejo, & J. Vassileva (Eds.), *International Workshop on Adaptation and Personalization for Web 2.0* (pp. 21–30). Aachen, Germany: CEUR.

Ardissono, L., Bosio, G., Goy, A., Petrone, G., & Segnan, M. (2009b). Managing context-dependent workspace awareness in an e-collaboration environment. In S. Yamada & T. Murata (Eds.), *International Workshop on Intelligent Web Interaction* (pp. 42–45). New York, NY: IEEE Press.

Ardissono, L., Bosio, G., Goy, A., Petrone, G., Segnan, M., & Torretta, F. (2010). Collaborative service clouds. *International Journal of Information Technology and Web Engineering*, 5(4), 23–39. doi:10.4018/jitwe.2010100102

Ardissono, L., Bosio, G., & Segnan, M. (2011). An activity awareness visualization approach supporting context resumption in collaboration environments. In A. Paramythis, L. Lau, S. Demetriadis, M. Tzagarakis, & S. Kleanthous (Eds.), *International Workshop on Adaptive Support Team Collaboration 2011* (pp. 15-25). Aachen, Germany: CEUR.

Ardissono, L., Goy, A., Petrone, G., & Segnan, M. (2009c). SynCFr: Synchronization collaboration framework. In M. Perry, H. Sasaki, M. Ehmann, G. Otiz Bellot, & O. Dini (Eds.), *5th Conference on Internet and Web Applications and Services* (pp. 18–23). New York, NY: IEEE Press.

Beck, K. (2000). *Extreme programming explained*. Reading, MA: Addison-Wesley.

Budinsky, F., DeCandio, G., Earle, R., Francis, T., Jones, J., & Li, J. (2004). WebSphere Studio overview. *IBM Systems Journal*, 43(2), 384–419. doi:10.1147/sj.432.0384

Cohen, U. (2004). *Inside GigaSpaces XAP - Technical overview and value proposition*. White Paper. New York, NY: GigaSpace Technologies Ltd.

Collanos. (2011). *Products overview: Team enabling professionals*. Retrieved September 8, 2011, from http://www.collanos.com/en/products

Corso, M., & Mainetti, S. (2008). *Enterprise 2.0: La rivoluzione che viene dal web. Technical Report*. Milano, Italy: Politecnico di Milano, School of Management.

Creeger, M. (2009). CTO roundtable: Cloud computing. *Communications of the ACM*, 52(8), 50–65. doi:10.1145/1536616.1536633

Cultured Code. (2011). *Things Mac*. Retrieved September 8, 2011, from http://culturedcode.com/things

Dikaiakos, M. D., Pallis, G., Katsaros, D., Mehra, P., & Vakali, A. (2009). Cloud computing. Distributed internet computing for IT and scientific research. *IEEE Internet Computing, 13*(5), 10–13. doi:10.1109/MIC.2009.103

DoIt.im. (2011). *Smart way to manage tasks*. Retrieved September 8, 2011, from http://www.doit.im

Dourish, P., & Bellotti, V. (1992). Awareness and coordination in shared workspaces. In M. Mantel & R. Baecker (Eds.), *1992 ACM Conference on Computer-Supported Cooperative Work* (pp. 107–114). New York, NY: ACM Press.

Dynamic Markets. (2008). *Corporate social networking in Europe. Independent market research report. Commissioned by AT&T (Technical Report)*. Abergavenny, UK: Dynamic Markets.

Feng Office. (2011). *Project management - Easy, powerful collaborative project management*. Retrieved September 8, 2011, from http://www.fengoffice.com

Google. (2011a). *Google Documents: Create documents, spreadsheets and presentations online*. Retrieved September 8, 2011, from http://www.google.com/google-d-s/tour1.html.

Google. (2011b). *Google Tasks: Keep track of what you need to do*. Retrieved September 8, 2011, from http://mail.google.com/mail/help/tasks

Google. (2011c). *Pubsubhubbub*. Retrieved September 8, 2011, from http://code.google.com/p/pubsubhubbub

Google. (2011d). *Google+*. Retrieved August 18, 2011, from http://www.google.com/+/learnmore/

Grimes, A., & Brush, A. (2008). Life scheduling to support multiple social roles. In M. Burnett, M. F. Costabile, T. Catarci, B. de Ruyter, D. Tan, M. Czerwinski, & A. Lund (Eds.), *26th Annual CHI Conference on Human Factors in Computing Systems* (pp. 821–824), New York, NY: ACM Press.

Gutwin, C., Greenberg, S., & Roseman, M. (1996). Workspace awareness in real-time distributed groupware: Framework, widgets, and evaluation. In Sasse, M. A., Cunningham, R. J., & Winder, R. L. (Eds.), *HCI 96 People and Computers XI* (pp. 281–298). Berlin, Germany: Springer-Verlag.

Haake, J., Hussein, T., Joop, B., Lukosch, S., Veiel, D., & Ziegler, J. (2010). Modeling and exploiting context for adaptive collaboration. *International Journal of Cooperative Information Systems, 19*(1-2), 71–120. doi:10.1142/S0218843010002115

Hislop, D., & Axtell, C. (2007). The neglect of spatial mobility in contemporary studies of work: The case of telework. *New Technology, Work and Employment, 22*(1), 34–51. doi:10.1111/j.1468-005X.2007.00182.x

InBox Foundry. (2010). *GTDInbox*. Retrieved March 22, 2010, from http://gtdinbox.com/better_inbox.htm

Mozilla Labs. (2010). *RainDrop*. Retrieved March 22, 2010, from http://mozillalabs.com/raindrop

Nielsen, J. (2009). *Social networking on intranets*. Retrieved March 22, 2010, from http://www.useit.com/alertbox/social-intranet-features.html

O'Reilly, T. (2007). What is Web 2.0: Design patterns and business models for the next generation of software. *Communications & Strategies, 1*(65), 17–37.

OrbiTeam Software GmbH & Co. KG. (2011). *Be smart - cooperative, worldwide*. Retrieved August 18, 2011, from http://public.bscw.de/en/about.html

Papadopoulos, G., & Arbab, F. (1998). Coordination models and languages. In Zelkowitz, M. (Ed.), *Advances in computers* (pp. 329–400). San Diego, CA: Academic Press.

Prinz, W., Löh, H., Pallot, M., Schaffers, H., Skarmeta, A., & Decker, S. (2006). ECOSPACE - Towards an integrated collaboration space for eProfessionals. In *2nd International Conference on Collaborative Computing: Networking, Applications and Worksharing* (pp. 39–45). New York, NY: IEEE Press.

Stylite. (2011). *eGroupware*. Retrieved September 8, 2011, from http://www.egroupware.org

Tsyrklevich, E., & Tsyrklevich, V. (2007). *Single sign-on for the Internet: A security story*. White Paper. Seattle, WA: Black Hat.

Turner, M., Budgen, D., & Brereton, P. (2003). Turning software into a service. *Communications & Strategies, 36*(10), 38–44.

Yahoo. (2011). *Pipes*. Retrieved August 18, 2011, from http://pipes.yahoo.com/pipes/

Chapter 2
Information Acquisition and Recall in Location–Aware and Search Engine Retrieval Systems

Sorin A. Matei
Purdue University, USA

Lance W. Madsen
Innervision Advanced Medical Imaging, USA

Robert Bruno
Purdue University, USA

ABSTRACT

This paper examines the potential cognitive impact of location aware information systems compared to that of search engines using a dual coding and conjoint retention theoretical framework. Supported by virtual reality or mobile devices, location aware systems deliver information that is relevant for a specific location. Research questions and hypotheses formulated under the assumption that location aware systems are better prepared to contextualize and make information memorable are explored using a planned comparison repeated measures 3 (2 treatment; 1 control) x 3 (pre-test, post-test, one week post-test) design. The results indicate that information acquisition in location-aware systems is just as powerful as that facilitated by search engines and that information recall (after 1 week) of facts is superior when using location-aware systems. The findings reinforce and extend dual coding theory suggesting that spatial and three-dimensional indexing can be one of the channels used in indexing and recalling information. The results also indicate that location-aware applications are a promising technology for distributing information in general and for learning in particular.

DOI: 10.4018/978-1-4666-0023-2.ch002

INTRODUCTION

New communication technologies capable of delivering information when and where we need it have emerged over the past decade (Yang, Okamoto, & Teng, 2008). These tools promise a radical shift in learning strategies by overcoming the current limitations imposed by search engines (Weiler, 2005). Location aware and augmented reality applications powered by mobile devices, such as Wikitude (http://wikitude.org), Layar (http://layar.com), Ubimark (http://ubimark.com), or Junaio (http://metaio.com) will one day provide details about any and all aspects of our physical and social environment (Sandor, Kitahara, Reitmayr, Feiner, & Ohta, 2009). Yet, despite the progress made, more needs to be done until our mobile phones would be able to answer in direct, timely, and accurate manner the question "What is that building with the tall red spire by the grocery store?" A great deal of contextual information would need to be added to the location aware systems, while mobile devices would need to be vastly improved in terms of usability, spatial sensitivity, and ability to discern which aspects of physical reality are of interest to the user.

Until then, researchers should make a case for the relevance and superiority of such technologies. Specifically, they need to provide a convincing answer to the question if location-aware delivery of information is more likely to foster learning than traditional, keyword-driven retrieval of information. Some researchers have already investigated some aspects of location aware learning and information retrieval, yet much more needs to be done. From the earlier studies on how location-aware systems can suggest points of interest in buildings, such as printers, elevators, or vending machines (Koo, Rosenberg, Chan, & Lee, 2003), we have advanced to massive, city, region, or worldwide projects or studies that focus on the role of space in delivering information (Anand, Harrington, & Agostinho, 2008; Armstrong & Bennett, 2005; Borriello, Chalmers, Lamarca, & Nixon, 2005;

Matei, Miller, Arns, Rauh, Hartman, & Bruno, 2007). In fact, an incipient and solid research program is budding out in the field of location aware information diffusion (Anand et al., 2008; Barbosa, Hahn, Rabello, & Barbosa, 2008; Yang, Okamoto, & Tseng, 2008)

Yet researchers interested in studying contextually-defined information delivery, especially information acquired by visitors and newcomers (Abowd et al., 1997), have primarily focused their attention on device usability, infrastructure architecture, and information delivery (De Jong, Specht, & Koper, 2008) rather than on understanding how location-aware systems can facilitate deeper cognitive processes. A search on Google Scholar for business, engineering, and social sciences papers that contain "location aware" in their title retrieved for 2009-2010 63 articles (http://bit.ly/locaware), of which only one dealt with spatial learning in an explicit way (Kim et al., 2009).

The present paper is an attempt to address this knowledge gap by providing answers to some basic questions related to how humans might learn in location-aware contexts. We are interested to learn not only if location-aware applications are feasible (Armstrong & Bennett, 2005; Borriello et al., 2005), but also whether or not these location-aware systems actually increase a person's ability to remember the information they receive in a location-aware context. Furthermore, the paper strives to bridge the gap between traditional research on spatial cognition with recent research related to the learning advantages of location-aware technology.

Previous research suggests that knowledge acquisition and retention are significantly improved when information is presented in a location-aware delivery system (Matei, Madsen, Arns, Bertoline, & Davidson, 2005; Mayer & Anderson, 1991). The present paper extends this line of research by comparing location-aware communication systems with other methods of information delivery. Consistent with a number of spatio-cognitive theories, it is anticipated that those who use a location-aware communication system might recall information better than those who learn through comparable technologies, such as search engines.

The basic assumption is that location-aware systems present three specific advantages over other, more traditional, methods of information delivery: 1) enhanced capacity to organize information focused on a specific physical space, 2) superior ability to convey this information while the user is in this specific physical space, and 3) unique ability to simultaneously deliver information through a number of channels (Anand et al., 2008; Barbosa et al., 2008). Of these three advantages the current paper focuses on the third, which will be explored in view of dual coding theory (Paivio, 1990) and the conjoint retention hypothesis (Kulhavy, Lee, & Caterino, 1985). According to dual coding theory, location-aware information should be easier to remember because it activates two cognitive systems simultaneously: the verbal and the imagery system. The interaction of these two systems enhances the ability to remember both types of information. The conjoint retention hypothesis suggests that the use of internalized geographical maps to organize information facilitates recall. Both theories are relevant for studying the impact of location-aware communication systems on learning since such systems facilitate organization of information as three-dimensional mental maps modeled after a specific physical environment.

Studying location-aware communication systems and their unique capacity to take advantage of the spatial dimension of human learning requires a comparative perspective. Common communication systems used today for searching and retrieving information are keyword-driven search engines, which can help individuals find relevant information by entering keywords about specific topics. However, the use of search engines might have more limited effects on learning and recall than location-aware systems. Search engines take only limited advantage of dual coding (they

usually retrieve and deliver information in textual format), and are in very small measure, if at all, influenced by the processes predicated by the conjoint retention hypothesis, which proposes that information organized around a spatial dimension is easier to remember. Search results are in most cases delivered as lists of links generated using abstract "relevance" criteria that typically have no spatial or visual implications. By comparing location-aware applications to the use of search engines it can be determined if the visual and spatial component inherent to location-aware applications provides an advantage in learning and retaining information compared with an existing communication technology.

A discussion of the literature about the theoretical implications of location-aware communication systems will synthesize these concerns into a number of research questions. They will be tested and validated by a multi-condition experimental design study in which information about a collection of spatial locations (landmarks situated on a university campus) was delivered with two treatment conditions: a simulation of a location-aware system and a web- based search engine.

LITERATURE REVIEW

Mental Mapping

The relationship between mental maps, spatial cognition, and memory is deeply intertwined. We create mental maps to arrange in space physical objects, which in turn help us to remember information about these objects. In view of this, mental mapping can be defined as the cognitive ability to make a mental abstraction of a physical environment in which information is organized, stored, and from which it can be later recalled (Dodds, 1982). This process is similar to those identified by William James (1890) and later developed by Piaget (Piaget & Inhelder, 1956), who found that the human mind has a tendency to search for and recognize patterns. We find patterns in how objects are physically situated and the recognition of these patterns allows us to remember their spatial orientation and information associated with them. Our recognition of these patterns is not discretely remembered as individual pieces of information, rather it is usually integrated into a holistic pattern, giving memories meaning and permanence (MacEachren, 1992). Radvansky and Copeland (2000) have found, for example, that recall is more likely to occur in situations where the various components of information to be remembered are in a functional (and often spatial) relationship with each other.

One way we use mental mapping is through the use of landmarks (Downs & Stea, 1977; Golledge & Stimson, 1997; Gould, 1975), which serve as anchor points for mental maps. Urban planner Kevin Lynch argued that certain urban space structural elements, such as landmarks, paths, and boundaries, can be used to facilitate the mental mapping and subjective understanding of urban environments (Lynch, 1960). There is also some evidence that we depend mostly on landmarks in order to navigate whenever they are available (Foo, Warren, Duchon, & Tarr, 2005). We also use landmarks to create personalized spatial information about where we are physically located; our memory infers the meaning and context of that location from architectural and environmental details (Parker, 1997).

Lloyd and Heivly (1987) have shown that landmarks, especially those closer to one's residence, have an inordinately large "gravitational pull" on our knowledge. The researchers found that key reference points distort the cognitive maps in their direction, so that landmarks farther afield appear to be closer than they are, especially when they are situated on a main transportation axis. In a related study, Lloyd (1994) also emphasized that spatial prototypes, which are organized around various patterns, are crucial in anchoring information in the human mind.

Mounin (1980) studied a cross-cultural sample of residents living in large cities in order to determine how prevalent landmarks are in socio-spatial orientation. He found that even people living in cities with numbered street grids (such as New York City or Brazil) preferred giving directions using landmarks and neighborhood names, rather than utilizing the more abstract grid convention. More importantly, Tuan (1975) has demonstrated that landmark anchored mental maps are rarely used as only simple navigational aids. They are most often used as mnemonic devices, allowing quick identification of a body of knowledge and constructing relationships with neighboring categories and ideas. This means that something like a tall water tower may function well as a navigational aid because it is a highly visible and recognizable landmark. However, the water tower itself may hold meaning due to stories involving the structure which can shape the perception of other landmarks around it. For example, if someone had died in an accident involving that water tower then people might have the impression that it is a dangerous location. Such perception might further lead to the idea that buildings near the water tower should also be avoided because of their close proximity to this perceived dangerous location. Clearly, landmarks can hold far more meaning than just navigational information.

Dual Coding Theory and Conjoint Retention Hypothesis

Spatial orientation, in general, and its use for mnemonic purposes, in particular, seems to be based on one's own life experience. We form mental representations about the things we personally find most important. This follows the classical distinction between perceptual and representational space, initially proposed by Kant and later developed by Piaget (Piaget & Inhelder, 1956).

Our ability to construct and use perceptual spatial representations is influenced by a number of information-acquisition mechanisms. Of these mechanisms the dual coding theory and the conjoint retention hypothesis are most relevant in the context of this research. Dual coding theory suggests that there are two independent but inter-connected cognitive systems that process and store information: an imagery system and a linguistic processing system (Paivio, 1990). The imagery processing system has the ability to simultaneously acquire, store, recall, and interpret multiple por-tions of a mental image. It is far more flexible than the linguistic processing system, which processes verbal information sequentially (Vekiri, 2002). When the imagery system and verbal system are used simultaneously, the flexibility of the imagery processing system can be leveraged for greater accuracy in recalling verbal information. Dual coding theory suggests that when an object or concept is explored both verbally and visually, the information about it will be "dually-coded." This means that visual cues can be used for recalling linguistic information and vice-versa. Moreover, the recall of either piece of information, visual or linguistic, will be better since more anchoring points for the recall mechanism are present. For example, if one were to see a picture of a building while simultaneously reading a brief historical summary, according to dual coding theory the individual would be able to later recall the histori-cal summary more easily than if he or she were presented with just the historical summary alone. Furthermore, when the individual is presented with only the picture of the building the historical verbal information will be more readily recalled.

Dual coding theory is particularly relevant for learning processes. Gellevij, Meij, Jong, and Pieters (2002) compared the effectiveness of two different presentational formats employed in learning how to use a complex computer program. One version of the instructional material used text and pictures, whereas the other version used text only. The use of the multimodal manual was found to lead to better performance, created a stronger mental model of the computer program, improved identification of program elements, and led to less training time than the text-only version of the instructional material. Other researchers (Mayer & Anderson, 1991) found similar results when comparing instructional media that included the use of words with pictorial animations. If words are simultaneously presented with a visual stimu-lus, then both pieces of information are easier to recall, ostensibly due to the mechanisms described by dual coding theory.

The conjoint retention hypothesis draws on dual coding theory, but its focus is directed at explaining the use of geographical (spatial) maps for storing and remembering textual information (Kulhavy et al., 1985). In essence, the hypothesis predicts that memorizing conceptual information is enhanced with simultaneous exposure to the spatial organization (map) of the information to be memorized. The theoretical mechanism behind the hypothesis is that the geographic map is acquired first, which creates a framework to store and fix-ate the verbal/conceptual information (Verdi & Kulhavy, 2002).

For example, in one study students with learning disabilities were asked to memorize a number of battles and relevant verbal/conceptual information related to each. The students were divided into two groups. In one group, learning the verbal/conceptual information was facilitated by presenting the information on a map (Brigham, Scruggs, & Mastropieri, 1995). The other group was exposed only to the verbal information. The students in the map group were better able to recall the names and locations of the battles than the students that were not exposed to the map. In a similar study, students who used maps with clear feature markers recalled information (sentences) about the events/places depicted on the map bet-ter than students who used maps without feature markers (Kulhavy, Caterino, & Melchiori, 1989). These insights were replicated by Webb, Thornton, Hancock, and McCarthy (1992) who found that verbal information was recalled better when it was spatially located and when the location was emphasized by icons. In addition, subjects were better at reconstructing features and locations on a map, effectively illustrating the link between spatial and verbal storage and retrieval (Webb & Saltz, 1994).

If two-dimensional maps can be used to store and retrieve verbal information, it is not inconceivable to extend this hypothesis to a three-dimensional situation. In addition, due to their perceptual dimension, three-dimensional maps derived from the environment we navigate every day will be particularly suitable for storing information. Such maps work more efficiently as information managers because of the personal and experiential meaning of the information stored in them.

This means that in a location-aware information context, where information is directly embedded within a three-dimensional map made of landmarks, users will map, store, and retrieve information about their environment more efficiently. The spatial dimension of the learning process will allow users to more effectively dually code and conjointly retain information, which should further facilitate its subsequent retrieval. In brief, dual coding and conjoint retention processes are particularly adept at explaining how information is acquired, stored, and recalled in a location-aware context.

Search Engines

Gauging how effective location-aware systems are in shaping learning and improving retention of information requires comparing such systems with other common methods of learning through communication technology. Search engine technology is identified by many as the central tool for retrieving information from the Internet (Weiler, 2005) and has become, especially in formal educational settings, an indispensable learning tool (Gerber & Shuell, 1998; Peterson & Merino, 2003).

Internet browsing and searching methods geared toward learning are constantly being evaluated and improved (Chen, Houston, Sewell, & Schatz, 1998; Mostafa, 2005). Yet, despite intrinsic advantages, especially ability to sift through vast amounts of data, search engines lack the ability to contextualize information spatially, or those that have this capacity are still in the early stages of development. Furthermore, the emphasis is on querying simple, preset information associated with a place, rather than more complex semantic-spatial associations. Although all search engines use relevance criteria, these are limited to keyword matching. Usually no context relationships are provided and even remoter is the presence of spatial contextual information. While information may be spatially organized on individual web pages, information across various web pages has no inherent spatial organization because these pages share the same physical space -- a computer monitor -- which is dreadfully inadequate for massive spatial representations.

In conclusion, comparing location-aware applications with web search engines can help elucidate if the spatial component inherent to location-aware applications provides a net advantage in learning and retaining information.

RESEARCH QUESTIONS AND HYPOTHESES

Since the knowledge delivered by location-aware communication systems has a strong spatial component, and since spatiality enhances acquiring and recalling information, it is possible that information acquired in a location-aware environment will be comparatively more easily learned and recalled.

Landmarks in a location-aware system are "soaked" with information and act like "mnemonic anchors" in our spatial cognitive maps. Learning in a location-aware communication environment will be facilitated by the cognitive processes described by the dual coding theory and by the conjoint retention hypothesis. Such learning will take place along a number of hypothesized dimensions, however, which vary with specific types of information content, such as factual, narrative, spatial, and landmark-specific (name and location).

Our study focuses on these four types of content, operationalizing its response variables into specific measurable indicators of information acquisition and recall for each of these dimensions. Indicators refer to subjects' ability to acquire and recall factual, narrative, location and name descriptors of specific landmarks. This information typology was constructed in light of the above reviewed literature, which makes important distinctions between narrative, factual, and geo-spatial information.

Our overall expectation was that location-aware systems would: 1) be an adequate means of information acquisition, 2) foster information recall, and most importantly that it would 3) lead to better information acquisition and recall than other technologies, such as search engines. To explore these theoretically-derived expectations we developed the following research questions and hypotheses, which will be tested through an experimental research design.

Research questions are organized into two clusters, which address the ability of location-aware systems to foster information acquisition and recall. Operationally, these questions will be addressed via within-subjects measures and tests.

RQ1

RQ1a: Does use of location-aware systems lead to context-defined landmark name information acquisition?

RQ1b: Does use of location-aware systems lead to context-defined landmark location information acquisition?

RQ1a: Does use of location-aware systems lead to context-defined landmark fact information acquisition?

RQ1a: Does use of location-aware systems lead to context-defined landmark story (narrative) information acquisition?

RQ2

RQ2a: Does use of location-aware systems lead to landmark name recall after longer periods of time (one week)?

RQ2b: Does use of location-aware systems lead to landmark location recall after longer periods of time (one week)?

RQ2c: Does use of location-aware systems lead to landmark factual recall after longer periods of time (one week)?

RQ2d: Does use of location-aware systems lead to landmark story (narrative) information recall after longer periods of time (one week)?

The study core hypotheses refer to expected positive and superior impact of the location-aware system, compared to search engines, on information acquisition and recall along the four information dimensions described above. Operationally, the hypotheses are tested via between-subjects comparisons.

H1a: Location-aware systems are more effective at facilitating landmark name acquisition and recall than search engine based systems.

H1b: Location-aware systems are more effective at facilitating landmark location information acquisition and recall than search engine based systems.

H1c: Location-aware systems are more effective at facilitating landmark factual information acquisition and recall than search engine based systems.

H1d: Location-aware systems are more effective at facilitating landmark story (narrative) information acquisition and recall than search engine based systems.

METHOD

Informational Places

In location-aware systems specific spatial points, arrays or containers are seeded with information. These locations can be as large as cities or merely single rooms in a building. While it is technically possible to create location-aware devices in the real world, testing them is costly and difficult and introduces many uncontrollable variables that would complicate practical research procedures. This research was conducted using a state-of-the-art virtual reality system which made possible precise control of environmental variables and simulated the functionality of a location-aware system (Davis, Scott, Pair, Hodges, & Oliverio, 1999).

A simulated environment of the Purdue University campus was utilized in this experiment, created by a FLEX-driven system and presented through a Virtual Reality Theater (FakeSpace Systems, 2005). The FLEX features three ten-foot by eight-foot panels for rear projection of large-scale three-dimensional images. These panels are arranged in a U-shaped configuration, surrounding the user on three sides. The landscape images are projected on each of the three panels and as well as one on the floor. The images projected on the panels are seamless, offering a panoramic, landscaped perspective of the simulated environment. Each panel has two projectors behind it that project an image slightly offset from one another, allowing the user to see the images stereoscopically (giving the image a simulated sense of depth) through the active LCD shutter glasses. These glasses also serve as movement trackers, allowing the projected image to be continuously adjusted to the physical perspective of the user. As the user moves his or her body or head while standing in a certain location the images on the panels and on the floor change, creating the illusion of movement through the virtual landscape.

The Virtual Reality Theater is equipped with an Intersense IS-900 tracking system, which allows correct perspective rendering and direct interaction with the virtual environment (Intersense, 2005), as well as a five-channel speaker system located in the corners of this facility that further contributes to the effect by adding surround-sound cues. For this research an audio cue was used to indicate to subjects when location information was available.

In addition to the use of the simulated environment, a tablet PC was employed to act as the location-aware delivery tool. A tablet PC is a specialized portable computer that utilizes a pen rather than a mouse for navigation. The tablet PC was mounted on a clear Plexiglas podium and the participant navigated the interface by using the pen. The use of the tablet PC and podium reduced the amount of cumbersome activity that the participants might have had to deal with and allowed them instead to simply focus on information acquisition through the device.

The simulated environment replicates a real-life experience in an outdoor space (in this case the Purdue University campus) while the tablet PC serves as a simulated location-aware portable device. As participants wandered around the "campus" they were informed by a brief audio cue that they were in close proximity to a building or landmark for which location information was available on the tablet PC. The user could then click on various navigation tabs to obtain additional written information about the location.

The information delivered to the respondents was entirely fictitious. This ensured that an effective baseline measurement could be established across all conditions and also greatly simplified the data analysis procedures; if participants reported learning this fictitious information they could have learned it through the treatment condition only and not through any other previous experience with the landmark. This helped to maximize the measurable effect of the treatment.

The time constraints of this research were set at 60 minutes to avoid participant fatigue, allowing 20 minutes for survey measures and 40 minutes for treatment time. Through trial runs it was found that 40 minutes would be a sufficient amount of time to see every landmark on campus. It was also expected that each participant would view all information available. However there remained the possibility that participants might spend too much of this set 40-minute time period on one part of the campus and not enough time at another. Therefore, it was decided that the treatment time would be divided into four 10- minute blocks of time and the tour design would focus on four specific campus vicinities. This increased the likelihood that the participants would view the four specific locations of interest. These vicinities were determined using a procedure that took into account the patterns of campus pedestrian traffic.

On the Purdue campus it is a common practice for event announcements and reminders for various organizations to be printed on color flyers and taped to sidewalks. There are locations on campus where it is not uncommon to find 50 or more flyers taped within a 15-foot radius. These locations are in areas where the flyers are likely to be seen by many passersby. It thus can be safely assumed that large congregations of flyers represent "informational places" where it is perceived that the traffic flow is high and the place is of significance in the campus geography. For this reason, we recorded flier density and determined what locations are most desirable for flier postings. The results revealed four main areas of concentration on campus with the highest level of flier density. Flier density was determined using the inverse distance weight method of density interpolation when the units of measurement are points. The maximum density points were determined to be the areas that were two standard deviations higher than the average flier density on campus. The density plotting and interpolation procedures were done in an ArcGIS environment.

RECRUITMENT

A total of 59 participants were recruited as part of a random-digit dialing survey of students currently registered at Purdue University (N=800). The survey included questions about awareness of the physical layout of the Purdue campus and through what channels they obtained information about the campus. Participants were then asked if they would like to participate in further research for which they would be compensated $20. Those who agreed to participate where randomly assigned to three groups: two treatments and one control group.

EXPERIMENTAL DESIGN

All three groups completed online evaluation instruments before and after their specific activities (virtual reality, search engine and placebo treatment). After an initial exposure to the instrument, participants were exposed to their treatments for a time period of 40 minutes. Immediately after this treatment they again took the online survey.

This study uses a longitudinal design to assess the degree to which subjects can store and retrieve information held in long-term memory. After one week, the subjects were invited to fill out the same questionnaire used before and after the treatment. A period of one week was chosen to test for long-term memory retention, a procedure that replicates similar methodologies used in other studies focused on information retention (Chun & Plass, 1996; Gagne, 1978; King, 1992; Lai, 2000; Wheeler, Ewers, & Buonanno, 2003). One week was found in the literature to be long enough for any traces of short-term memory storage to fade away.

Due to attrition a total of 38 participants completed all three phases of research.

Location-Aware Treatment Group

Participants were introduced to the location-aware virtual environment and instructed to learn as much as they could about each building or landmark they encountered by reading the information delivered on the location-aware device (tablet PC). They were allowed to explore each section for 10 minutes. Each section was centered on a specific landmark, although the respondents were not informed of this. The respondents were encouraged to "walk" around the environment and to approach any landmark or location available, seeking information through the location-aware device. After 10 minutes spent at a specific location the respondents were "teleported" to another area of campus.

Search Engine Treatment Group

Participants in this condition were shown four web pages on a desktop computer, each containing an image of one of the four general campus vicinities that were explored by the location-aware condition respondents in the virtual reality environment. The images depicted prominent landmarks and their surroundings at the four campus locations. The respondents were asked to use a custom-made search engine that indexed the same information about the campus locations as the one delivered through the virtual campus environment. They were asked to explore the area they saw in each of the four pictures utilizing the search engine and were told to think of keywords about the vicinity and to enter them into the search engine. No specific keywords or information about the locations were provided. Participants were simply asked to perform searches, explore the hyperlinks and to learn as much as possible from each web page retrieved. The web pages retrieved information identical to that delivered through the location-aware system in the virtual campus environment. The search activity was divided into four blocks of 10 minutes for a total search time of 40 minutes.

Control Group

This group played a three-dimensional computer game unrelated to the campus environment or to the information presented to the other groups using an ordinary desktop situated in a computer lab. The game consisted of a training session which taught respondents how to navigate an underground maze and how to use various virtual objects. Respondents played the game for 40 minutes, the same amount of time spent by the treatment participants on their respective activities. They were encouraged to visualize themselves in the game and to learn as much as they could to further their progress. This control group was included to rule out random guessing of the new information introduced through the location- aware system or via the search engine.

MEASURES

The cognitive impact of the location-aware system was measured through a web-based questionnaire administered to all conditions once before and twice after stimuli were presented. The post-intervention questionnaires were completed in two steps: immediately after completing the in-lab activity and one week later. All surveys were administered online.

The questions asked were related to four prominent landmarks or locations found in the four distinct sections of the campus the treatment groups were asked to explore. To tap into the four different dimensions of information acquisition via search engines or location-aware systems described previously respondents were asked questions about landmark names and their locations or about specific facts and stories related to them. As previously described, these "facts" and stories were fictitious. The fact-based questions referred to various fictitious details about the architectural style, size, or construction dates of landmarks. The story-based questions related to the narratives offered for each landmark. The answers to the fact-related questions were multiple-choice, while the story, location, and name questions required open-ended answers.

All answers, for all landmarks, were translated into binary variables. For the name question the answer was coded as positive (1) if the respondent mentioned at least partially (through a non-generic moniker) the name of the landmark. For example, for Founder's Fountain answers that mentioned at least "founder" where coded as 1. Answers that only mentioned fountain were coded as 0. For the story question the answer was coded as 1 if the respondent used at least one non-generic sentence or narrative that indicated correctly the actor and the action involved in the story. For the fact and location questions, which utilize objective measures (multiple-choice answers and correct map grid identification), answers were coded as positive (1) if respondents indicated at least one correct map grid and if they chose the correct answer to the multiple choice questions.

The binary variables thus obtained were combined by summation into synthetic, time-specific variables. The points in time were defined as T1=pre, T2=post, and T3 = one week post treatment. Since information was collected on four dimensions (name, location, fact, and story), 12 final variables were generated (3 times x 4 dimensions). Each variable measured recognition of a specific kind of information (NAME, LOCATION, FACT, STORY) for all 4 target landmarks combined. Consequently, values ranged from 0 to 4, where 0 indicated that no information was recognized or recalled during that time period for any landmark and 4 indicated that relevant information was identified during that time period for all 4 landmarks. These 12 variables constitute the key dependent variables.

RESULTS

Data was analyzed using a planned comparison repeated measures 3 (2 treatment; 1 control) x 3 (pre-test, post-test, one week post-test) ANOVA[1] for each of the four types of information requested (landmark NAME, landmark LOCATION, landmark FACT, and landmark STORY). The assumption of sphericity was violated in the tests for the NAME, (Mauchly's W =.286, p <.001), LOCATION (Mauchly's W =.255, p <.001), STORY (Mauchly's W =.815, p =.031 and FACT (Mauchly's W =.768, p =.022) variables. The Greenhouse-Geiser corrected F was used in these cases to compensate for this violation.

The overall results of this analysis suggest that the location-aware simulation had a similar, if not better, effect on information acquisition and recall as the search engine information approach for three out of the four variables (NAME, LOCATION, STORY). The respondents learned equally well new information about their environment using either channel of communication. Learning FACT-based information seems, however, to have been favored in a significant way by location-aware systems. In what follows we will discuss the answers to the research questions and hypotheses advanced in this study in view of the data analysis results.

RQ1: Information Acquisition through Location-Aware Systems

The first cluster of research questions asked if the use of location-aware systems lead to the acquisition of context-defined information, in general. Thus, our goal is to investigate if there is a net information gain after exposure to the location-aware system (T1 compared to T2). The data suggest that three of the four questions (RQ1a,c,d) should be answered yes, since for

*Table 1. Condition * Point in Time (NAME)*

Condition	Point in Time	Mean	Std. Error	N	95% Confidence Interval	
					Lower Bound	Upper Bound
Control	T1	3.268 [a]	0.263	10	2.729	3.808
	T2	3.369 [a]	0.133	10	3.096	3.641
	T3	3.462 [a]	0.115	10	3.226	3.699
Location-Aware	T1	3.155 [a]	0.236	12	2.671	3.639
	T2	3.926 [a]	0.119	12	3.681	4.171
	T3	3.938 [a]	0.104	12	3.726	4.15
Search Engine	T1	3.573 [a]	0.23	13	3.103	4.044
	T2	4.016 [a]	0.116	13	3.778	4.254
	T3	3.932 [a]	0.101	13	3.726	4.138

a. Covariates appearing in the model are evaluated at the following values: Number of Years on Campus = 2.7143, Familiarity = 3.2857, Gender =.6857, Job on campus = 1.54.

each of these information dimensions (NAME, FACT, STORY) we had a net information within subject gain. The increase was more modest for LOCATION, proving not to be significant between T1-T2. However, the gain has been consolidated over time, the within-subject test across all three stages becoming significant (see next section for details). Table 1, Table 2, Table 3, and Table 4 provide mean scores for each dependent variable across time and condition.

In more specific terms, there was found to be a within-subjects main effect over time for the NAME variable, Greenhouse-Geisser corrected $F(1.17, 37.33) = 12.256, p = .001$, Partial $R^2 = .277$). A within-subjects planned comparison for the location-aware condition over time on the NAME variable (i.e., -1, .5, .5) also showed a statistically significant difference between T1 and T2, $F(1,11) = 9, p = .012$, Partial $R^2 = .450$. (Notably, a similar comparison was not statistically significant for the search engine condition at T1 and T2, $F(1,12)$

*Table 2. Condition * Point in Time (LOCATION)*

Condition	Point in Time	Mean	Std. Error	N	95% Confidence Interval	
					Lower Bound	Upper Bound
Control	T1	3.420 [a]	0.296	10	2.815	4.026
	T2	3.606 [a]	0.197	10	3.201	4.01
	T3	3.718 [a]	0.174	10	3.362	4.073
Location-Aware	T1	3.564 [a]	0.265	12	3.02	4.107
	T2	3.827 [a]	0.177	12	3.464	4.189
	T3	3.828 [a]	0.156	12	3.509	4.147
Search Engine	T1	3.156 [a]	0.258	13	2.628	3.685
	T2	3.771 [a]	0.172	13	3.418	4.124
	T3	3.838 [a]	0.152	13	3.527	4.148

a. Covariates appearing in the model are evaluated at the following values: Number of Years on Campus = 2.7143, Familiarity = 3.2857, Gender =.6857, Job on campus = 1.54.

*Table 3. Condition * Point in Time (FACT)*

Condition	Point in Time	Mean	Std. Error	N	95% Confidence Interval	
					Lower Bound	Upper Bound
Control	T1	1.270 [a]	0.235	11	0.786	1.753
	T2	.418 [a]	0.219	11	-0.033	0.868
	T3	.574 [a]	0.247	11	0.066	1.082
Location-Aware	T1	1.803 [a]	0.247	10	1.294	2.312
	T2	3.850 [a]	0.231	10	3.376	4.324
	T3	3.902 [a]	0.26	10	3.367	4.436
Search Engine	T1	1.417 [a]	0.22	12	0.965	1.869
	T2	3.409 [a]	0.205	12	2.988	3.831
	T3	2.972 [a]	0.231	12	2.497	3.447

a. Covariates appearing in the model are evaluated at the following values: Number of Years on Campus = 2.6667, Familiarity = 3.2727, Gender =.6667, Job on campus = 1.55

= 4.596, p =.053, Partial R^2 =.277, or for the control condition $F(1,9) = 1$, $p = 343$, Partial R^2 =.1). This difference is effectively demonstrated in Figure 1 showing a sharp increase in name recognition from T1 to T2 for the location-aware condition.

For the LOCATION variable there was a within-subjects main effect over time, Greenhouse-Geisser corrected $F(1.15,36.69) = 8.357$, $p =.006$, Partial R^2 =.207[1]. The within-subjects planned comparison (i.e. -1,.5,.5) for the location-

aware condition also captures a slight increase in recall between T1 and T2, yet this is not statistically significant $F(1,11) = 1.94$, $p =.191$, Partial R^2 =.180 (see also Figure 2).

For the STORY variable the data had a much different pattern. There was a within-subjects main effect over time for the STORY variable, Greenhouse-Geisser corrected $F(1.688,59.072) = 91.463$, $p <.001$, Partial R^2 =.723. A within-subjects planned comparison for the location-aware condition at T1 and T2 showed a statisti-

*Table 4. Condition * Point in Time (STORY)*

Condition	Point in Time	Mean	Std. Error	N	95% Confidence Interval	
					Lower Bound	Upper Bound
Control	T1	.000 [a]	0	13	0	0
	T2	-.009 [a]	.185	13	-.387	.368
	T3	.013 [a]	.253	13	-.504	.529
Location-Aware	T1	.000 [a]	.000	12	.000	.000
	T2	2.675 [a]	.191	12	2.285	3.066
	T3	1.720 [a]	.262	12	1.186	2.254
Search Engine	T1	.000 [a]	.000	13	.000	.000
	T2	2.694 [a]	.185	13	2.317	3.070
	T3	1.323 [a]	.252	13	.808	1.837

a. Covariates appearing in the model are evaluated at the following values: Number of Years on Campus = 2.6316, Familiarity = 3.2368, Gender =.6579, Job on campus = 1.55.

Figure 1. Landmark name recognition over time

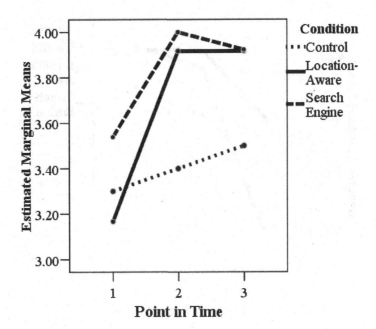

cally significant difference, $F(1,11) = 108.3$, p <.001, Partial R^2 =.908. This difference is effectively demonstrated in Figure 3 showing a sharp increase from T1 to T2.

Likewise, for the FACT variable there was a within-subjects main effect over time, Greenhouse-Geisser correct $F(1.62, 48.71) = 38.462$, p <.001, Partial R^2 =.562. A within-subjects planned comparison for the location-aware condition at T1 and T2 showed a statistically significant difference, $F(1,9) = 60$, p <.001, Partial R^2 =.87. This difference is effectively demonstrated in Figure 4 showing a sharp increase in factual information learned between T1 and T2.

RQ1: Information Recall through Location-Aware Systems

The second cluster of research question asked if the use of location-aware systems lead to information recall after a longer period of time (one week). The answer to this question is yes for the same variables as above (RQ1a,c,d). A within-

subjects planned comparison for the location-aware condition over time on the NAME variable (i.e., -1,.5,.5) showed a statistically-significant difference between T1 and T3 (Figure 1), $F(1,11)$ = 9, p =.012, Partial R^2 =.450. (A within-subjects planned comparison for the control condition was not statistically significant between T1 and T3, F $(1,9) = 2.25$, p =.168, Partial R^2 =.2 nor was the search engine condition $(1,12) = 3.26$, p =.096).

The location-aware condition did not show a statistically-significant difference for a within-subjects planned comparison between T1 and T3 for the LOCATION variable, $F(1, 11) = 1.94$, p =.191, Partial R^2 =.150 or for the control condition at T1 and T3 $F(1,9) = 1.976$, p =.193, Partial R^2 =.180. (The search engine condition on the other hand did show a statistically-significant difference for a within-subjects planned comparison at T1 and T3 $F(1, 12) = 6.94$, p =.022, Partial R^2 =.367 for LOCATION. While this test was not significant for the location-aware condition it is again important to note that the mean at T2 and

Figure 2. Landmark location recognition over time

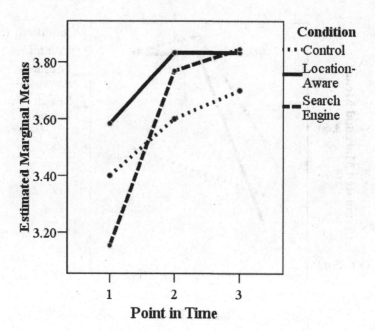

T3 for the location-aware condition remained constant (M = 3.83, SD =.577)).

For the STORY variable the location-aware condition had a statistically- significant within-subjects planned comparison at T1 and T3 $F(1,11) = 25$, $p <.001$, Partial $R^2 =.694$ as well as for the FACT variable at T1 and T3, $F(1,9) = 60$, $p <.001$, Partial $R^2 =.87$. While the location-aware condition did show a decline in recall at T3 for the STORY variable (Figure 3) the amount of information retained was still statistically significant. In addition, the mean for the FACT variable remained constant from T2 to T3 (M = 3.8, SD =.42) as show in Figure 4.

H1: Location-Aware vs. Search Engine Acquisition and Recall

The central hypothesis of this study proposes that location-aware systems are more effective at facilitating information acquisition and recall than search engine-based systems. The same data indicates that the hypothesis is supported only by

the FACT variable (H1c). In other words, subjects recalled factual information better when it was delivered via the location-aware system compared to the search engine.

The FACT variable indicates a clear between-subjects main effect and a statistically-significant difference between the location-aware condition and the search engine condition, $F(1,20) = 4.78$, $p =.041$, Partial $R^2 =.193$. This statistical difference is due to the mean difference of the location-aware condition at T3 (M = 3.8, SD =.95) and the search engine condition at T3 (M = 3, SD = 1.54). While there is not a difference between location-aware systems and search engine-based systems for information acquisition (as measured at T2), there is evidence that location-aware systems are more effective at facilitating recall of information for factual information (as measured at T3).

There was no between-subjects main effect for condition on the NAME and LOCATION variables. There was a between-subjects main effect of condition for the STORY variable; however, the difference between the location-aware con-

Figure 3. Landmark story recognition over time

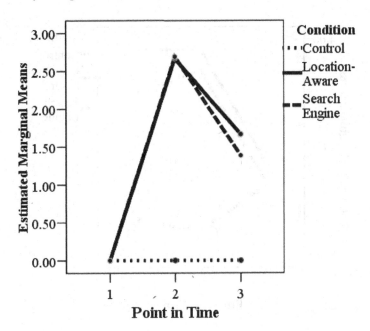

dition and the search engine condition was not statistically significant, $F(1,23) =.158$, $p =.694$, Partial $R^2 =.007$. It is important to note that while this difference was not statistically significant the mean for the location-aware condition at T3 (M = 2.69, SD =.89) was in the expected direction compared to the search engine condition at T3 (M = 1.38, SD = 1.12).

DISCUSSION

The results from this analysis are promising for the future of location-aware applications. Although their information recall effects are not spectacularly greater than those of search engines, there is an indication that compared to search engines location-aware applications enhance our capacity to recall factual information after a moderate amount of time.

The greater capacity to recall information of those who used location aware information is likely due to the effects predicted by our theo-

retical framework. More specifically, we believe that participants in the search engine condition dually-code the verbal information with the visual stimulus of the landmark. When it came to recall the information learned they could use the picture of the landmark to assist in remembering the verbal information they coded into visual memory. However, as most effectively shown in Figure 1, this may not be enough to recall everything learned. Participants in the location-aware condition on the other hand, were able not only to dually-code this verbal information with the visual stimulus of the landmark, but they were also able to then use another mental mechanism to situate this information in relation to the other landmarks within their mental map of the environment. This mnemonic device for recall was mental mapping created during the physical experience they had in the three-dimensional environment. Participants moved from one landmark to the next, learning new verbal information about each one. As they moved from landmark to landmark the information learned was associated with a physical

Figure 4. Landmark fact recognition over time

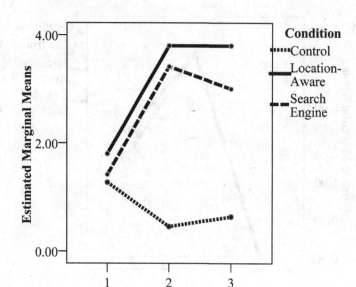

location that could then be translated into pertinent information situated within the individual's mental map. Subsequently, when it came time to recall the information, they were able to use the visual stimulus of the picture they saw in the survey to recall the information, just like those in the search engine condition. However, if they were unable to recall the information based off of the visual stimulus they would also be able to use their mental map of the environment to help facilitate the recall of the needed information. For example, if a participant was unable to recall the verbal information for landmark 3 they could "retrace their steps" to landmark 2 in their mental map and use the information they remembered about landmark 2 to trigger a recall mechanism of the neighboring landmark 3. In view of this we believe that dual coding and conjoint retention hypothesis should be extended to include not only visual but also spatial dimensions.

While the data does indicate that location-aware systems lead to greater long-term recall than search engine systems, this difference was only statistically significant for one variable, and the

differences at T2 were not as large as expected. There are several possible reasons for these results.

First, due to attrition factors, there were an overall low number of participants in each cell. This attrition was due to participants not responding to the one-week follow up survey at T3, which affected the statistical power of the study.

Second, the chosen dependent variables had low variability as their values only ranged from 0-4. This created a ceiling effect, preventing higher variability in the dependent variables which was further compounded by a floor effect, particularly with the NAME and LOCATION variables. Because participants already knew, on average, three out of the four names and locations of the landmarks prior to treatment, finding a statistically-significant difference for these variables was more difficult.

Third, because participants were all treated in the same way, they all shared the same starting locations at the same intervening time periods for discovering the campus. Because of this there may have been an ordering effect that affected the results. However, no participant took the same path from these starting points and so ordering effects may have been diminished by the participants' ability to choose their destinations from those initial starting points.

Last, because the environment was in a virtual reality setting many participants may have learned the new information merely because it was a new and exciting experience. Learning and recalling would, in this scenario, be influenced by the Hawthorne Effect (the respondents reacted to the novelty of the stimulus rather than to its actual content). The study design allowed, however, for an introductory period to the location-aware application, which was meant to ease the participant into the environment and to reduce the excitement associated with it. However, this might not have completely eliminated the effect. An alternative and better way of ameliorating the Hawthorne Effect would have been to reserve a prolonged time block for acclimatization purposes, but this was severely constrained by time limitations.

Some of these limiting factors could have been handled by using several variations of the basic design. Unfortunately, most of these variations were not implemented due to material or time limitations. The most obvious limitation that could have been addressed is the small sample size. A larger number of respondents would have increased the statistical power. The time and funding available limited the number of respondents to 20 per condition.

Regarding the low variability of the dependent variables, it would have been better to have participants learn as much as they could about the environment and then report as much new information as they could remember about every landmark in the entire environment, rather than just about four key landmarks. However, this method would have been extremely time-consuming, both in terms of administering the surveys and of coding the results. In addition, this would have led to even higher attrition rates.

Finally, the fact that the virtual environment replicated a known geographic location could have had an impact on learning and recall. Superior visual acquaintance with the landmarks might have positively influenced information acquisition and recall. Yet, when utilizing covariates that measure respondents' familiarity with the campus the results were not influenced in a major way. However, a better method would have been to use a completely unknown virtual environment. This design would likely have impacted not only the three-dimensional experience but also the effectiveness of the search engine condition, as participants would have been severely limited in their ability to search for information about an environment that they had little to no information about. In view of this, we should point out that the search engine effect on information acquisition and recall is likely overestimated. We intend to re-estimate these effects by creating a virtual reality environment of an ancient location (The Roman Forum), which will be used in the context of an undergraduate history class (Matei et al., 2007). The environment will be new to all students and the subject information will be largely unknown to them.

In addition, we have developed a mobile location aware environment, in which information seeded on a central website is delivered to an iPhone through a dedicated app (http://ubimark.com/in/link/88). Two projects, Ubimark (http://ubimark.com) and Visible Past (http://visiblepast.net) will employ the platform for measuring the cognitive impact of mobile location aware systems.

The knowledge gained through this study can be used more broadly by web engineers to shape their own future location aware and mobile search applications. One of the findings that is of significance for web and mobile application designers is the need to connect factual information to images and to embed both in spatial context through mapping applications. Mobile search applications tend to replicate the desktop experience, search results being listed sequentially as plain text. With some exceptions, most mobile search applications do not make a conscious effort to connect information to maps or to spatial or imagistic representations of the information retrieved. All mobile applications need to match information to be served in a mobile manner with a map location and at least one picture that is recognizable from the immediate environment experienced by the user. Furthermore, maps and images need to be organically integrated in the search results. Better yet, they could be served directly through map applications, the way the mobile Google Maps handles searches. Connecting information to searcher location also needs to take advantage of easily visible landmarks. This creates a perceptual information map that is easy to recognize and utilize. Landmark dominated maps can better assist users to position themselves in space and are more conducive to dual coding and conjoint retention of information. Landmarks will connect images, a felt sense for where things are, and information in a holistic pattern, which facilitates information retrieval.

CONCLUSION

Despite some limitations, the present study provides an important glimpse into the possibilities created by location-aware systems. It study lays the groundwork for further research related to the learning effects of location-aware communication systems and points to the need to study how physical space and mental maps merge through the use of this communication technology. This is an important research issue, as the emergence of location-aware applications has the potential to greatly impact many facets of the information landscape, to say nothing of the world at-large.

For generations children have gone to museums and learning centers to learn something in a way not possible in a classroom. They learn by "being there" and by directly experiencing both the information and environment. Location-aware learning experiences can move such experiences out of museums and into university campuses and other locations. The days of static plaques at exhibits are numbered as a large computerized database of knowledge could be used for each exhibit in a museum and a location-aware device could be programmed to deliver information pertinent not only to that physical space but also to the individuals using the device. This could tailor an age-appropriate experience for each user and could allow for superficial or deeper modes of inquiry, according to the circumstances. More importantly, if these findings are confirmed by further study, what is learned in a location-aware environment is more likely to be retained.

Yet, to transform location-aware systems from experimental into actual learning tools more research has to be done in a number of directions. Researchers interested in the learning effects of location-aware technology should try to further isolate and investigate the cognitive mechanisms involved. This could be done by looking at the role physical attributes of landmarks and their relationships with other landmarks play in constructing mental maps and shaping the memorization process. The working hypothesis here would be that more concrete mental maps of the environment might lead to even greater acquisition and recall of the information embedded in the landmarks. It may also be useful to have a comparison condition that would have participants use a printed pamphlet instead of a location-aware device to further isolate the influence of this new communication technology on learning and retention.

Research in location-aware systems also needs to focus on determining if location-aware applications increase the bounds of mental capacity. Can location-aware systems increase the net capacity of humans to remember information? This can be explored with new and appropriate experimental designs.

Overall, location-aware systems have a promising future. The hope is that the theoretical insights gathered through this study will drive further research into discovering the underlying theoretical processes that shape the effect of location-aware systems on mental mapping and learning.

REFERENCES

Abowd, G. D., Atkeson, C. G., Hong, J., Long, S., Kooper, R., & Pinkerton, M. (1997). Cyberguide: A mobile context-aware tour guide. *ACM Wireless Networks*, *3*(5), 421–433. doi:10.1023/A:1019194325861

Anand, P., Herrington, A., & Agostinho, S. (2008). Constructivist-based learning using location-aware mobile technology: an exploratory study. In *Proceedings of World Conference on Educational Multimedia, Hypermedia and Telecommunications 2008* (pp. 2312-2316). Chesapeake, VA: AACE.

Armstrong, M. P., & Bennett, D. A. (2005). A manifesto on mobile computing in geographic education. *The Professional Geographer*, *57*(4), 506–515. doi:10.1111/j.1467-9272.2005.00495.x

Barbosa, J., Hahn, R., Rabello, S., & Barbosa, D. (2008). Local: a model geared towards ubiquitous learning. In *Proceedings of the 39th SIGCSE technical symposium on Computer science education* (pp. 432-436).

Borriello, G., Chalmers, M., Lamarca, A., & Nixon, P. (2005). Delivering real-world ubiquitous location systems. *Communications of the ACM*, *48*(3), 36–41. doi:10.1145/1047671.1047701

Brigham, F. J., Scruggs, T. E., & Mastropieri, M. A. (1995). Elaborative maps for enhanced learning of historical information: Uniting spatial, verbal, and imaginal information. *The Journal of Special Education*, *28*(3), 440–460. doi:10.1177/002246699502800404

Chen, H., Houston, A. L., Sewell, R. R., & Schatz, B. R. (1998). Internet browsing and searching: User evaluations of category map and concept space techniques. *Journal of the American Society for Information Science American Society for Information Science*, *49*(7), 582–603.

Chun, D. M., & Plass, J. L. (1996). Effects of multimedia annotations on vocabulary acquisition. *Modern Language Journal*, *80*(2), 183–198. doi:10.2307/328635

Davis, E. T., Scott, K., Pair, J., Hodges, L. F., & Oliverio, J. (1999). *Can audio enhance visual perception and performance in a virtual environment?* Paper presented at the Proceedings of The Human Factors and Ergonomics Society 43rd Annual Meeting, Houston, TX.

De Jong, T., Specht, M., & Koper, R. (2008). A reference model for mobile social software for learning. *International Journal of Continuing Engineering Education and Lifelong Learning*, *18*(1), 118–138. doi:10.1504/IJCEELL.2008.016079

Dodds, A. G. (1982). The Mental Maps of the Blind: The Role of Previous Visual Experience. *Journal of Visual Impairment & Blindness*, *76*(1), 5–12.

Downs, R. M., & Stea, D. (1977). *Maps in minds: Reflections on cognitive mapping*. New York: Harper & Row.

FakeSpace. (2005). *FLEX™ and reFLEX™. Innovative designs allow for fast reconfiguration for fully detached module capabilities*. Retrieved September 28, 2005, from http://www.fakespace-systems.com/pdfs/solutions/Fakespace-FLEX-reFLEX.pdf

Foo, P., Warren, W. H., Duchon, A., & Tarr, M. (2005). Do humans integrate routes into a cognitive map? Map versus landmark-based navigation of novel shortcuts. *Journal of Experimental Psychology*, *31*(2), 195–215.

Gagne, E. D. (1978). Long-term retention of information following learning from prose. *Review of Educational Research, 48*(4), 629–665.

Gellevij, M., Meij, H. V. D., Jong, T. D., & Pieters, J. (2002). Multimodal versus unimodal instruction in a complex learning context. *Journal of Experimental Education, 70*(3), 215–239. doi:10.1080/00220970209599507

Gerber, S., & Shuell, T. J. (1998). Using the internet to learn mathematics. *Journal of Computers in Mathematics and Science Teaching, 17*(2), 113–132.

Golledge, R. G., & Stimson, R. J. (1997). *Spatial behavior: A geographic perspective*. New York: Guilford Press.

Gould, P. (1975). Acquiring spatial information. *Economic Geography, 51*(2), 87–99. doi:10.2307/143066

Intersense. (2005). *InterSense IS-900 Precision Motion Tracker*. Retrieved September 28, 2005, from http://www.isense.com/products/prec/is900/

James, W. (1890). *The principles of psychology*. New York: Holt.

Kim, G., Han, M., Park, J., Park, H., Park, S., Kim, L., & Ha, S. (2009). An OWL-Based Knowledge Model for Combined-Process-and-Location Aware Service. In *Proceedings of the Symposium on Human Interface 2009 on Human Interface and the Management of Information. Information and Interaction. Part II: Held as part of HCI International 2009* (p. 167).

King, A. (1992). Comparison of self-questioning, summarizing, and note-taking review as strategies for learning from lectures. *American Educational Research Journal, 29*(2), 303–323.

Koo, S., Rosenberg, C., Chan, H. H., & Lee, Y. C. (2003, March). *Location-based e-campus web services: from design to deployment.* Paper presented at the IEEE International Conference on Pervasive Computing and Communications (PerCom), Dallas, TX.

Kulhavy, R. W., Caterino, L. C., & Melchiori, F. (1989). Spatially cued retrieval of sentences. *The Journal of General Psychology, 116*(3), 297–304.

Kulhavy, R. W., Lee, J. B., & Caterino, L. C. (1985). Conjoint retention of maps and related discourse. *Contemporary Educational Psychology, 10*, 28–37. doi:10.1016/0361-476X(85)90003-7

Lai, S. (2000). Increasing associative learning of abstract concepts through audiovisual redundancy. *Journal of Educational Computing Research, 23*(3), 275–289. doi:10.2190/XKLM-3A96-2LAV-CB3L

Lloyd, R. (1994). Learning spatial prototypes. *Annals of the Association of American Geographers. Association of American Geographers, 84*(3), 418–440. doi:10.1111/j.1467-8306.1994.tb01868.x

Lloyd, R., & Heivly, C. (1987). Systematic distortions in urban cognitive maps. *Annals of the Association of American Geographers. Association of American Geographers, 77*(2), 191–207. doi:10.1111/j.1467-8306.1987.tb00153.x

Lynch, K. (1960). *Image of the City*. Cambridge, MA: The MIT Press.

MacEachren, A. M. (1992). Application of environmental learning theory to spatial knowledge acquisition from maps. *Annals of the Association of American Geographers. Association of American Geographers, 82*(2), 245–274. doi:10.1111/j.1467-8306.1992.tb01907.x

Matei, S. A., Madsen, L., Arns, L., Bertoline, G., & Davidson, D. (2005, October 5-9). *Socio-spatial cognition and community identification in the context of the next generation of location-aware information systems.* Paper presented at The Association of Internet Researchers Annual Conference, Chicago.

Matei, S. A., Miller, C. C., Arns, L., Rauh, N., Hartman, C., & Bruno, R. (2007). *Visible Past: Learning and discovering in real and virtual space and time* (*Vol. 12*). First Monday.

Mayer, R. E., & Anderson, R. B. (1991). Animations need narrations: An experimental test of a dual-coding hypothesis. *Journal of Educational Psychology, 83*(4), 484–490. doi:10.1037/0022-0663.83.4.484

Mostafa, J. (2005). Seeking better web searches. *Scientific American, 292*(2), 66–73. doi:10.1038/scientificamerican0205-66

Mounin, G. (1980). The semiology of orientation in urban space. *Current Anthropology, 21*(4), 491–501. doi:10.1086/202498

Paivio, A. (1990). *Mental representations: a dual coding approach.* New York: Oxford University Press.

Parker, R. D. (1997). The architectonics of memory: On built form and built thought. *Leonardo, 30*(2), 147–152. doi:10.2307/1576426

Peterson, R. A., & Merino, M. C. (2003). Consumer information search and the internet. *Psychology and Marketing, 20*(2), 99–121. doi:10.1002/mar.10062

Piaget, J., & Inhelder, B. (1956). *The child's conception of space.* London: Routledge & K. Paul.

Radvansky, G. A., & Copeland, D. E. (2000). Functionality and spatial relations in memory and language. *Memory & Cognition, 28*(6), 987–992.

Sandor, C., Kitahara, I., Reitmayr, G., Feiner, S., & Ohta, Y. (2009). Let's go out: Research in outdoor mixed and augmented reality. In *Proceedings of the 2009 8th IEEE International Symposium on Mixed and Augmented Reality* (p. 229). Washington, DC: IEEE Computer Society.

Tuan, Y.-F. (1975). Images and mental maps. *Annals of the Association of American Geographers. Association of American Geographers, 65*(2), 205–213. doi:10.1111/j.1467-8306.1975.tb01031.x

Vekiri, I. (2002). What is the value of graphical displays in learning? *Educational Psychology Review, 14*(3), 261–312. doi:10.1023/A:1016064429161

Verdi, M. P., & Kulhavy, R. W. (2002). Learning with maps and text: An overview. *Educational Psychology Review, 14*(1), 27–46. doi:10.1023/A:1013128426099

Webb, J. M., & Saltz, E. D. (1994). Conjoint influence of maps and auded prose on children's retrieval of instruction. *Journal of Experimental Education, 62*(3), 195–208.

Webb, J. M., Thornton, N. E., Hancock, T. E., & McCarthy, M. T. (1992). Drawing maps from text: A test of conjoint retention. *The Journal of General Psychology, 119*(3), 303–313.

Weiler, A. (2005). Information-seeking behavior in generation y students: Motivation, critical thinking, and learning theory. *Journal of Academic Librarianship, 31*(1), 46–53. doi:10.1016/j.acalib.2004.09.009

Wheeler, M. A., Ewers, M., & Buonanno, J. F. (2003). Different rates of forgetting following study versus test trials. *Psychology Press, 11*(6), 571–580.

Yang, S. J., Okamoto, T., & Tseng, S. S. (2008). Context-aware and ubiquitous learning. *Educational Technology & Society, 11*(2), 1–2.

ENDNOTES

[1] Repeated measures 3 X 3 ANCOVA was originally used to analyze the data, the procedure including covariates concerned with one's familiarity with the environment. It was thought that having a better familiarity with the environment would lead to a greater capacity to incorporate new information into a predefined mental map. The number of years that the participant had been attending the university, whether or not they worked on campus, and their own subjective measurement of their familiarity of the campus on a 1-4 Likert-like scale were used. However, none of these variables were not found to be statistically significant and we employed the more direct, ANOVA procedure. The ANCOVA results were: NAME (number of years, $F(1, 26) = 1.854$, $p = .185$, Partial $R^2 = .067$; worked on campus, $F(1, 26) = .237$, $p = .631$, Partial $R^2 = .009$; familiarity, $F(1, 26) = 1.002$, $p = .326$, Partial $R^2 = .037$); LOCATION (number of years, $F(1, 28) = 2.52$, $p = .620$, Partial $R^2 = .009$; worked on campus, $F(1, 28) = .721$, $p = .403$, Partial $R^2 = .025$; familiarity, $F(1, 28) = .038$, $p = .844$, Partial $R^2 = .001$); STORY (number of years, $F(1, 31) = .371$, $p = .547$, Partial $R^2 = .012$; worked on campus, $F(1, 31) = 2.121$, $p = .155$, Partial $R^2 = .064$; familiarity, $F(1, 31) = .576$, $p = .454$, Partial $R^2 = .018$); FACT (number of years, $F(1, 26) = 1.854$, $p = .185$, Partial $R^2 = .067$; worked on campus, $F(1, 26) = .237$, $p = .631$, Partial $R^2 = .009$; familiarity, $F(1, 26) = 1.002$, $p = .326$, Partial $R^2 = .037$).

[2] Different degrees of freedom are due to different N values resulting from participants neglecting to respond to all questions (Tables 1, 2, 3, and 4).

This work was previously published in International Journal of Information Technology and Web Engineering, Volume 5, Issue 2, edited by Ghazi I. Alkhatib, pp. 32-52, copyright 2010 by IGI Publishing (an imprint of IGI Global).

Chapter 3
Applying Ontology Similarity Functions to Improve Software Agent Communication

Jairo Francisco de Souza
Federal University of Juiz de Fora, Brazil

Sean Wolfgand Matsui Siqueira
Federal University of the State of Rio de Janeiro, Brazil

Rubens Nascimento Melo
Pontifical Catholic University of Rio de Janeiro, Brazil

ABSTRACT

In order to perform its tasks on the Semantic Web, software agents must be able to communicate with other agents using domain ontologies, even when considering different ontologies. Thus, it's necessary to address the semantic interoperability issue to enable agents to recognize common concepts and misunderstandings. This work proposes the use of GNoSIS, a tool for composing ontology similarity functions, and specific modules in Goddard agent architecture in order for software agents to negotiate meanings of terms not defined in its ontology.

INTRODUCTION

Dealing with systems interoperability has been a research issue for some time, but the use of a knowledge structure to allow system interoperability - whether in communication between agents, in database integration or still in other scenarios – has still several problems. Interoperability is compromised when different knowledge structures

are used and overlapping domain concepts can become a computing issue.

According to O'Hara (2004), the highest layers of the Semantic Web architecture contain social phenomena that cannot be overlooked in computational solutions (such as the trust layer). As the structuring of knowledge is present in the upper layers of the Semantic Web, a genuinely social phenomenon that can be observed is related to

DOI: 10.4018/978-1-4666-0023-2.ch003

the achievement of consensus for the creation and compatibilization of these knowledge structures (in our case, ontologies).

To execute their tasks, software agents need to be interactive and adaptive, that is, they should be capable of receiving and sending messages to other agents or to the environment and should be capable of understanding these messages. The understanding of the messages takes place through a standardization of the vocabulary of the agents. The attaining of such compatibility can be made with the use of domain ontologies. In open environments, however, software agents are subjected to receiving messages from agents that do not share the same standardized vocabulary, which characterizes one of the challenges in this area. The software agents will be responsible for dealing with the harmonization of ontologies (Breitman et al, 2007), discovering similarities between concepts or the wrong interpretation of some concept during the communication with other agents, to execute some task that requires interaction between agents.

However, harmonizing ontologies is a hard task and stills an unsolved issue. The ontological divergences can be divided into (1) divergences on the level of language (differences caused by the use of different formalisms) and (2) divergences on the level of conceptualization (differences related to the structuring of the concepts in the ontology) (Klein, 2001).

Divergences on the level of language are solved with the changing the formalism of one or of the two ontologies. The changing of the formalism also generates new problems, such as those caused by the difference in expressiveness of a formalism in relation to the other but even then this is the most adequate solution to solve this type of divergence. In this work, we adopted OWL language as the standard for ontology description and thus do not deal with divergence issues of this level.

Divergences on the level of conceptualization occur, amongst other cases, due to a difference in coding, use of synonyms, use of distinctive generic ontologies, difference in granularity between the ontologies, etc. These cases demand a comparison of the structure of the concepts and of the context, that is, a semantic comparison. Syntactic comparisons can add good results to semantic comparisons by finding semantic relations between terms, as it happens in many algorithms that mine text corpus (Chakrabarti, 2000; Faatz & Steinmetz, 2002).

This work presents an approach that allows the harmonization of ontologies during the communication of software agents. The approach addresses how agents must encapsulate similarity functions to harmonizing ontologies during communication. This approach was implemented in GNoSIS system. The system evaluation (Souza et al, 2010) shows that the system can be used to reach similarity degrees very close to 1 (that is, close kinship) between concepts even with syntactic or structural differences.

The GnoSIS algorithm uses resemblance functions and calculates the degree of similarity between the concepts in a recursive manner, calculating the degree of similarity between two concepts based on the degree of total similarity between the concepts that have close kinship.

ADDING MEANING NEGOTIATION SKILLS TO MULTIAGENT SYSTEMS

Software agents communication languages, such as KQML, allow describing domain ontologies used in the message content. The negotiation of meanings for software agents takes place through the alignment of their respective domain ontologies. This negotiation of meanings is done immediately after the identification of (1) a term that does not exist in the ontology of the agent that receives the message or (2) homonymous terms. As domains are denoted in ontology languages as *namespaces,* the terms described in ontologies as classes and that are in distinct *namespaces* are considered possible homonyms. Figure 1 shows two messages sent by two agents, A and B, us-

Figure 1. Possible homonymous concept. The message MSG1 has an term "process" from ontology opencyc as its content and the message MSG2 has an homonymous term "process" from ontology SUMO as its content.

```
[Message MSG1]
(evaluate
  :sender A
  :receiver B
  :language KIF
  :ontology opencyc
  :reply-with q1
  :content (val (process m1))
)
CYC namespace:      http://opencyc.sourceforge.net/daml/cyc
[Message MSG2]
(reply
  :sender B
  :receiver A
  :language KIF
  :ontology SUMO
  :in-reply-to q1
  :content (= (process m1) (scalar 12))
)
SUMO namespace:   http://reliant.teknowledge.with/daml/sumo
```

ing the standard KQML agent communication language (Finn et al, 1994), where agents use the "process" concept of two ontologies (CYC and SUMO) with distinct *namespaces*.

In the MSG1, agent (*:sender*) A request the value of a variable process to agent B (*:receiver*). The agent A sends the message (*:content*) that has the term process to agent B. This term is described by the high-level ontology (*:ontology*) opencyc (Lenat & Guha, 1990). Agent B replies to the message (MSG2) answered the query received also using a process term, albeit described in the SUMO high-level ontology (Niles & Pease, 2001). As the ontologies have distinct *namespaces*, the terms can be constructed as possible homonymous concepts. This approach to identify homonymous concepts is very simplistic but quite useful as distinct ontologies have no indication of correlation between concepts with the same name (providing this correlation is not explicit in the ontology via some type of relation, such as mappings).

The identification of homonymous terms is carried out to identify a need for alignment between the terms. The alignment process should be capable of confirming if the terms truly denote

concepts that are different, supplementary, or equal. In some cases, however, the terms can denote equal concepts but with distinct restrictions, such as cardinality restrictions. The problem with the harmonization of concepts is its complexity and therefore it is not a problem that can be fully automatic, requiring human intervention (Klein, 2001). However, it is not reasonable to think of direct and constant human intervention in a multi-agent environment to solve compatibility problems in ontology concepts. Thus, it is possible to see that this work, as well as related works on the automatic alignment of concepts in ontologies has a positive outcome, in their majority, in simpler structures and with alignments that are not too complex.

Once a need of alignment is identified, the agents should start this procedure through a process of negotiation defined in the negotiation protocol. The negotiation protocol sets the types of messages that can be sent and how messages are related.

Once an agent A identifies a non-existing term in its domain ontology, it replies to a message (KQML performative: ask-about) to the agent B, sender of the message requesting the definitions of the term contained in the message, that is, its set of restrictions and their hierarchical and non-hierarchical relationships. Agent B, in its turn, replies to the message (KQML performative: reply) to agent A with the term definition (i.e., an excerpt from agent ontology with the concept definitions). If the definition has new, non-identified terms, agent A repeats the process for the new terms until agent B has no more definitions to reply about (ending the negotiation without alignment) or until agent A has no more queries (defining the alignment). This process is based on the technique of negotiation of meanings as described by Teresa Pica (1987), which describes the process through which people try to reach an understanding of words that are not understood, as used during a dialogue. A simple example of this process for the negotiation of meanings is given in Figure 2.

In the example above, sender A sends a message. Receiver B does not know one of its terms and sends a message querying the meaning from B. The sender and receptor exchanges messages to define the concept received until receiver B understands the meaning of the term. Firstly, the sender sends the concepts that are closest in the hierarchy to the concept being queried. This way, sender A sent the definition of Triangle using its Polygon superclass. Polygon is a concept known to receiver B and this, in its turn, sends a confirmation message. Sender A then forwards the definition of the Triangle concept based on its attributes. Receiver B confirms the sending and creates its definition of Triangle. Figure 3 shows the definitions exchanged by the players as description logic sentences that are described in the OWL language by the agents and interpreted by logical reasoning agent for description logics.

The alignment process, however, most of the time needs to use more sophisticated techniques to find the correct alignment between concepts, such as in distinct domains, as the difficulty to

recognize synonyms and homonyms is a challenge. In this case, techniques for the calculation of similarity are used. The techniques for the calculation of similarity usually use the names of concepts and properties (syntactic techniques) and the ontological structure and instances (semantic techniques) to recognize the concepts with a higher degree of similarity (see "Concept Similarity Analysis" section).

To provide compatibility to the concepts during the process of conversation of agents in a meaning negotiation approach, independently from the similarity techniques that are used, we proposed the agent development architecture from Figure 4. An extension was built of the Goddard agent architecture (Truszkowski, 2006), with the adding of highlighted boxes to the agent communication module (Agent Communication Perceptor/Effector).

The Goddard agent architecture subdivides the agent in 8 modules and defines the relationships between these modules. The main module for the previously described problem is the communication module (Agent Communication Perceptor/Effector), which is responsible for interacting with the environment, receiving messages in an agent communication language. These messages are sent to a reasoning agent, responsible for interpreting the message and processing the contents of the message using the plan defined in the Planning and Scheduling module and the current state of the agent (Modelling and State module). Once the decision for the next step of

Figure 2. Example of negotiation of meanings

```
A: it's the triangle.
B: triangle?
A: the triangle is the polygon.
B: ok, the triangle is the polygon.
A: the triangle has exactly three sides.
B: ok, the triangle is the polygon with exactly
   three sides.
```

Figure 3. Definitions in description logics

Agent A:

$Triangle \sqsubseteq Polygon$

$Triangle \sqsubseteq (\geq 3\ hasSide) \sqcap (\leq 3\ hasSide)$

Agent B:

$Triangle \equiv Polygon \sqcap (\geq 3\ hasSide) \sqcap (\leq 3\ hasSide)$

the agent is made, it executes the action (Execution) and alters the environment (Effector).

The communication module was expanded, being detailed into 4 sub-modules. When receiving a message, the Receptor/Sender sub-module is responsible for forwarding the concepts that exist in the message to the sub-module Ontology Reasoning, which represents the logic inference machine that is used to process the definition of the concepts of the ontologies. This sub-module checks if the concepts contained in the message are defined in the agent ontology through a syntactic comparison between the canonical names of the concepts as described in OWL (*namespace#concept*). If it has

been defined, the Receptor/Sender sub-module forwards the message to the agent reasoning module which is responsible for the understanding of the message and for decision-making pursuant to the BDI (Belief-Desire-Intention) model (Bratman, 1999). On the other hand, if the concept is not defined in the domain ontology of the agent, the Ontology Reasoning sub-module consult the negotiation plan in the Negotiation Planning sub-module, which stores the techniques that will be used to attempt the alignment of the concepts and the reserve value for the negotiation of the agent. The alignments considered as valid by the Ontology Reasoning are stored in the Mapping

Figure 4. Agent architecture showing the negotiation of meaning components added in agent communication perceptor/effector component

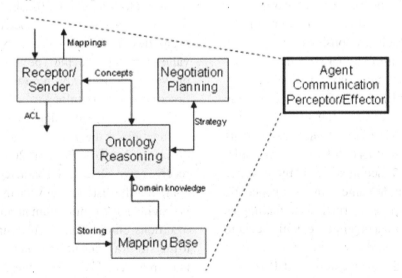

Base sub-module and used in the future to avoid repeated negotiations and to enrich the inference process in new negotiations.

If the agent cannot align the concepts of the ontologies, the negotiation is considered as non-satisfactory. The agent's plan, described in the Negotiation Planning sub-module, defines the best alternative to an unsuccessful negotiation. In this case, the agent should try to use the plan to implement an action that may have a relevant outcome. If the agent is making a consultation in another agent to present to the user, for example, the agent can present the data as probable but not reliable results.

CONCEPT SIMILARITY ANALISYS

This section presents the method to calculate the similarity of ontology concepts that we use during software agent meaning negotiation process. This method is available as alignment method in Negotiation Planning module (Figure 4). The method receives pairs of ontologies (or subontologies) as input and returns a table with the most similar concepts. To fill this table, the module analyzes the ontologies in a syntactic manner (names of concepts and properties) and semantic manner (hierarchy, relationships and property constraints). The value for total similarity between two concepts is recursively calculated based on the hierarchy of concepts and their relationships.

The calculation of similarity is done by using unsymmetrical similarity functions $F(a, b)$, where $F_i(a, b)$, $i=1,k$; $F: A \times B \rightarrow [0,1]$, A is the set of ontology elements of the agent reasoning the ontology and B is the set of ontology elements of the agent that sent its ontology. For example, $F_1(a, b)$ can be a function such as "the quantity of properties with the same name of a concept", where value 1 can be returned if the ontologies have the same number of properties with the same name, and 0 if they do not have any identical property. Intermediate values represent the degree of similarity between the concepts and, the highest the value for $F(a, b)$, the greater the similarity between the concepts compared will be.

For each ontological element (class, property, constraints, etc.), a set of similarity functions is applied. At first, all the constructors used to create an ontology can be useful in the calculation of similarity and thus have one or more associated similarity functions. We used similarity functions in this approach applied to concept names, properties, and relationships.

Each property of a concept of the ontology has similarity functions that analyze its name and type. In OWL, the properties can be data type or object properties, represented by constructors owl:DatatypeProperty and owl:ObjectProperty, respectively. Data type properties are those that are intrinsic to the concept, that is, that do not relate to other concepts. For example, data type properties for the "Person" concept can be: name, age, size, etc. In its turn, object properties are those that are extrinsic to the concept, i.e., that relate to other objects. For the same "Person" concept, the examples of object properties can be: "works in" or "is the son of".

Data type properties have primitive data types and object properties have complex data types, that is, other concepts of the ontology. Thus, for example, property "name", as mentioned before, allows values as strings (or, in OWL, the type xsd:string) as it is a data property. Similarly, the properties "is the son of" and "works in" allow values as instances of the type "Person" and "Place of Work", respectively; both types are concepts described in the ontology.

The first of the techniques we used as similarity functions is the *edit distance* or *Levenshtein's distance* (Shvaiko & Euzenat, 2005). This distance receives two chains of characters as input and computes the distance between the strings, which is provided by the minimum number of character insertions, eliminations or substitutions as needed to transform a string into another. The edit distance is normalized. The greater the editing distance, the smallest the similarity between the chains of characters is. Thus, we declare a function of similarity in equation 1.

$$F(a,b) = 1 - \frac{editDistance(a,b)}{\max(length(a), length(b))} \quad (1)$$

The edit distance is used to compare concept names and property names. We also calculated the similarity between the property types. The similarity function applied to property names is identical to that applied to the name of object and data type properties and, although the similarity function applied to the types of properties is different for the two. Should the properties have primitive data types as a value (round, floating point, etc.) that is identical, they will then be given value for similarity 1. Otherwise, they will be given value for similarity 0. If the properties are relationships between concepts (object property), we then compare the degree of similarity for the property type by calculating the edit distance between the domains of these properties. We then have it that, if two properties relate to each other with concepts that are similar, then these properties have a certain degree of similarity and consequently the two concepts that have these properties also have a certain degree of similarity.

To give an example of the above two similarity functions consider the concepts "Vehicle" from ontology A and "Car" from ontology B. Concept "Vehicle" has a data type property named "Year of Manufacture" that receives integer values (xsd:int in RDF notation) and an object type property named "has" that relates to the concept "Wheel". In its turn, the concept "Car" has a data type property named "Date of Manufacture" that receives string values (xsd:string in RDF notation) and an object type property named "contains" that also relates to a concept named "Wheel", as shown in Figure 5. We can calculate the similarity function for property name analyzing "Year of Manufac-

ture" and "Date of Manufacture". When applying the edit distance to the name of these properties we will find value $(1 - 4 / 19) = 0.79$; see Equation 1. When applying the similarity function for property type also to "Year of Manufacture" and "Date of Manufacture", we will find value zero (different data types). The same similarity function when applied to object type properties "has" and "contains" will produce value 1 as it is the return of the edit distance between the two concepts Wheel.

The hierarchy of the concepts is used when comparing their children and parents, i.e., two non-leaf concepts are structurally similar if the set of their immediate children is highly similar. The same idea is also used for the immediate parents of the concepts. This similarity function of hierarchy analyzes the context of the concept, that is, to calculate the degree of similarity of two concepts A and B, it is necessary to calculate the degree of similarity of their immediate parents. In this function, concepts that exist in the leaves, that is, concepts that have no children (subclasses) are computed with a degree of similarity 1. Similarly, root concepts, that is, concepts that have no super-classes, are computed with a degree of similarity equal to 1.

The total similarity degree of two concepts is calculated through equation 2 as proposed in (Souza, 1986). The equation is used to assess the similarity function F_x as the summation of the m similarity functions F_q applied to the pairs of ontological elements "a" and "b" with their associated weights W_q.

$$F_x(a,b) = \sum_{q=1}^{m} F_q(a,b) \times W_q \quad (2)$$

Figure 6 shows an example of similarity functions (in the boxes) and the standard weights used in an comparison between concepts. The weights are used to adjust the algorithm and are normal-

Figure 5. Section of ontologies containing a concept and its properties

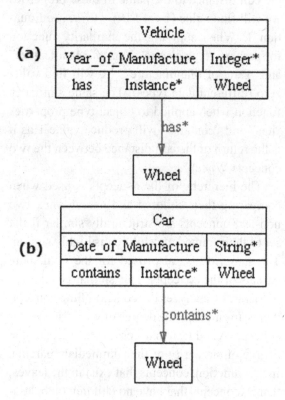

For the cases where more than one related element exists, to be used in the similarity function, that is, concepts that have several properties or concepts with more than a sub-class, the average from the result of the functions is used in the cases where both sides have the same number of elements. If some information is missing in one of the ontologies, for example, if the number of properties that the two concepts under comparison have is different, we use a technique to produce an average of the weights. The goal is to introduce a penalty (equivalent to a negative weight) that is calculated as in equation 3, where L_{min} and L_{max} represent the minimum and maximum number of concept properties (or children or parents of the concept), respectively.

$$\rho = \frac{\sum_{k=0}^{n} F_k(a,b)}{L_{min} + Penalty \times (L_{max} - L_{min})} \qquad (3)$$

The penalty ranges from 0 (when the difference in the number of elements of a concept is not important, resulting into a simple arithmetical average) to 1 (when the difference in the number of elements is important).

When analyzing the result of the function applied to two concepts where each concept has more than one property or sub-class, we were faced with the problem of choosing which relationships should be chosen as relevant. For example, imagine concepts A and B, with respective subclasses [a1, a2] and [b1, b2, b3]. The similarity function will be applied to the entire subclass combination to generate, for example, a matrix such as that shown in Table 1, ordered by degree of similarity.

The return of this function can happen through the choice of the first combination found, without repeating elements, where we would have [a1, b2, 0.8] and [a2, b1, 0.4] totalling a 1.20 similarity. An algorithm that checks all the combinations to choose the best response possible would select combinations [a2, b2, 0.7] and [a1, b3, 0.6] total-

ized to sum 1. The similarity functions return 1 to a perfect similarity and a smaller positive value for pairs with smaller similarities. The equation above is used to aggregate the information on the sub-similarities from the bottom up in the tree shown in the figure below, until the root. That is, the final similarity between two distinct ontology concepts is provided by the sum of their similarity functions.

Suppose that two concepts under comparison have one single property and that similarity functions "Name" and "Type" have produced 0.5 and 1, respectively. This information will contribute to the result of the similarity function "Property" with weights 0.6 and 0.4, respectively. Thus, the similarity function "Property" will be calculated as: 0.5*0.6 + 1 * 0.4 = 0.7.

Figure 6. Similarity functions and their respective weights

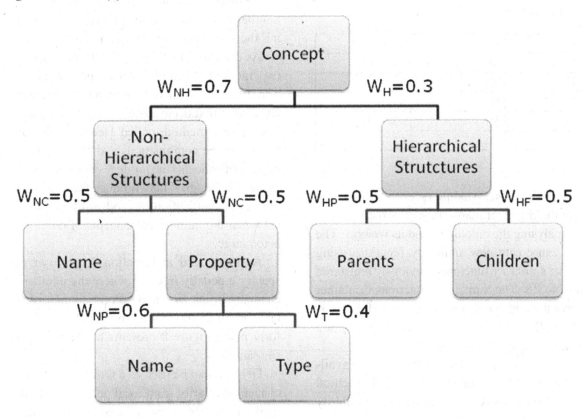

ling 1.30. Even with some cases producing a lower value, we propose the algorithm uses the first approach, as (1) our main concern is to find the biggest similarities between the concepts and find a very similar element is more interesting than to find two less similar elements, even if the sum total of the functions is bigger; apart from that, (2) to find the response that produces the highest total value is a NP-complete problem (Souza, 1986) and can prove to be impractical to calculate the response when the number of elements is too large.

After the calculation of the similarities, the concepts that meet equation 4 are listed to the user as highly similar.

$$\sum_{q=1}^{m} F_q(a,b) \times W_q \geq \Omega \qquad (4)$$

This method was evaluated in (Souza et al, 2010) and shows that the algorithm can be used to reach similarity degrees very close to 1 (that is, close kinship) between concepts even with syntactic or structural differences.

GNOSIS IMPLEMENTATION

The GNoSIS has been developed in Knowledge Engineering Research Group from Federal University of Juiz de Fora in partnership with Federal University of the State of Rio de Janeiro and Pontifical Catholic University of Rio de Janeiro.

GNoSIS allows similarity calculation from two distinct ontology concepts through the specification of several similarity functions. Figure 7 shows the system architecture.

Table 1. Matrix with combinations of similarity [Souza, 1986]

a1	b1	0.8
a2	b2	0.7
a1	b1	0.6
a2	b2	0.4
a1	b1	0.3
a2	b2	0.3

The Composer Module receives a XML file with the list of similarity functions the user sets for analysing the ontologies and its weights. The user can create new similarity functions using atomic similarity functions previously registered in GNoSIS. The Similarity Functions Container stores the registered atomic similarity functions. The XML file represents a function composition such as exemplified in Figure 6.

Some atomic similarity functions are available in GNoSIS, mainly syntactical and structural ones. The user can compose syntactical similarity functions using edit distances metrics such as Levenshtein, Damerau-Levenshtein, Hamming, Jaro-Winkler, and so on. These edit distances metrics can be applied on syntactical similarity functions that compares ontology terms as concept names, property names, labels, descriptions, etc. In the same way, atomic structural similarity functions for comparing concept hierarchy, property range and domain, and individuals. GNoSIS allows that other similarity functions can be implemented and registered. Thus, the system can be useful to assess distinct similarity functions proposed in literature in regards with performance and similarity results.

The Similarity Engine is responsible for processing the composite functions on the ontologies and populate a table with the similarity degrees from each pair of concepts. The user can specify which concepts will be compared, or in case of any concept is chosen, the system will compare all the concepts of two ontologies. In the latter case, the similarity table has mn lines, where m is the number of concepts from the first ontology and n is the number of concepts from the second one.

As discussed in Table 1, the user can specify two methods to combinating multiple results. The first one, called FirstMatchCombination, checks the first occurrence of a pair of concepts. The second method, called DeepCombination, checks all the combinations and chooses the set of pairs of concepts the produces better similarity degree. Although it may produce the best result, the DeepCombination method can be impractical when the number of concepts of ontologies is too large.

As mentioned in Equation 3, the user can specify a penalty in cases where the number of elements in the first and second column of the table are different, that is, some elements do not form pairs. Figure 8 presents a GNoSIS XML file sample.

GNoSIS can be used as a tool for calculating concept similarity from local ontologies. In this work, though, we use GNoSIS within Agent Strategy Planning module (Figure 4). Figure 9 shows a agent communication process in our approach.

As mentioned earlier, we propose adding meaning negotiation skills to agents by altering Goddard agent architecture. However, the proposal aims to minimize the impact in traditional agent communication. Thus, our agent communication process starts and ends with traditional request and response messages. In case of receiving messages with terms not defined in the receiver ontology, the usual agent communication stops and waits for the meaning negotiation ends.

In agent communication above, the receiver verifies if the message received contains terms not found in the agent ontology. The list of terms not found is sent to sender. Each term definition is sent to the receiver, which calculates the similarity degree using a predefined similarity function composition from GNoSIS. The proposed map-

Figure 7. GNoSIS architecture

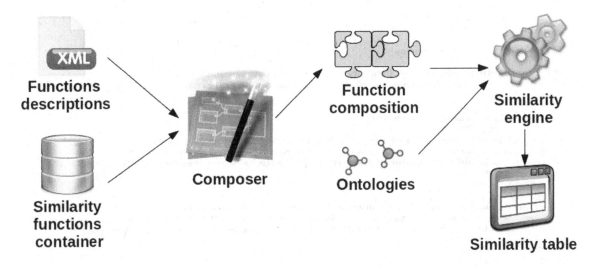

ping is sent to the sender that verify the similarity degree with its predefined similarity function composition. The sender can send a message with mapping agreement, refusal or a weak agreement. In case of agreement, the agents store the mapping in its Mapping Base module. In case of refusal, the meaning negotiation fails and the communication breaks. A weak agreement means that the similarity degree calculated by sender did not reach a upper threshold, but did not reach a lower threshold as well. In case of weak agreement, the agreement depends on the receiver Negotiating Planning module configuration, since the sender is not sure about the correctness of the mapping.

RELATED WORK

Multi-agent systems that use negotiation concepts are proposed in the literature (Chen et al, 1999; Collins & Gini, 2008; Huang & Sycara, 2002), although each agent has the capacity to consult a single, global domain ontology to avoid communication problems, which makes these approaches applicable only to environments that are more controlled than the Web.

A negotiation agent is considered an agent capable of exchanging offers to reach reasonable consensus to the parties. However, systems that use negotiation agents are usually systems that aim at the object of the negotiation, be it the acquisition of a product or a mutual action that will be implemented.

On the other hand, agents for the negotiation of meaning have been proposed in the literature (Packer et al, 2009; Aschoff et al, 2004, Elst & Abecker, 2002; Tijerino et al, 2004), where the goal of the agent is to solve conflicts with the understanding of concepts. This scenario also differs from this work, for the same reason mentioned earlier, as the goal of these agents is the object of the negotiation. This work, in its turn, uses the negotiation capacity of agents independently from its goals. The negotiation of meanings is part only of the agent communication process, ensuring an understanding of the concepts involved in the message received by the agent.

In the articles mentioned in the previous paragraph, the negotiation of meanings is made to achieve the compatibility of domain ontologies as evolved by human users. In (Aschoff et al, 2004, Elst & Abecker, 2002), negotiation performatives are defined and the negotiation

Figure 8. GNoSIS XML file sample

```
<?xml version="1.0" encoding="ISO-8859-1" ?>
<!DOCTYPE similarityCalc (View Source for full doctype...)>
- <similarityCalc>
  - <container name="Funcoes principais">
    - <function weight="0.3">
        <class>br.ufjf.ontology.gnosis.similarity.structure.ConceptNameSimilarity</class>
        <strategy>br.ufjf.ontology.gnosis.similarity.editdistance.DamerauLevenshteinEditDistance</strategy>
        <penalty>0.0</penalty>
      </function>
    - <container name="Funcoes Secundarias" weight="0.7">
      - <function weight="0.35">
          <class>br.ufjf.ontology.gnosis.similarity.structure.DirectSuperClassSimilarity</class>
          <combination>br.ufjf.ontology.gnosis.similarity.combination.DeepCombination</combination>
          <strategy>br.ufjf.ontology.gnosis.similarity.editdistance.DamerauLevenshteinEditDistance</strategy>
          <penalty>0.0</penalty>
        </function>
      - <function weight="0.35">
          <class>br.ufjf.ontology.gnosis.similarity.structure.DirectSubClassSimilarity</class>
          <combination>br.ufjf.ontology.gnosis.similarity.combination.DeepCombination</combination>
          <strategy>br.ufjf.ontology.gnosis.similarity.editdistance.DamerauLevenshteinEditDistance</strategy>
          <penalty>0.0</penalty>
        </function>
      + <function weight="0.1">
    - <container name="Funcoes Propriedades" weight="0.2">
      - <function weight="0.2">
          <class>br.ufjf.ontology.gnosis.similarity.structure.DirectDataTypePropertybyRangeSimilarity</class>
          <penalty>0.0</penalty>
```

protocol, as in our work, describes the process through which agents can communicate on the meaning of the concepts. However, the need of human intervention in the negotiation process of the agents is a point of difference in the application scenario that is described at that work. Apart from that, our proposal extends an architecture for the construction of agents (Truszkowski, 2006) and introduces sub-modules into the communication module to deal with concepts that are typical to negotiation with the objective of allowing the process of communication between agents to be made without the interference in the intention of the agent. Apart from that, the intention is to see this proposal used by agents that communicate between themselves in open environments such as the Web, where constant human intervention is not reasonable.

Approaches to calculate similarity between concepts have been studied by several scholars (Euzenat & Shvaiko, 2007). Some proposals try to solve the problem with the use of semantic comparisons (Bouquet et al, 2003) or syntactic ones. As RiMON (Yi Li et al, 2006), our approach uses different similarity functions in a set. The main contribution of GNoSIS is in the separation of the similarity functions in each group of elements of the ontology in a recursive manner over the concepts and the possibility of creating new composite similarity functions. Once mapping discovery technique depends on some ontologies characteristics and the better technique can change from one to another pair of ontologies, each agent can use distinct composite functions.

CONCLUSION

Two active (and related) research topics in the Semantic Web are (1) finding solutions for the semantic interoperability between the growing number of ontologies that, one imagines, will exponentially increase as the area matures, and (2) allowing agents to communicate with other agents using distinct domain ontologies. This work contributes to these two research topics

Figure 9. Example of an agent communication process

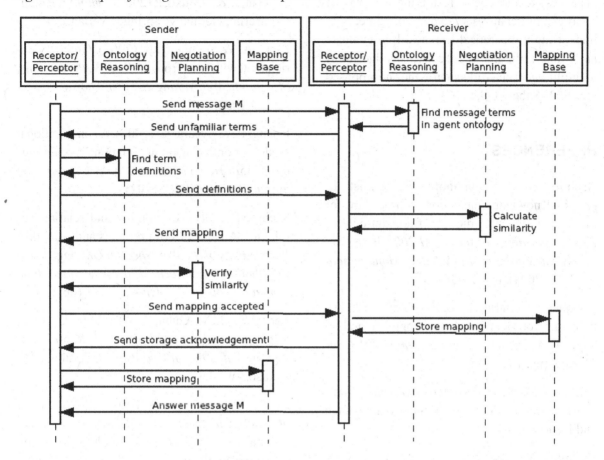

when proposing a negotiation approach to assist in obtaining consensus during the harmonization between knowledge structures.

The work presented an extension to the Goddard agent architecture to carry out a treatment of the terms defined in ontologies as received during the communication between agents that use distinct domain ontologies. The treatment of the terms is made through a process of negotiation of meanings that results in an alignment between the concepts of the ontologies.

Together to agent architecture, this article presented the GNoSIS, tool for combining different approaches for the alignment of ontologies: hierarchical, lexical and relational. GNoSIS carries out a recursive calculation on the concepts of the ontologies where the partial function for the

similarity between subclasses or super-classes of a concept depends on the calculation of the total similarity function between these two concepts. GNoSIS was implemented as a module in agent architecture proposed. In spite of the higher degree of complexity of the algorithm, the evaluations showed in (Souza et al, 2010) that this approach can be used in an efficient way in ontological structures with conceptual divergences of a median degree of complexity.

As future work, an analysis of different methods for the alignment of ontologies that could be better applied during the communication between the agents is suggested, considering similarity calculations that are more specific to the structure of ontologies (that is, considering concepts, properties and rules, as well as possibly instances), to allow better results that can provide a more significant reserve value to the negotiations using a theoretical framework to evaluate algorithms in the analysis of consensus algorithms for multi-agent networks such as the one proposed in (Olfati-Saber et al, 2007).

REFERENCES

Aschoff, F. R., Schmalhofer, F., & Elst, L. (2004). Knowledge mediation: A procedure for the cooperative construction of domain ontologies. *Proceedings of the ECAI-2004 Workshop on Agent-Mediated Knowledge Management* (AMKM-2004), (pp. 29-38).

Bouquet, P., Serafini, L., & Zanobini, S. (2003). Semantic coordination: A new approach and an application. In *International Semantic Web Conference*, (pp. 130-143).

Bratman, M. E. (1999). *Intention, plans, and practical reason*. Center for the Study of Language and Informatics.

Breitman, K., Casanova, M. A., & Truszkowski, W. (2007). *Semantic Web: Concepts, technologies and applications*. London, UK: Springer-Verlag.

Chakrabarti, S. (2000). Data mining for hypertext: A tutorial survey. *Proceedings of the ACM SIGKDD Explorations: Newsletter of the Special Interest Group (SIG) on Knowledge Discovery & Data Mining, 1*(2), 1-11.

Chen, Y., Peng, Y., Finin, T., Labrou, Y., & Cost, S. (1999). Negotiating agents for supply chain management. *Proceedings of the AAAI Workshop on Artificial Intelligence for Electronic Commerce.*

Collins, J., & Gini, M. (2008). Scheduling tasks using combinatorial auctions: The MAGNET approach. In Adomavicius, G., & Gupta, A. (Eds.), *Handbooks in Information Systems series: Business computing*. Elsevier.

Euzenat, J., & Shvaiko, P. (2007). *Ontology matching*. Berlin, Germany: Springer-Verlag.

Faatz, A., & Steinmetz, R. (2002). Ontology enrichment with texts from the WWW. *Proceedings of Semantic Web Mining 2nd Workshop at ECML/PKDD-2002.*

Huang, P., & Sycara, K. (2002). A computational model for online agent negotiation. *Proceedings of the 35th Hawaii International Conference on System Sciences* (HICSS'02).

Klein, M. (2001). Combining and relating ontologies: An analysis of problems and solutions. *Proceedings of the Workshop on Ontologies and Information Sharing at the 17th International Joint Conference on Artificial Intelligence*, (pp. 53-62).

Lenat, D. B., & Guha, R. V. (1990). *Building large knowledge-based systems: Representation and inference in Cyc Project*. Boston, MA: Addison-Wesley.

Li, Y., Li, J., Zhang, D., & Tang, J. (2006). *Result of ontology alignment with RiMOM at OAEI'06*. International Workshop on Ontology Matching collocated with the 5th International Semantic Web Conference.

Niles, I., & Pease, A. (2001). Towards a standard upper ontology. *Proceedings of the International Conference on Formal Ontology in Information Systems.*

Olfati-Saber, R., Fax, J. A., & Murray, R. M. (2007). Consensus and cooperation in networked multi-agent systems. *Proceedings of the IEEE, 95*(1), 215–233. doi:10.1109/JPROC.2006.887293

Packer, S. H., Gibbins, N., & Jennings, N. R. (2009). *Ontology evolution through agent collaboration*. Workshop on Matching and Meaning: Automated Development, Evolution and Interpretation of Ontologies.

Pica, T. (1987). Second-language acquisition, social interaction and the classroom. *Applied Linguistics, 8*(1), 3–21. doi:10.1093/applin/8.1.3

Shvaiko, P. (2004). A classification of schema-based matching approaches. *Proceedings of the Meaning Coordination and Negotiation workshop at International Semantic Web Conference (ISWC).*

Shvaiko, P., & Euzenat, J. (2005). A survey of schema-based matching approaches. *Journal on Data Semantics, 4,* 146–171.

Souza, J. F., Melo, R. N., & Siqueira, S. W. (2010). Improving software agent communication with structural ontology alignment methods. *International Journal of Information Technology and Web Engineering, 5,* 49–64. doi:10.4018/jitwe.2010070103

Souza, J. F., Paula, M., Oliveira, J., & Souza, J. M. (2006). Meaning negotiation: Applying negotiation models to reach semantic consensus in multidisciplinary teams. *Group Decision and Negotiation,* 297–300.

Souza, J. F., Siqueira, S. W., & Melo, R. N. (2009). Adding meaning negotiation skills in multiagent systems. *Proceedings of the IEEE International Conference on Intelligent Computing and Intelligent Systems,* (pp. 663-667).

Souza, J. M. (1986). *Software tools for conceptual schema integration.* University of East Anglia.

Tijerino, Y. A., Al-Muhammed, M., & Embley, D. W. (2004). Toward a flexible human-agent collaboration framework with mediating domain ontologies for the Semantic Web. *Proceedings of the ISWC '04 Workshop on Meaning Coordination and Negotiation,* (pp. 131-142).

Truszkowski, W. F. (2006). What is an agent? And what is an agent community? In Rouff, C. A., Hinchey, M., Rash, J., Truszkowski, W., & Gordon-Spears, D. (Eds.), *Agent technology from a formal perspective* (pp. 3–24). London, UK: Springer-Verlag. doi:10.1007/1-84628-271-3_1

Section 2
Web Engineering Discoveries

Chapter 4
A Pattern Language for Knowledge Discovery in a Semantic Web Context

Mehdi Adda
University of Quebec at Rimouski, Canada

ABSTRACT

Ontologies are used to represent data and share knowledge of a specific domain, and in recent years they tend to be used in many applications such as database integration, peer-to-peer systems, e-commerce, semantic web services, bioinformatics, or social networks. Feeding ontological domain knowledge into those applications has proven to increase flexibility and inter-operability and interpretability of data and knowledge. As more data is gathered/generated by those applications, it becomes important to analyze and transform it to meaningful information. One possibility is to use data mining techniques to extract patterns from those large amounts of data. One challenging general problem in mining ontological data is taking into account not only domain concepts, properties and instances, but also hierarchical structures of those concepts and properties. In this paper, the authors research the specific problem of extracting ontology-based sequential patterns.

INTRODUCTION

Web usage mining (WUM) is about the discovery of trends in user navigational behavior within a Web application (Mobasher, Jin, & Zhou, 2004; Pierrakos, Paliouras, Papatheodorou, & Spyro-poulos, 2003; Eirinaki, Lampos, Vazirgiannis, & Varlamis, 2003). A WUM task is typically motivated by the need to understand the strategies adopted by the user (e.g., within an e-Learning context) or to personalize the content served by the application (e.g., in e-Commerce). The standard approach for WUM is association rule extraction

DOI: 10.4018/978-1-4666-0023-2.ch004

from the application logs (Mobasher, Jin, & Zhou, 2004; Melville, Mooney, & Nagarajan, 2002; Jin & Mobasher, 2003).

With the emergence of the Semantic Web and the associated use of ontologies to describe the semantic content of web documents, the emphasis has been shifted on the ability of WUM engines to incorporate large amounts of domain knowledge (Eirinaki, Lampos, Vazirgiannis, & Varlamis, 2003; Dai & Mobasher, 2004; Liao, Chen, & Hsu, 2009). Although the topic of combining semantics and WUM has been largely covered in the literature, the specific impact of using an ontology in the mining process has not been clearly stated. In fact, most studies assume the domain ontology to be a mere conceptual hierarchy, thus reducing the mining problem to, at worst, a generalized sequential pattern extraction.

It is noteworthy that the full-blown ontologies that may underlay sophisticated Web information systems such as portals, virtual museums, Web e-learning and e-commerce environments, etc., have more structure than a mere conceptual *is-a* tree, in particular, they often comprise a collection of inter-concept relations, possibly organized into a separate *is-a* hierarchy. At the individual level, the relations materialize as inter-object links that belong to object descriptions and hence may be relevant for mining.

Such links induce higher-order dependencies between the content objects in a sequence, leading to a secondary graph structure (Bandyopadhyay, Maulik, Holder, & Cook, 2005; Mineau, Moulin, & Sowa, 1993; Huan, Wang, Prins, & Yang, 2004). While the mining of graph-shaped data is now a relatively well understood topic, its enhancement to graphs with hierarchically organized labels on both vertices (objects) and edges (links) as provided by a full-blown ontology, is far less studied (Washio & Motoda, 2003). The goal of this study is therefore to define a mining discipline for this new pattern type.

The work presented in this paper is an enhancement and completeness of a framework previously proposed to mine this new category of patterns (Adda, Valtchev, Missaoui, & Djeraba, 2007). It is composed of a pair of languages (a data and a pattern one) provided with a generality relationship as well as a level-wise mining method, *xPMiner*, based on a set of specialization operators.

RELATED WORK

Semantic Web is designed to let users and machines describe resources, share that data in a distributed manner and enable interpretation and processing of the related data (Lee, Hendler, & Lassila, 2001; Lee, 2001). In recent years, ontologies are widely used to realize the data layer of Semantic Web. In this paper the specific problem of mining onlogy-powered systems is addressed. One reason for focusing in the knowledge discovery aspect of a Semantic Web environment is that the concepts and relations of an ontology have an impact on the quantity and quality of the patterns that may be extracted from such graph-based data (Di-Jorio, Bringay, Fiot, Laurent, & Teisseire, 2008; Liao, Chen, & Hsu, 2009; Rajapaksha & Kodagoda, 2008).

Mining data in the context of ontology-powered systems involves different fields such as ontology engineering, knowledge discovery and pattern mining. Hereafter, the background knowledge related to this study on both pattern mining and ontologies is presented.

Pattern Mining

Given a universe of objects, or items, O, a database D made of records combining items from O, and a frequency, or *support*, threshold σ, frequent pattern mining amounts to extracting the family F_σ of item collections, or *patterns* that are present in at least σ records. Two languages, the pattern language Γ and the data one Δ, and two binary relations underlay the problem: the *generality* between patterns $\sqsubseteq$ and *instantiation* between a data record and a pattern $\prec$. Generality follows instantiation as given a pattern $f \in \Gamma$ and a super-pattern thereof $\bar{f}$ ($f \sqsubseteq \bar{f}$), each record $d \in \Delta$ instantiating $d \prec f$ instantiates $\bar{f}$ as well.

In the simplest settings, both data records and patterns are sets of items (*itemsets*), i.e., $\Delta = \Gamma = 2^\circ$, while $\sqsubseteq$ and $\prec$ boil down to set-theoretic inclusion. Hence the mining goal amounts to finding all the frequent subsets of a family of sets $D \subseteq \Delta$.

More elaborate record structures have been studied including sequences (Agrawal & Srikant, 1995; Di-Jorio, Bringay, Fiot, Laurent, & Teisseire, 2008) Δ, $\Gamma = w^\circ$) and graphs (Inokuchi, Washio, & Motoda, 2000) with equally more complex generality and instantiation (subsequence and sub-graph relations, respectively). A somewhat orthogonal research axis researches *generalized* pattern languages which are built on top of a set C_Ω of abstract concepts from the domain underlying O, i.e., Γ derived from 2^{C_Ω} instead of 2°. Concepts typically form a taxonomy, $H = <C_\Omega, \leq >$, ($\leq$ is the is-a relationship), which is a highly simplified case of a full-scale domain ontology (Ding, Kolari, Ding, & Avancha, 2007).

Pattern mining methods search for F_σ through the pattern space $<\Gamma, \sqsubseteq>$ with frequency checks, by typically exploring the monotony in frequency w.r.t. $\sqsubseteq$ (a super-patterns of a frequent pattern is frequent). Apriori (Agrawal & Srikant, 1994) is the prototypical pattern miner which, basically, performs a level-wise top-down traversal of $<\Gamma, \sqsubseteq>$. On itemsets, it examines patterns at level k, i.e., of size k ($k = 1, 2, 3,...$) on two points: frequency of *candidate* patterns is computed by matching them against the records in D, while $k + 1$ candidates are generated by combining pairs of frequent k-patterns. As scanning the entire database is expensive, frequency scores are exploited for a priori invalidation of infrequent

candidates. The level-wise mining approach has been successfully adapted to both structured and generalized patterns, but less used in ontology-based environments.

Feeding a Full-Scale Ontology

A better trade-off between flexibility, interpretability and precision can be achieved by feeding into the mining process knowledge about semantic links between objects (Adda, Valtchev, Missaoui, & Djeraba, 2007). The basic assumption is that co-occurrences between objects often reflect the existence of a link between them (e.g., clicks on the page of *Paris* are usually followed by clicks on pages of Parisian hotels). Hence, manipulating links explicitly can increase the focus of concept-based patterns while preserving its flexibility.

As ontologies constitute the standard way of representing domain concepts and relations, thus one has to assume that an ontology powers the target system on which the mining task will be performed. In this work, it is made of two *is-a* taxonomies, on concepts and relations, respectively, whose elements are connected by an attribution relation. Figure 1 (on the left) provides a partial view of the *Travel* ontology[1]. Formally, an ontology is a four-tuple, $\Omega = < C_\Omega, R_\Omega, \sqsubseteq, \rho>$, where C_Ω is the set of concepts, or *classes*, and R_Ω a set of domain relations, or *properties*. In Figure 1, these are depicted as two-section rectangles and ovals, respectively (*e.g.*, *LuxuryHotel* is a class and *hasAcccommodation* a property). The generality order $\sqsubseteq$ holds both for concepts and properties $\sqsubseteq \subseteq (C_\Omega \times C_\Omega) \bigcup (R_\Omega \times R_\Omega)$. Generality links are depicted by plain arrows (e.g., *LuxuryHotel* $\sqsubseteq$ *Hotel* and *hasHotel* $\sqsubseteq$ *hasAcccommodation*). Attribution is ternary relation $\rho \subseteq (C_\Omega \times R_\Omega \times C_\Omega)$ which connects a property r to classes c and $\bar{c}$ whose instances are, respectively, origin and destination of the links of type r. In Figure 1, attribution is expressed through *domain*- and *range*-labeled dashed arrows pointed to a prop-

Figure 1. (A) Partial view of Travel ontology and sample sequences: (B) object sequence, extended object sequence and (C) two patterns. (©2009, mehdi adda)

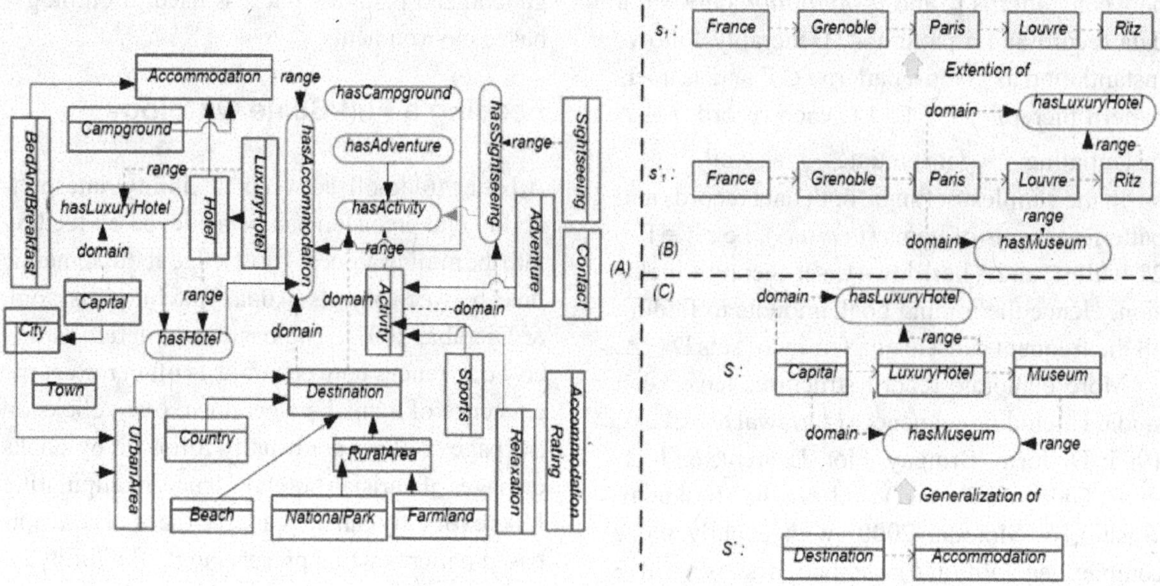

erty (e.g., *Destination* is the domain and *Accommodation* the range class *hasAcccommodation*).

Descriptions of class and property instances are assumed organized into a separate knowledge base. Formally, let

$$K = <O, \preccurlyeq, \rho_o> \quad \rho \subseteq (C_\Omega \times R_\Omega \times C_\Omega)$$

be such a base where O stands for objects and $\preccurlyeq$ for the instantiation relationship. In Figure 1 (on the right), single-section rectangles represent objects while instantiation remains implicit (e.g., *Paris* clearly instantiates *Capital*). Table 6 provides a set of $\preccurlyeq$ links. Furthermore, ρ_o expresses inter-object links $(\rho_o \subseteq (O \times R_\Omega \times O))$ in a way similar to ρ hence its identical visualization in Figure 1 (e.g., a *hasLuxuryHotel* link between *Paris* and *Ritz*) (Table 1). A set of attribution links is provided in Table 2.

In the following, the definitions of both data and pattern languages are consolidated and the relationships between their respective elements are formalized. An optimized method for pattern generation is proposed and new evidence about its behavior on click-stream data is presented.

ONTOLOGY-BASED FREQUENT SEQUENTIAL PATTERNS

The ontological knowledge encoded in Ω and K induces a characteristic pattern space with a specific mining discipline composed of two languages: the first to represent data and the second to represent patterns. As in this work only sequential patterns are looked for, initial data structures are ordered sets (lists) of objects. The two languages are presented below.

Descriptive Languages and Pattern Space

The data language Δ_Ω is derived from raw data records of sequences of objects. Those objects could be simple page URIs or products/items visited or bought together. In the running example of this

Table 1. Partial extensions for some travel classes

Class	Instances
Country	France, Australia, Italy
Capital	Paris, Roma
City	Grenoble, Cairns, Sydney
Museum	Louvre, Augustus_Mausoleum
Beach	Clifton_Beach, Bondi_Beach
Safari	Cape_York_Safari
Surf	Bondi_Beach_Surf

Table 2. A subset of inter-object links

Property	Object pairs
hasMuseum	(Paris, Louvre), (Roma, Augustus_Mausoleum)
hasLuxury-Hotel	(Paris, Ritz), (Roma, St_George), (Sydney, Swissotel)
hasSafari	(Cairns, Cape_York_Safari)
hasSurf	(Sydney, Bondi_Beach_Surf)

work objects correspond to page URIs representing destinations, activities and accommodations of an e-tourism website. These are readily translated into object IDs from K. The set of raw object sequences composing the example is given in Table 5 whereas s_1 is depicted in Figure 1.

Each sequence is further extended with all the links from ρ_O involving exclusively objects from the sequence. The resulting structures are labelled *digraphs* where vertices are objects and edges either represent semantic links or belong to the initial sequential structure. For instance, the completion of s_1, denoted s'_1, is depicted beneath it in Figure 1 (Table 3).

Formally, the language Δ_Ω is made of pairs $s = (\varsigma, \theta)$ where $\varsigma \in w^o$ is an object sequence, and $\theta \in E(\rho_O)^2$ is a set of links. For conciseness reasons, the triplets in each $s.\theta$ will be simplified from (o_d, r, o_r) to $r(l, m)$ where l and m are natural numbers denoting the ranks of o_d and o_r, respectively, within $s.\theta$. Table 4 illustrates the θ completions of the sequences in Table 5. Currently, it is

required that all links in $s.\theta$ be co-linear with the sequence $s.\varsigma$, *i.e.*, $\forall r\,(l,m) \in s.\theta,\ l \leq m$.

The language Γ_Ω has a structure identical to that of Δ_Ω: A pattern $S \in \Gamma_\Omega$ is a pair of a sequence and a triplet set, (ς, θ). However, here $\varsigma \in w_\Omega^c$ and $\theta \in E(\rho_O)^2$ follow the ontology rather than the knowledge base. For instance, a pattern S is depicted in Figure 1, on the right, whereby its textual form spells: (*<Capital, LuixuryHotel, Museum1>, {hasLuxuryHotel(1,2), hasMuseum(1,3)}*) Further patterns are given in Tables 7 to 10.

Interpretation and Generality Relationship

A pattern $S \in \Gamma_\Omega$ is interpreted by a subset of Δ_Ω. The interpretation is based on two elements: (i) domain of interpretation (Δ_Ω), and (ii) an interpretation function which associates to each pattern a set of elements of the interpretation domain. This function is based on the instantiation relationship between a pattern and a data record. Furthermore, the generality order among patterns that follows instantiations is defined.

Instantiation Relationship

Instantiation between data records and patterns relies on graph morphism. Intuitively, $s \in \Delta_\Omega$ instantiates $S \in \Gamma_\Omega$ whenever a sub-structure of s exists that may be homomorphically mapped *onto* S. Formally, the instantiation relationship between a pattern and a data record is defined as follows:

Definition 3.1. *Instantiation relationship*

An instantiation between s and S, denoted by $s \prec S$, is witnessed by a partial integer map $\phi : [1, |\, S.\varsigma\,|]^3 \to [1, |\, s.\varsigma\,|]$ satisfying:

- ϕ is an order preserving and injective function,

Table 3. Initial object sequences

Id	Object sequences ($\mathbf{\grave{o}}$)
s_1	*<France, Grenoble, Paris, Louvre, Ritzi>*
s_2	*<Australia, Cairns, Clifton_Beach, Cape_York_Sa-far>*
s_3	*<Italy, Roma, Augustus_Mausoleum, Grand_Ho-tel_Plaza>*

Table 4. Relational completions of the object sequences

Id	Relation triplet sets (è)
s_1	*{hasLuxuryHotel(3,5), hasMuseum(3,4)}*
s_2	*{hasSafari(2,4)}*
s_3	*{hasLuxuryHotel(2,4), hasMuseum(2,3)}*

- For each $i \in [1.. \mid S.\varsigma \mid]$, the object $s.\varsigma[\phi(i)]$ is an instance of the concept $S.\varsigma[i] (s.\varsigma[\phi(i)] \in [S.\varsigma[i]]_\Omega)$,

- For each triplet r'(i_1, i_2) in $S.\theta$ exists an instantiating triplet $r(j_1, j_2)$ in $s.\theta$, such that (j_1, j_2) in $([1.. \mid s.\varsigma \mid])^2 2$ and $\phi(i_1) = j_1$, $\phi(i_2) = j_2$ ($r \sqsubseteq_\Omega$ r' *and* $(s.\varsigma[j_1], s.\varsigma[j_2]) \in [r'(i_1, i_2)]_\Omega)$.

For example, in Figure 2, S' is instantiated by s'_1 (via $\phi = \{(1, 2), (2, 4)\}$), but not S (classes in reverse order).

Based on the definition above, the formalization of the interpretation is depicted bellow.

Definition 3.2. *Interpretation relationship*

The interpretation of the pattern S, denoted $[S]_{\Gamma_\Omega}$, is a set of object records s such that $s \ll S$, which means $\forall S \in \Gamma_\Omega$: $[S]_{\Gamma_\Omega} = \{s \in \Delta_\Omega \mid s\ S\}$.

Generality Order

The generality order on Γ_Ω follows instantiations as a pattern S_1 is a super-pattern of another one, S_2, whenever every data sequence that instantiates S_2 also instantiates S_1. For example, given the Travel ontology and the knowledge base composed of Tables 6 and 2, one could check that each sequence instantiating S from Figure 1, also instantiates S_0. However, the inverse is not true as a counter-example was given above. Hence S_0 is strictly more general than S. Formally, it is defined as follows:

Definition 3.3. *Generality relationship*

Let S_1 and S_2 be two patterns from $\Gamma_\Omega S_1$, is said more general than S_2, denoted by $S_2 \leq S_1$, if and only if $[S_2]_{\Gamma_\Omega} \subseteq [S_1]_{\Gamma_\Omega}$.

The instantiation-based generality test is unpractical. In fact, only part of the domain of interpretation of a pattern is available. An alternative computation mechanism is designed that mimics the homomorphic mapping in instantiation tests. Hence, a pattern relationship, called pattern subsumption, is proposed. This relationship relays on the syntactical structure of patterns to replace instantiation test.

Table 5. Object sequences

Id	Object seauence
s_1	*<France, Grenoble, Paris, Louvre, Ritz>*
s_2	*<Australia,Cairns,Clifton_Beach,Cape_Y ork_Safari>*
s_3	*<Italy,Roma, Mausoleum_of_Augustus, Grand_Hotel_Plaza>*

Table 6. Class/instance mapping

Class	Instances
Country	France, Australia, Italy
Capital	Paris, Roma
City	Grenoble, Cairns
Museum	Louvre, Augustus_Mausoleum
LuxuryHotel	Ritz, Grand_Hotel_Plaza
Safari	Cape_York_Safari
Beach	Clifton_Beach

Table 7. 1-Frequents and their associated support

Pattern	Support
$S_1 = <<Accommodation>,\{\}>$	2
$S_2 = <<AccommodationRating>,\{\}>$	0
$S_3 = <<Activity>,\{\}>$	3
$S_4 = <<Contact>,\{\}>$	0
$S_5 = <<Destination>,\{\}>$	3

Pattern Subsumption

The generality order on Γ_Ω follows instantiations as a pattern S_1 is a super-pattern of another one, S_2.

Similarly to instantiation, subsumption reflects the existence of a sub-structure of the more specific pattern homomorphic to the more general one (in the above sense, by simply replacing ontology instantiation by is-a links). The only difference with instantiation resides in the use of inter-class specialization instead of object-to-class instantiation links. Within the example from Figure 1, the function ψ for S and S' is $\{(1,1),(2,2)\}$.

Definition 3.4. *Pattern Subsumption*

Let S_1 and S_2 two patterns from Γ_Ω, S_2 subsumes S_1, denoted by $S_1 \sqsubseteq_{\Gamma_\Omega} S_2$, if it exists an injective and monotonously increasing function ψ: $[1,|S_2.\varsigma|] \rightarrow [1,|S_1.\varsigma|]$ such that:

- For each $i \in [1..|S_2.\varsigma|]$, $S_1.\varsigma[\psi(i)] \sqsubseteq_{\dot{U}} S_2.\varsigma[i]$,
- For each triplet $r(i_1, i_2)$ in $S_2.\theta$ exists (j_1,j_2) in $([1..|S_1.\varsigma|])^2 : \psi(i_1)=j_1$, $\psi(i_2)=j_2$ and it exists $r'(j_1,j_2)$ in $S_1.\theta$, such that $r \sqsubseteq_\Omega r'$.

Figure 3 is an illustration of pattern subsumption with regards to the ontology *Travel*. In this figure, it is noteworthy that for the patterns S' and S, the conditions of *Definition 3.4* are satisfied, and hence conclude that S' subsumes S since *Capital* and *LuxuryHotel* of S are subclasses of *Destination* and *Accommodation* of S' in *Travel*, respectively. Additionally, the relation *hasLuxuryHotel* of S is mapped to the more general relation *hasAccommodation* of S'. However, S'' does not subsume S because of the existence of a concept in S'' (namely *Accommodation*) which do not have an antecedent in S.

Remark 3.5. For each pattern S of Γ_Ω: $S \sqsubseteq_{\Gamma_\Omega} \varnothing_{\Gamma_\Omega}$, which means that the empty pattern $\varnothing_{\Gamma_\Omega}$ is the root of the pattern space Γ_Ω.

Subsumption relationship is a partial order (see *Proposition 1*). Thus, for two patterns S_1 and S_2 from Γ_Ω, $S_1 = S_2$ if and only if $S_1 \sqsubseteq_{\Gamma_\Omega} S_2$ and $S_2 \sqsubseteq_{\Gamma_\Omega} S_1$.

Proposition 1. *Partial Order*

The relation $\sqsubseteq_{\Gamma_\Omega}$ is a partial order over the pattern universe Γ_Ω.

Now that the subsumption relationship is defined, it can be stated that it effectively replace the generality relationship between two patterns (see *Proposition 2*).

Proposition 2. *Subsumption relationship to test Pattern Generality*

Figure 2. Example of mapping with structure preservation. (©2009,mehdi adda)

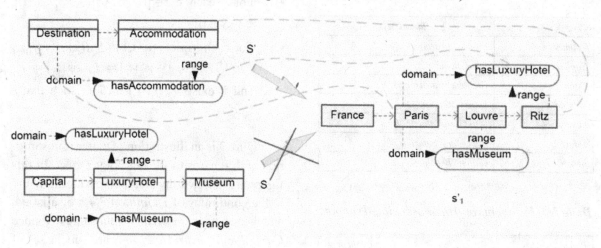

Let S_1 and S_2 two patterns from Γ_Ω such that $S_1 \sqsubseteq_{\Gamma_\Omega} S_2$ then $S_1 \leq S_2$

Pattern Space Traversal

To insure precise traversal of the generality-based pattern hierarchy, the precedence relationship behind generality needs to be followed (corresponding to moving from a set of size k to its $k + 1$ supersets in *Apriori*). Moreover, a level in the new space is defined as the set of patterns having the same generality degree. The level is independent of the pattern size since two equally-sized elements from Γ_Ω may well present a generality relationship. Moving to a finer notion that reflects the generality degree of pattern components, classes and properties, within the ontology, the generality rank measure (rank: $\Gamma_\Omega \rightarrow N$) is presented. It is based on the depth of classes/properties ($h : C_\Omega \cup R_\Omega \rightarrow N$), *i.e.*, the length of a maximal path from a component to a root of its respective *is-a* hierarchy (e.g., h (*LuxuryHotel*) = 3 in Figure 1). The rank of a pattern is the sum of its component depths (in Figure 1, rank(S') = 3 and rank(S) = 16). The level k in the pattern space is thus made of all patterns of rank k.

Furthermore, the level-wise descent relies on generation of $k + 1$-level patterns from k-level

ones which amounts to the computation of the precedence $\prec_{\Gamma_\Omega}$ on Γ_Ω (transitive reduction of $\sqsubseteq_{\Gamma_\Omega}$). The target operations clearly increase the pattern rank by 1, hence are easily translated in terms of component precedence. In summary, four unit operations are defined for a pattern S:

- **Class insertion (AddCls(S, c, j))**: insertion of a root class c from C_Ω into a pattern S at a position j within the range of the sequence size:
  - $1 \leq j \leq |S.\varsigma| + 1$,
  - $c \in \max_{\sqsubseteq_\Omega} (C_\Omega)$.

- **Class specialization (SplCls(S, c, j))**: substitution of a non-leaf class at position j of a pattern $S = (\varsigma, \grave{e})$ ($S.\varsigma[j]$) with c, immediate subclass of $S.\varsigma[j]$ from C_Ω, provided c is comparable with the actual ranges of all the properties r involved in triplets $r(,j)$ from θ:
  - $S.\varsigma[j] \notin \min_{\sqsubseteq_\Omega} (C_\Omega)$,
  - $c \prec_\Omega S.\varsigma[j]4$,
  - $\forall r(i, j) \in S.\theta : \exists c' \in ran_\Omega (r, S.\varsigma[i])5$

  such that: $c \sqsubseteq_\Omega c'$.

Figure 3. Pattern subsumption illustrated. (©2009, mehdi adda)

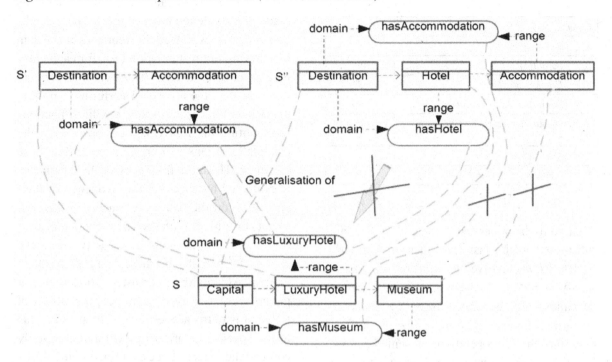

- **Relational triplet insertion (AddRel(S, r, i, j)):** insertion of a relational triplet involving the positions i and j and a root relation r from R provided that the corresponding sequence members are at least sub-classes of a valid pair of domain and range classes for r in the ontology Ω, and that in θ there is no triplet $r(i, j)$ such that r' is equal or more specific that r:

  ◦ $r \in \max_{\sqsubseteq_{\Omega}}(R_{\Omega})$,

  ◦ $\exists c \in dom_{\Omega}(r)$ 6 such that: $S.\varsigma[i] \sqsubseteq_{\Omega} c$,

  ◦ $\exists c \in ran_{\Omega}(r, S.\varsigma[i])$ such that: $S.\varsigma[i] \sqsubseteq_{\Omega} c$,

  ◦ $\nexists r' \in S.\theta$ such that: $r'(i, j) \sqsubseteq_{\Omega} r(i, j)$.

- **Relational specialization (SplRel(S, r, i, j, r')):** substitution of a non-leaf relation r by an immediate sub-relation r' from R_{Ω} in a triplet $r(i, j)$ from $S.\theta$ provided that the range and domain constraints are respected:

  ◦ $r' \prec_{\Omega} r$,

  ◦ $\exists c \in dom_{\Omega}(r)$ such that: $S.\varsigma[i] \sqsubseteq_{\Omega} c$,

  ◦ $\exists c \in ran_{\Omega}(r, S.\varsigma[i])$ such that: $S.\varsigma[i] \sqsubseteq_{\Omega} c$.

For instance, in Tables 9 and 10, S_{14} and S_{18} can be obtained by S_9 by, respectively, adding a triplet labelled by a root property (*hasAccommodation*), and substituting a class (*Destination*) by an immediate specialization thereof (*UrbanArea*).

Although the above operations constitute an effective generation mechanism, their unrestricted use is redundancy-prone and hence would harm the efficiency of the global mining process. As an illustration, take S_{16} in Table 10 which can be obtained from both S_9 and S_6 (given in Table 9) by class substitution and insertion, respectively. Although impossible to completely eliminate, this kind of redundancy can be greatly decreased by fixing the position where operations apply. Thus, only canonical operations, i.e., involving the last class of $S.\varsigma$, are admitted. In other terms, class-

Table 8. 2-candidates and their associated support

Pattern	Support
$S_6 = <<Beach>,\{\}>$	1
$S_7 = <<RuralArea>,\{\}>$	0
$S_8 = <<UrbanArea>,\{\}>$	3
$S_9 = <<Destination, Accommodation>,\{\}>$	2
$S_{10} = << Destination, AccommodationRating >,\{\}>$	0
$S_{11} = <<Destination, Contact>,\{\}>$	0
$S_{12} = << Destination, Activity >,\{\}>$	3

es can only be appended at the end of $S.\varsigma$, and substituted to the last class thereof. Similarly, triplets $r(l, m)$ involved in insertion/substitution satisfy $m = | S.\varsigma |$. In the new settings, S_{16} cannot be canonically generated from S_9. Canonical operations are named by appending the suffix C to each of the four operations presented bellow.

Mining sequential patterns from object sequences D, amounts to finding all the members of the pattern space Γ_Ω having support greater than *minsup*, a user provided threshold. The target set is thus $\Gamma_\Omega^s = \{S \in \Gamma_\Omega | \sup p(S) \geq \min sup\}$ where the support of S is the size of the subset of D_Ω^0 that passes the instantiation test for S.

The mining algorithm, called *xPMiner*, is a top-down level-wise miner similar in spirit to *Apriori*. This means that at the $(k + 1)$-*th* level, *xPMiner* uses the frequent patterns generated at level k to compose candidates and then checks their frequency in order to establish which ones are to be kept. The pseudo-code of *xPMiner* is given in Figure 4.

The algorithm works on an input made of an ontology Ω and a set of object sequences D. At its initialization step, it transforms the sequences into extended object sequences, i.e., it adds all the links from the ontology that connect two objects in the same order (a source object followed by a target object) as they appear in the sequence (primitive *iSeqTrans(D)7*). Moreover, the structure carrying the candidate patterns is initialized

with all patterns of rank one, i.e., the patterns made of a single top most concept. These candidates are then evaluated for frequency in the data by carrying out instantiation tests for each entry in D (primitive *candTest()*). At each subsequent iteration step, the algorithm performs two main operations. First, using the four auxiliary functions representing the four canonical operations, the candidates of rank $k+1$ are generated from those of rank k. Then, the frequency of the candidates is evaluated by checking the frequency of their apparition on the dataset by means of the *candTest()* primitive. With such a procedure, a question must be asked about its capacity to retrieve exactly the target pattern set, i.e., all frequent patterns. Although both consistency (no infrequent patterns are retrieved) and completeness (all frequent patterns are retrieved) are at stake, due to the specific nature of the computation, only completeness needs to be examined. The following *Proposition 3* states the completeness of *xPMiner*.

Proposition 3. *xPMiner Completeness*

For an arbitrary natural number k, if patterns of rank k+1 exist, then any of them can be generated by at least one pattern of rank k by means of the canonical operations.

An example of the application of the proposed method is provided in the next section which is dedicated to its application as a recommendation tool.

Patterns Extracted with xPMiner

Table 5 shows a set of three object sequences, while Table 6 shows the correspondence between the individual objects and their membership classes in Travel. Moreover, the minimum (absolute) support is set to 2 (*minsup* = 2).

At the first step, the candidate patterns S_1, S_2, S_3, S_4 and S_6 have only one concept. Then, the candidate test procedure calculates for each

Figure 4. Algorithm xPMiner. (©2009, mehdi adda)

```
Algorithm 1 xPMiner
1: Input:
2: Ω, D, σ:        ▷ Domain ontology, set of object sequences and the minimal
   support

3: Output:                                        ▷ Set of frequent patterns
4: ∪ₖ𝓕ₖσ:                                   ▷ Set of all frequent patterns

5: Initialization:
6: D_Ω ← ISEQTRANS(D, Ω):              ▷ Set of extended object sequences
7: 𝓕⁰σ ← ∅_Γ_Ω:

8: Method:
9: for k = 2; 𝓕ᵏ⁻¹σ ≠ ∅; k + + do
10:     𝓕ᵏσ ← ∅:
11:     for all 𝒮 ∈ 𝓕ᵏ⁻¹σ do
12:         𝓕ᵏσ ← 𝓕ᵏσ ∪ ADDCLSC(𝒮) ∪ SPLCLSC(𝒮) ∪ ADDRELC(𝒮) ∪
        SPLRELC(𝒮)
13:     end for
14:     𝓕ᵏσ ← CANDTEST(𝓕ᵏσ, σ):
15: end for
16: return (∪ₖ𝓕ᵏσ):
```

candidate the number of its instances in the object sequences. For example, S_1 succeeds the instantiation test with the first and third sequences but fails the test with the second sequence. The support of the five patterns is presented in Table 7. From this table, it can be observed that S_2 and S_4 have a support inferior to the fixed *minsupp*. In fact, it is easy to see that neither *AccommodationRating* in S_2 nor Contact in S_4 have instances in the set of object sequences. At this point, the anti-monotony property of non-frequency ensures that all patterns that can be generated from a non-frequent pattern are non-frequent. However, S_1, S_3, and S_5 have a frequency greater than *minsup* and are marked as frequent and will be considered as a seed set for the next pass.

Table 9. 2-Frequents and their associated support

Pattern	Support
S_8 =<<UrbanArea>,{}>	3
S_9 =<<Destination, Accommodation>,{}>	2
S_{10} =<< Destination, Activity >,{}>	3

Table 10. 3-Frequents and their associated support

Pattern	Support
S_{13}=<<City>,{}>	2
S_{14} =<<Destination, Accommodation>,{hasAccommodation(1,2)}>	2
S_{15} =<<UrbanArea, Activity>,{}>	3
S_{16} =<< UrbanArea, Accommodation>,{}>	2
S_{17} =<< Destination, Activity >,{has Activity(1,2)}>	3
S_{18} =<<Destination, Activity, Accommodation>,{}>	2

Each *1-frequent* pattern is tentatively extended by means of canonical operations thus yielding the set of *2-candidates* presented in Table 8. For instance, concept specialization on S_5 produces the following set of *2-candidates*: S_6, S_7 and S_8.

It is noteworthy that operations involving relations are not applicable until the step two of the process. Now, the *2-candidates* are filtered through the instantiation test in order to determine *2-frequent* patterns. It can be observed that only patterns presented in Table 9 are frequent. These patterns, having rank 2, are then used to generate 3-frequent patterns.

As an illustration, consider the extension of the pattern S_{12}. Candidates generated by concept specialization (on *Activity*) are: <<Destination, *Adventure*>, {}>, <<Destination, *Relaxation*>, {}>, <<Destination,Sightseeing>,{}>, <<Destination,Sports>,{}>, <<Activity, Accommodation>,{}>. Observe that several concept insertions generate the candidates: <<Destination, Activity, AccommodationRating>, {}>, <<Destination, Activity, Contact>, {}>, and <<Destination, Activity, Destination>, {}>. The pattern S_{17} is the only one obtained by relation insertion. Table 10 provides the frequency of the candidates at the third step.

At the fourth iteration, after the generation and the subsequent instantiation test a set of eight 4-frequent patterns is obtained (see Table 11). Actually, the process continue until the 16-th step where the frequent pattern S_{30} = <<*Capital, Museum, LuxuryHotel>, {hasMuseum(1; 2), hasLuxuryHotel(1; 3)}*> having a support equal to 2 is discovered. Then, the process stop because there is no new candidate to be generated.

EFFICIENCY STUDY

The feasibility of the ontology-based pattern mining approach has been examined through a comprehensive performance study using synthetic object records related to an e-tourism application. For this purpose, a Java implementation of *xPMiner* has been integrated as a plug-in into the *Protégé*8 ontology manipulation platform. Such an architecture enables Protégé to be relied on for ontology exploration, e.g., moving along the *is-a* hierarchy and attribution network, and knowledge base querying.

The experimental settings include the *Travel* ontology (36 classes and 44 properties) and a knowledge base of 150 objects. On top of these, several collections of raw object sequences have been generated, using a multi-level randomization model adapted from (Agrawal & Srikant, 1994) that embodies a particular understanding of how user navigation unfolds.

Experiments have been executed on a CentOs virtual machine with 1024 MB RAM and the equivalent of a 2 Giga-Hertz Pentium 4 processor.

A first experiment tested the scale-up of *xPMiner*. Its dataset comprised sequences of average length of 6 and were mined with two σ values: 0.7 and 0.5. The evolution of the CPU time upon an

Table 11. 4-Frequents and their associated support

Pattern	Support
S_{19}=<<*City, Activity*>,{}>	2
S_{20}=<< UrbanArea, *Accommodation*>,{hasAccommodation(1,2)}>	2
S_{21}=<<*UrbanArea, Hotel*>,{}>	2
S_{22}=<< UrbanArea, *Sightseeing*>,{}>	2
S_{23}=<< UrbanArea, Activity >,{has Activity(1,2)}>	3
S_{24}=<< UrbanArea, Activity, Accommodation>,{}>	2
S_{25}=<< Destination, Hotel>,{hasAccommodation(1,2)}>	2
S_{26}=<<Destination,Activity,Accommodation>,{hasAccommodation (1,3)}>	2
S_{27}=<<Destination,Activity,Accommodation>,{hasActivity (1,2)}>	2
S_{28}=<<Destination,Activity,Hotel >,{ }>	2
S_{29}=<<*Capital*>,{}>	2

Figure 5. Scale-up: number of object sequences. (©2010, mehdi adda)

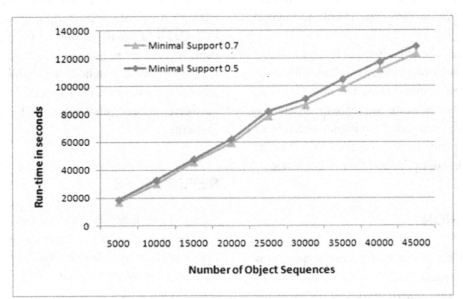

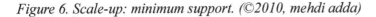

Figure 6. Scale-up: minimum support. (©2010, mehdi adda)

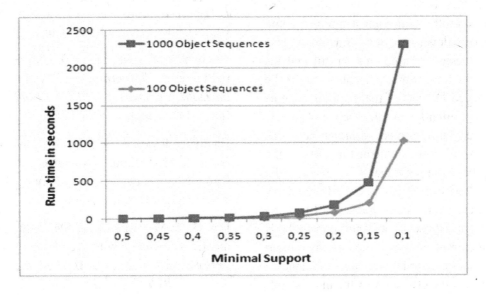

increase in the dataset size (from 1000 to 4500) is shown in Figure 5: both graphs seem to represent a roughly linear dependency. The impact of minimum support variation - within value range of 0.1 to 0.5 - on *xPMiner* performances was examined next.

As shown in Figure 6, the performance of *xPMiner* rapidly deteriorates as σ decreases. Nevertheless, this is consistent with other studies on sequential pattern mining and reflects the sharp increase in the frequent pattern number. As an illustration, with 200 sequences of average size 6, the decrease of the minimal support from 0.1

to 0.5 makes the pattern number grow from several dozen (76 in one experiment) to several thousands (4582).

To sum up, the experimental results indicate that the behavior of *xPMiner* is in line with those of comparable method from the literature, i.e., *Apriori*-like methods for sequential pattern mining. It is noteworthy that the implementation of *xPMiner* depends heavily on the *OWL Protégé API* which in not optimized for mining tasks.

CONCLUSION

In this paper, the problem of mining frequent patterns from sequences of objects characterized as instances within a domain ontology is investigated and a couple of languages to describe the data and the target patterns with its syntax and semantics are described. The structure of the underlying pattern space was carefully studied and a mining method to explore it in a level-wise manner, *xPminer*, was defined. Its design is based on a careful analysis of the *Apriori* framework, whereas some of the most advanced features of that method, such as the a priori elimination of infrequent candidates, remain to be implemented. *xPminer* performs a complete and non-redundant traversal of the pattern space in the sense that it discovers all the frequent patterns while generating all candidates only once. The overall performance profile of the method matches those of similar mining methods whereas its current implementation, *i.e.*, as a *plugin* of *Protégé*, has a clear impact on its efficiency.

To improve the efficiency of the method, additional optimization techniques from *Apriori* are currently translated into the pattern description and mining framework. The next step is to design a depth-first method for pattern mining, since this category of mining methods are knowingly more efficient than level-wise ones, especially for short sequences.

At a further step, it would be interesting to examine how the current pattern model could be extended by allowing multiplicity marks on concepts of the pattern (saying how many instances in a row may be matched by the concept) or by admitting chains instead of direct links (relations) between objects (concepts) in the sequences (patterns).

REFERENCES

Adda, M., Valtchev, P., Missaoui, R., & Djeraba, C. (2007). Toward Recommendation Based on Ontology-Powered Web-Usage Mining. *IEEE Internet Computing Journal*, 45-52.

Agrawal, R., & Srikant, R. (1994). Fast algorithms for mining association rules. *VLDB '94, Proceedings of 20th International Conference on Very Large Data Bases* (pp. 487-499). Santiago de Chile, Chile: Morgan Kaufmann.

Agrawal, R., & Srikant, R. (1995). Mining Sequential Patterns. *Eleventh International Conference on Data Engineering, IEEE Computer Society* (pp. 3-14). Taipei, Taiwan: IEEE.

Bandyopadhyay, S., Maulik, U., Holder, L.-B., & Cook, D.-J. (2005). Advanced Methods for Knowledge Discovery from Complex Data. 95-121. Springer Berlin Heidelberg.

Dai, H., & Mobasher, B. (2004). Integrating Semantic Knowledge with Web Usage Mining for Personalization. *The AAAI 2004 Workshop on Semantic Web Personalization (SWP '04)*, (pp. 276-306). San Jose, California, USA.

Di-Jorio, L., Bringay, S., Fiot, C., Laurent, A., & Teisseire, M. (2008). Sequential Patterns for Maintaining Ontologies over Time. *OTM '08: Proceedings of the OTM 2008 Confederated International Conferences, CoopIS, DOA, GADA, IS, and ODBASE 2008. Part II on On the Move to Meaningful Internet Systems* (pp. 1385-1403). Berlin, Heidelberg: Springer-Verlag.

Ding, L., Kolari, P., Ding, Z., & Avancha, S. (2007). A Survey. In Sharman, R., Kishore, R., & Ramesh, R. (Eds.), *Ontologies A Handbook of Principles, Concepts and Applications in Information Systems* (pp. 79–113). Springer, US: Using Ontologies in the Semantic Web.

Eirinaki, M., Lampos, H., Vazirgiannis, M., & Varlamis, I. (2003). Sewep: Using site semantics and a taxonomy to enhance the web personalization process. *The Ninth ACM SIGKDD International Conference on Knowledge Discovery and Data Mining* (pp. 99-108). Washington, DC, USA: ACM Special Interest Group on Knowledge Discovery in Data.

Huan, J., Wang, W., Prins, J., & Yang, J. (2004). SPIN: Mining Maximal Frequent Subgraphs from Graph Databases. *10th ACM SIGKDD International Conference on Knowledge Discovery and Data Mining*, (pp. 581-586).

Inokuchi, A., Washio, T., & Motoda, H. (2000). An Apriori-Based Algorithm for Mining Frequent Substructures from Graph Data. *PKDD '00: Proceedings of the 4th European Conference on Principles of Data Mining and Knowledge Discovery* (pp. 13-23). Lyon, France: Springer.

Jin, X., & Mobasher, B. (2003). Using semantic similarity to enhance item-based collaborative filtering. *The 2nd IASTED International Conference on Information and Knowledge Sharing.* Scottsdale, AZ, US.

Knublauch, H. (n.d.). Retrieved September 12, 2008, from http://protege.cim3.net/: http://protege.cim3.net/file/pub/ontologies/travel/travel.owl

Lee, T., Hendler, J., & Lassila, O. (2001, May). *The semantic web. Scientific American.* Retrieved 02 2010, from http://www.sciam.com/article.cfm?id=the-semantic-web

Lee, T. B. (2001). *W3C Semantic Web Activity.* Retrieved 2010, from http://www.w3.org/2001/sw: http://www.w3.org/2001/sw

Liao, S.-H., Chen, J.-L., & Hsu, T.-Y. (2009). Ontology-based data mining approach implemented for sport marketing. *Expert Systems with Applications: An International Journal*, 11045-11056.

Melville, P., Mooney, R. J., & Nagarajan, R. (2002). Content-boosted collaborative ltering for improved recommendations. *Eighteenth National Conference on Artificial Intelligence (AAAI-2002)* (pp. 187-192). Edmonton, Alberta, Canada: ACM Special Interest Group on Artificial Intelligence.

Mineau, G. W., Moulin, B., & Sowa, J. F. (1993). Conceptual graphs for knowledge representation. *Lecture Notes in Artificial Intelligence 699.*

Mobasher, B., Jin, X., & Zhou, Y. (2004). Web Semantically enhanced collaborative filtering on the web. *Web Mining: From Web to Semantic*, 57-76.

Pierrakos, D., Paliouras, G., Papatheodorou, C., & Spyropoulos, C. (2003). Web usage mining as a tool for personalization: A survey. *User Modeling and User-Adapted Interaction, Kluwer Academic Publishers*, 311-372.

Rajapaksha, S., & Kodagoda, N. (2008). Internal Structure and Semantic Web Link Structure Based Ontology Ranking. *ICIAFS 2008. 4th International Conference on Information and Automation for Sustainability*, (pp. 86-90). Colombo.

Washio, T., & Motoda, H. (2003). State of the art of graph-based data mining. *ACM SIGKDD Explorations Newsletter*, 59-68.

ENDNOTES

[1] Adapted from (Knublauch).
[2] E(X): the powerset of the set X.
[3] |X|: cardinality of the collection X.
[4] $\prec_{\Omega}$: concept (relation) subsumption relationship in the ontology Ω.
[5] $ran_{\Omega}(r,c)$: set of co-domain concepts of the relation r in the ontology Ω.
[6] $dom_{\Omega}(r)$: set of domain concepts of the relation r in the ontology Ω.
[7] Pseudo-code skipped since straightforward.
[8] http://protege.stanford.edu.

This work was previously published in International Journal of Information Technology and Web Engineering, Volume 5, Issue 2, edited by Ghazi I. Alkhatib, pp. 16-31, copyright 2010 by IGI Publishing (an imprint of IGI Global).

Chapter 5
Deep Web Information Retrieval Process:
A Technical Survey

Dilip Kumar Sharma
G.L.A. Institute of Technology and Management, Mathura, India

A. K. Sharma
YMCA University of Science and Technology, Faridabad, India

ABSTRACT

Web crawlers specialize in downloading web content and analyzing and indexing from surface web, consisting of interlinked HTML pages. Web crawlers have limitations if the data is behind the query interface. Response depends on the querying party's context in order to engage in dialogue and negotiate for the information. In this paper, the authors discuss deep web searching techniques. A survey of technical literature on deep web searching contributes to the development of a general framework. Existing frameworks and mechanisms of present web crawlers are taxonomically classified into four steps and analyzed to find limitations in searching the deep web.

INTRODUCTION

The versatile use of the internet has proved a remarkable revolution in the history of technological advancement. Accessibility of web pages starting from zero in 1990 has reached more than 1.6 billion during 2009. It is like a perineal stream of knowledge. The more we dig the more thirst can

DOI: 10.4018/978-1-4666-0023-2.ch005

be quenched. Surface data is easily available on the web. Surface web pages can be easily indexed through conventional search engine. But the hidden, invisible and non-indexable contents which cannot be retrieved through conventional methods used for surface web and whose size is estimated to be thousands of times larger than the surface web is called deep web. The deep web consist of a large database of useful information such as audio, video, images, documents, presentations and various other types of media. Today people really heavily depend on internet for numerous applications such as flight and train reservations, to know about new product or to find any new locations and job etc. They can evaluate the search result and decide which of the bits or scraps reached by the search engine is most promising (Galler, Chun, & An, 2008).

Unlike the surface web, the deep web information is stored in searchable databases. These databases produce results dynamically after processing the user request (BrightPlanet.com LLC, 2000). Deep web information extraction first uses the two regularities of the domain knowledge and interface similarity to assign the tasks that are proposed from users and chooses the most effective set of sites to visit by ontology inspection. The conventional search engine has limitations in indexing the deep web pages so there is a requirement of an efficient algorithm to search and index the deep web pages (Akilandeswari & Gopalan, 2008). Figure 1 shows the barrier in information extraction in the form of search form or login form.

Contributions: This paper attempts to find the limitations of the current web crawlers in searching the deep web contents. For this purpose a general framework for searching the deep web contents is developed as per existing web crawling techniques. In particular, it concentrates on survey of techniques extracting contents from the portion of the web that is hidden behind search interface in large searchable databases with the following points.

- After profound analysis of entire working of deep web crawling process, we extracted qualified steps and developed a framework of deep web searching.
- Taxonomic classification of different mechanisms of the deep web extraction as per synchronism with developed framework.

Figure 1. Query or credentials required for contents extraction

- Comparison of different algorithms web searching with their advantages and limitations.
- Discuss the limitations of existing web searching mechanisms in large scale crawling of deep web.

CURRENT DEEP WEB INFORMATION RETRIEVAL FRAMEWORK

After exhaustive analysis of existing deep web information retrieval processes, a deep web information retrieval framework is developed, in which different tasks in deep web crawling are identified, arranged and aggregated in sequential manner. This framework is useful for understanding entire working of deep web crawling mechanisms as well as it enables the researcher to find out the limitations of present web crawling mechanisms in searching the deep web. The taxonomical steps of developed framework can be classified into following four major parts.

- **Query Interface Analysis:** First of a crawler will request for any web server to fetch a page. After fetching process, it parses and process the form to build an internal representation of web page based on the model developed.
- **Values Allotment:** It provides appropriate value to each and every input element by using different combinations of keywords, which will be allocated by using some string matching algorithms by analyzing the form labels using knowledge base.
- **Response Analysis and Navigation:** Crawler analyzes the response web pages to check if the submission yielded valid search results. Crawler uses this feedback to tune the values assigned and crawl the hypertext links iteratively, received by response web page to some pre-specified depth.
- **Relevance Ranking:** Relevance ranking means the order in which search engine should return the URLs, produced in response to a user's query, to show more relevant pages on priority basis. During this step, the deep web is a completely different from traditional web. In deep web there is none of those <A href> links to content and no association of links to be followed. So in deep web retrieval process, quality of a page cannot be predicted with its reference. This need is definitely highly demanded in the future framework to be developed to increase the quality of deep web contents.

These above steps are represented in following Figure 2 which demonstrates the flow of control over the deep web contents extraction mechanism.

Exhaustive literature analysis of the above taxonomically classified steps are given below.

QUERY INTERFACE ANALYSIS

Query interface analysis can be taxonomically partitioned into the following steps.

Detection of Hidden Web Search Interfaces

Pages for search interfaces are commonly HTML forms which is filled and submitted by users and server respond appropriately according to filled forms. But every form is not search interfaces. The problem is to identify a form which is a search interface. Search interface identification can be taxonomically categorized in the following three ways:

Based on Heuristic Rules

One of the simplest methods of search interface identification is done by using heuristic rules. "Heuristic" here refers to a general problem-solving rule or set of rules that do not guarantee the best solution or even any solution, but serves as a useful guide for interface matching. Automatic search interface detection was first defined by (Raghavan & Garcia-Molina, 2001), whose crawling system use heuristic rules to detect the hidden databases. The paper (Lage et al., 2004) use two heuristic rules and utilizes a pre existing data repository to identify the contents of deep web. This paper exploits the advantage of some patterns that is available in websites to find out the navigation path to be followed.

Decision Trees Classification

One of the approaches for search interface identification is based on decision trees classification models to detect search interface. One of such example is random forest algorithm. A random forest

Figure 2. Contemporary deep Web content extraction and indexing framework

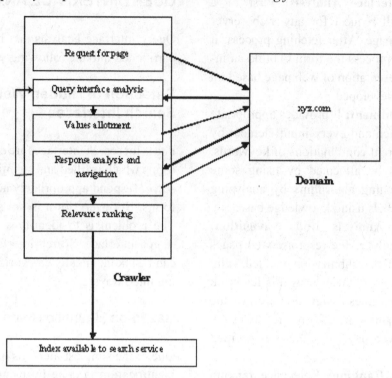

model contains a group of decision tress defined by bootstrapping the training data. An improved version of random forest algorithm (RFA) known as improved random forest algorithm (IRFA) is proposed by Deng et al. In IRFA the original RFA is extended with a weighted feature selection method to select more representative subset of features for building each decision trees. These can be vulnerable to changes in the training dataset. IRFA eliminates the problem of classification of high dimensional and sparse search interface data through the ensemble of decision trees (Deng, Ye, Li & Huang, 2008). Future work in this regard is to identify other techniques for feature waiting for the generation of random forest. Currently this paper uses the features available in the search form themselves. The paper (Jared Cope et al., 2003) defined a novel technique to automatically detect search interface from a group of HTML forms. Future work in this regard will be to develop a technique to eliminate false positives.

Best-Effort Parsing Framework

It identifies the search interface by continuously producing fresh instances by applying productions until attaining a fix-point, when no fresh instance can be produced. An example is shown in Figure 3 (Zhang, He & Chang, 2004) in which, the parser starts from a group of tokens to iteratively generate fresh instances and finally generates parse trees. A complete parse tree related to a unique instance of the start symbol QI that take cares of all tokens. But, due to the significant ambiguities and incompleteness, the parser may not derive any complete parse tree and only conclude with multiple incomplete parse trees. Best effort parser technique minimizes wrong interpretation as much as possible in a very fast manner. It also understands the interface to a large extent.

Table 1 depicts the comparison of different hidden web search interfaces detection mechanism.

Figure 3. Best-Effort (Fix point) parsing process

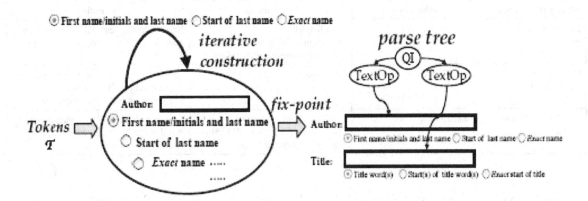

Search Form Schema Mapping

After detection of hidden web search interface, the next task is to identify accurate matching for finding semantic correspondences between elements of two schemas. Schema extraction of query interface is one of the very prime research challenges for comparing and analysis of an integrated query interface for the deep web. Many algorithms are suggested in last few years which can be classified in two groups.

Heuristic

Heuristic techniques for Search form Schema Mapping are based on guessing relations which may consider similar labels or graph structures and can be further taxonomically classified in two groups.

- **Element-level explicit techniques:** Explicit techniques use the semantics of labels such as used in precompiled dictionary (Cupid, COMA), Lexicons (S-Match, CTXmatch) e.g. sedan: Car is a hypernym for Four Wheeler, therefore, Car ⊆ Four Wheeler.

- **Structure-level explicit techniques:** These are based on taxonomic struc-

ture format (Anchor-Prompt, NOM) e.g. DEPTT and Department can be found as an appropriate match.

Formal

Formal techniques are based on model-theoretic semantics which is used to justify their results.

- **Element-level explicit techniques:** It uses the OWL properties (NOM) e.g. same Class as constructor explicitly states that one class is equivalent to the other such as hybrid car = Car or CNG based car.

- **Structure-level explicit techniques:** The approach is based on to translate the matching problem, namely the two graphs (trees) and mapping queries into propositional formula and then to check it for its validity.

- **Modal SAT (S-Match):** The idea is based on to enhance propositional logics with modal logic operators. Therefore, the matching problem is translated into a modal logic formula which is further checked for its validity using sound and complete search procedures.

Table 1. Comparison of different hidden web search interfaces detection techniques

Search Techniques	Authors Name	Advantages	Limitations
Based on Heuristic Rules	S. Raghavan et al.	Effective label extraction technique with high submission efficiency.	No auto-leaning capability, unstable, unscalable to diverse hidden web databases.
Decision Tree Classification Algorithm	Jared Cope et al.	Performed well when rule are generated on the same domain.	Long rules, large size of feature space in training samples, Over fitting, Classification precision is not very satisfying.
Best-Effort Parsing Framework	Zhen Zhang et al.	Very simple and consistent, No priority among preferences, Handling missing elements in form.	Critical to establish single global grammar that can be interacted to the machine globally.

Many automatic or semi-automatic search form schema mapping systems meticulous in a simple 1:1 matching are proposed such as Cupid method (Madhavan, Bernstein & Rahm, 2001). This method proposed a new technique by including subtential linguistic matching step and by biasing matches by leaves of a schema. Future work in this regard include intergating cupid transparently with an off-the-shelf thesaurus using schema notations for the linguistic matching and automatic tuning of control parameter.

Do and Rahm (2002), develops the COMA schema matching system to combine multiple matchers. It uses COMA as a framework to evaluate the effectiveness of different matcher and their combination for real world schemas. Future work in this regard is to add other match and combination algorithm in order to improve match quality. LSD method (Doan, Domingos & Levy, 2000), proposes a initial idea for automatic learning mappings between source schemas and mediated schemas. (Doan, Domingos & Halevy, 2001) describe LSD a system that employs and extends current machine learning techniques to semi-automatically find semantic mappings between the source schema and mediated schema.

The paper (Melnik, Garcia-Molina & Rahm, 2002) present a matching algorithm based on a fixed point computation. It uses two graphs such as schemas catalogs as input and produces as output a mapping between corresponding nodes of the graph. The paper (Kaljuvee, Buyukkokten,

Molina & Paepcke, 2001) proposed a technique for automatically and dynamically summarize and organize web pages for displaying on a small devices such as PDA. This paper proposed eight algorithms for performing label-widget matches in which some algorithms based on n-gram comparisons and others based on common form layout specifications. Results can be improved by using syntactic and structural feature analysis.

For schema extraction, (He, Meng, Yu & Wu, 2005) consider the non-hierarchical structure of query interface assuming that a query interface has a flat set of attributes and the mapping of fields over the interfaces is 1:1, which neglects the grouping and hierarchical relationships of attributes. So the semantics of a query interface cannot be captured correctly.

Literature (Wu, Yu, Doan & Meng, 2004) proposed a hierarchical model and schema extraction approach which can group the attributes and improve the performance of schema extraction of query interface, but they show the poor clustering capability of pre-clustering algorithm due to the simple grouping patterns and schema extraction algorithm and possibly outputs the subsets inconsistent with those grouped by pre-clustering algorithm. Semantic matching is based on two ideas: (i) To discover an alignment by computing semantic relations (e.g., equivalence, more general); (ii) To determine semantic relations by analyzing the meaning (concepts, not labels). Although this paper provides a good accuracy but it can be improved by investigating the interaction to help break ties when the ordering based strategy does not work. Another improvement can be done by investigating the use of automatic interface model procedure into the proposed approach.

Most of the proposed search form schema mapping techniques require human involvement and not suitable for dynamic large scale data sets. Other approaches such as DCM framework (He & Chang, 2006) and MGS framework (He et al., 2003) pursues a correlation mining approach by exploiting the co-occurrence patterns of attributes and proposes a new correlation measures while other (Zhong, Fu, Liu, Lin & Cui, 2007) hypothesizes that every application field has a hidden generative model and can be viewed as instances generated from models with possible behaviors. There are certain issues in this algorithm that can be improved such as how to select the appropriate measure to filter out false matching and how to design a dynamic threshold to apply it to all domains. Table 2 depicts the comparison of different schema mapping techniques.

New schema extraction algorithm Extr (Qiang, Xi, Qiang &, Zhang, 2008), which is based on three metrics (LCA) precision, (LCA) recall, and (LCA) F1 are employed to evaluate the performance of schema extraction algorithm by pre-clustering of attributes P by using MPreCluster than all the subsets in P are clustered once again according to spatial distance. Finally all singleton clusters are merged according to n-way constrained merging operation as algorithm Ncluster. So the result is a hierarchical clustering H over the attributes of query interface on the deep web. The experimental results indicate that proposed algorithm can obviously improve the performance of schema extraction of query interfaces on the deep web and avoid resulting inconsistencies between the subsets by pre-clustering algorithm and those by schema extraction algorithm.

The last but not least correlated-clustering framework works in four phases. In first phase, it finds frequent attributes in the input attribute groups. In second phase, it discovers group where positively correlated attributes to form potential attribute groups, according to positive correlation measure and defined threshold. In third phase, it partitions the attributes into concepts, and cluster the concepts by calculating the similarity of each

Table 2. Comparison of different schema mapping techniques

Name	Techniques	Advantages	Limitations
Artemis	Affinity-based analysis and hierarchical clustering	Effective label extraction technique, with high submission efficiency.	Falls into the alignments as likeness clues category.
Cupid	Structural schema matching techniques	Emphasize the name and data type similarities present at the finest level of granularity (leaf level).	Lack of integrating cupid transparently with an off-the-shelf thesaurus using schema notations for the linguistic matching and automatic tuning of control parameter.
DCM	Hybrid n:m	Identifies and clusters synonym elements by analyzing the co-occurrence of elements. DCM framework can find complex matching in many domains.	Ineffective against 'noisy' schema. DCM cannot differentiate frequent attributes from rare attributes. DCM tries to first identify all possible groups and then discover the matching between them.
GLUE	Composite n:m	It uses a composite approach, as in LSD, but does not utilize global schema.	Accuracy of the element similarity depends on training.
LSD	Corpus-based Matching	It provides a new set of machine-learning based matchers for specific types of complex mappings expressions. It provides prediction criterion for a match or mismatch.	Performance depends upon training data.

two concepts. At last, it ranks to discover matching and then use a greedy matching selection algorithm to select the final matching results.

Komal Kumar Bhatia et al. (Bhatia & Sharma, 2008) presented in his research literature where mapping is done by using domain specific interface mapper in which search interface repository will work for matching purpose. It includes extensible domain specific matcher library. The multi-strategy interface matching is done in three steps: parsing, semantic matching and semantic mapping generation, in step one SI parser is used to extract interface schema. In second step each tuple has mapped by fuzzy matching, domain specific thesaurus and data type matching. Finally SVM generator creates matrices of mapping that are identified by the matching library. DSIM also used mapping knowledge base for avoiding repetition in map effort.

Future work may include testing the schema extraction algorithm on real world data and testing the efficiency of schema matching and schema merging over variety of query interfaces.

Domain Ontology Identification

Ontology is a formal specification of a shared conceptualization (Niepert, Buckner & Allen, 2007). This step is required for analyzing area or specialization of web page so that in further steps appropriate data set will be efficiently placed in query part of the page. This can be taxonomically classified into four different groups.

RDF Annotations Based Ontology Identification

Deitel et al. (Deitel, Faron & Dieng, 2001) present an approach for learning ontology from resource description framework (RDF) annotations of web resources. To perform the learning process, a particular approach of concept formation is adopted, considering ontology as a concept hierarchy, where each concept is defined in extension by a cluster of resources and in intension by the most specific common description of these resources. A resource description is a RDF sub graph containing all resources reachable from the considered resource through properties. This approach leads to the systematic generation of all possible clusters of descriptions from the whole RDF graph incrementing the length of the description associated to each particular concept in the source graph.

Metadata Annotations Based Ontology Identification

Stojanovic et al. (Stojanovic, Stojanovic & Volz, 2002) presents an approach for an automated migration of data-intensive web sites into the semantic web. They extract light ontologies from resources such as XML Schema or relational database schema and try to build light ontologies from conceptual database schemas using a mapping process that can form the conceptual backbone for metadata annotations that are automatically created from the database instances.

Table Analysis Based Ontology Identification

The paper (Tijerino et al., 2005) presents an approach Table Analysis for Generating Ontologies (TANGO) to generate ontology based on HTML table analysis. TANGO discovers the constraints, match and merge mini-ontology based on conceptual modeling extraction techniques. TANGO is thus a formalized technique of processing the format and content of tables that can aim to incrementally build a appropriate reusable conceptual ontology.

DOM Based Ontology Identification

Zhiming Cui et al. published his research (Cui, Zhao, Fang & Lin, 2008) which works in following steps.

- Use the query interfaces information to generate a mini-ontology model.
- To draw the instances from the intermediate result pages.
- Use of various sources to generate ontology mappings and merging ontology.

In first step it employs vision-based approach to extract query interface (Zhao et al., 2007), and in second step data region discovery is done by employing a DOM parser to generate the DOM parsed trees from the result pages. Based on parent length, adjacent and normalized edit distance, extraction of hierarchical data is done by vision-based page segmentation algorithm. In final step merging is done by label instances pairs mined from the result pages into the domain ontology. They also give an idea, of absent attribute annotation for finding absent attributes in the data records. The next generation semantic web framework is required to be able for handling knowledge level querying and searches. Main area to be focused for research is concept relations learning to increase the efficiency of the system.

VALUES ALLOTMENT

Values allotment techniques can be taxonomically classified in the following groups.

Integrating the Databases for Values Mapping

Integration of the databases with the query interfaces is done in this process. The search form interface brings together the attributes and this step will analyze appropriate data values by structure characteristics of the interface and the order of attributes in the area as much as possible. The integration of query interfaces can provide a unified access channel for users to visit the databases which belong to the same area. For integrating interfaces, the core part is dynamic query translator, which can translate the users' query into different form (Meng, Yin & Xiao, 2006) (He, Meng, Yu & Wu, 2003). There is also some scope for improvement by using open directory hierarchy to detect more hypernymy relationship.

Fuzzy Comprehensive Evaluation Methods

In this mapping is done by fuzzy comprehensive evaluation (Chen, Wen, Hu & Li, 2008) which map the attribute of the form to the data values. First it analyzes a form mapping with a view the data range mapping is the key issues and then it select the optimum matching result. Further scope of work in this area includes finding a model to detect the optimal configuration parameter to produce the results with high accuracy.

Query Translation Technique

Query translation technique is used to get query across different deep web sources to translate queries to sources without primary knowledge. The framework takes source query form and target query form as input and output a query for target query.

Some methods can be concerned such as type-based search-driven translation framework by leveraging the "regularities" across the implicit data types of query constraints.

Patterns Based

(He & Zhang et al., 2005) found that query constraints of different concepts often share similar patterns i.e. same data type (page title or author name etc.) and encode more generic translational knowledge for each data type. This indicates explicit declaration of data type that localities the translatable patterns. Therefore, getting translation knowledge for each data types are more generic rather identifying translation based on source of information. For translation purpose it uses extensible search-driven mechanism which uses type based translation.

Hierarchical Relations Based

Other approach published by Hao Liang et al (Liang, Zuo, Ren & Sun, 2008) will map on the basis of three constraints:

1. The same word in different schemas of query forms generally has the same semantics, with or without the same formalizations.
2. The words of two different forms may have same meaning. For this purpose use of the-saurus or dictionaries is required. In addition to this for dealing some special subjects like computer and electronics, some specialized dictionaries related to that subject is required.
3. There may be some hierarchical relations between the words, e.g., X is a hypernym of Y if Y is a kind of X, and on the other hand Y is hyponym of X. For example, bus is a vehicle it indicates that vehicle is hypernym of bus, vehicle is also equivalent to automo-bile. Thus bus is equivalent to automobile semantically.

Domain ontology is widely used in different areas. There is a scope of lot of work to be done for building ontology. One of the issues is to make automatic domain ontology inspection for some particular domain to gain information about the domain.

Type-Based Predicate Mapping

Type-based predicate mapping method (Zhang et al., 2005) proposed by Z. Zhang focusing on text type attribute with some constraint. The constraint is restricted to the query condition, for example, any, all or exactly. The minimum search space is computed but the cost of the query is not considered.

Cost Model Based

Another method of query translation is proposed by Fangjiao Jiang, Linlin Jia, Weiyi Meng, Xiaofeng Meng which is based on a cost model for range query translation (Jiang et al., 2008) in deep web data integration. This paper proposes a multiple regression cost model based on statisti-cal analysis for global range queries that involve numeric range attributes. It works on the basis of following concepts.

1. Using a statistical-based approach for trans-lating the range query at the global level after proposing a multiple-regression cost model (MrCoM).
2. For selecting significant independent vari-ables into the MrCoM, a pre-processing-based stepwise algorithm is defined.
3. Global range queries are classified into three types and different models are proposed for each of the three types of global range query.
4. Experimental process is done to verify the efficiency of the proposed method.

After going through the above process conclu-sion is that MrCoM has good fitness and query strategy selection of MrCoM is highly accurate.

RESPONSE ANALYSIS AND NAVIGATION

Techniques for Response analysis and navigation can be taxonomically categorized in the follow-ing parts.

Data Extraction Algorithm

Data extraction is another important aspect of deep web research, which involves in extracting the information from semi-structured or unstructured web pages and saving the information as the XML document or relationship model. The paper (Crescenzi, Mecca, & Merialdo, 2001) have done a lot of work in this field. Additionally, in some papers, such as (Arlotta, Crescenzi, & Mecca et al., 2003) and (Song, Giri, & Ma, 2004), researchers have paid more attention to the influence of semantic information on deep web.

Jufeng Yang et al. (Yang, Shi, Zheng & Wang, 2007) has published his literature on data extraction in which web page is converted into a tree, in which the internal nodes represent the structure of the page and the leaf nodes preserve the information and compared with configuration tree. Moreover, structure rules are used to extract data from HTML pages and the logical and application rules are applied to correct the extraction results. The model has four layers, among which the access schedule, extraction layer and data cleaner are based on the rules of structure, logic and application. Proposed models are tested to three intelligent system i.e. scientific paper retrieval, electronic ticket ordering and resume searching. The results show that the proposed method is robust and feasible.

Iterative Deepening Search

Generally the dynamic web search interface generates some output if they are given with some input. The generation of output by some input can be visualized as a graph which is based on the keyword relationship. After getting the interface information about the targeted resource, primary query keywords are applied to generate new keywords, which can be used for further extraction. Iterative deepening search (Ibrahim, Fahmi, Hashmi & Choi, 2008) is proposed by Ahmad Ibrahim et al. His work is based on probability, iterative deepening search and graph theory. It has two phases. The first phase is about classification or identification of a resource behind search interface into some certain domain and in the second phase, each resource is queried according to its domain. Even if the deep web contains a large database but there is need of efficient technique for extracting information from deep web in relatively short time. Presently most of the techniques do not work on real time domain and a lot of time is consumed in processing to find the desired result.

Object Matching Method

Object matching process has vital role for integration of deep web sources. For integration of database information, a technique was proposed in (Hernandez et al., 1995) to identify the same object from variety of sources using well defined specific matching rules. This paper gives the solution of the merge/ purge problem i.e. the problem of merging data from multiple sources in an effective manner. The sorted neighborhood method is proposed for the solution of merge/ purge problem. An alternative technique based on clustering method is also proposed with the comparison with sorted neighbor method.

String Transformation Based Method

A technique to compare the same parameters of similar objects was proposed in literature (Tejada, Knoblock & Minton, 2002) through string transformation which is independent of application domain but uses application domain to gain the knowledge about attributes of weights through a very little user interaction. There are several future research areas with regard to this algorithm such as how to minimize the noise in the labels given by the user.

Training Based Method

A technique PROM was defined in (Doan, Lu, Lee & Han, 2003) to increase the accuracy of matching by using the constraint among the attributes available through training procedure or expert domain. It uses the objects from different sources having different attributes. Using the segmentation of pages into small semantic blocks defined on basis of HTML tags. The future work in this regard to implement the profilers generated in matching task to other related matching task to see the effect of transferring such knowledge.

Block Similarity Based Method

A technique proposed by (Ling, Liu, Wang, Ai & Meng, 2006),which segment pages into small semantic blocks based on html tags and change the problem of object matching into problem of block similarity. The method is based on high accuracy record and attributes extraction. Due to the limitation of existing information extraction technology, extracted object data from html pages is often incomplete.

Text Based Method

A new method of object matching is proposed by (Zhao, Lin, Fang & Cui, 2007) which is text-based standard TF/IDF cosine-similarity calculation method to calculate the object similarity, and further expended his framework to record-level object matching model, attribute-level object matching model and hybrid object matching model, which considers structured and unstructured features and multi-level errors in extraction. This paper compare the performance of the unstructured, structured and hybrid object matching models and concludes that hybrid method has the superior performance.

RELEVANCE RANKING

Surface web crawlers normally do the page-level ranking but this does not fulfill the purpose of vertical search for entity oriented. The need of entity level ranking for deep web resource has initiated a large amount of research in the area of entity-level ranking. Previously most of the approaches concentrate on the ranking the structured entities based on the global frequency of relevant documents or web pages. Method of relevance ranking can be taxonomically categorized in the following parts.

Data Warehouse Based Method

Many researchers such as (Nie, Ma, Shi, Wen & Ma, 2007) initiated the use of web data warehouse to pre-store all of entities having the capability of handling structured queries. This paper proposed various language models for web object retrieval such as an unstructured object retrieval model, structural object retrieval model and hybrid model having both structured and unstructured features. This paper concludes that hybrid model is the superior one with extraction errors at changing labels.

Global Aggregation and Local Scoring Based Method

A data integration methods based on the local uncertainties of entities is proposed in various literature. But nearly all of the method does not have the capability for local scoring of entities or aggregation of variety of web sources in a global environment. Literature survey indicates that various search engines are built for focusing on clear indication of entity type and context pattern in user request as illustrated in reference (Cheng, Yan & Chang, 2007). This paper concentrates on the ranking of entities by extracting its underline theoretical model and producing a probabilistic ranking framework that can be able to smoothly integrate both global and local information in ranking.

One of the latest techniques named as LG-ERM proposed by (Kou, Shen, Yu & Nie, 2008) for the entity-level ranking based on the global aggregation and local scoring of entities for deep web query purpose. This technique uses large number of parameters affecting the rank of entities such as relationship between the entities, style information of entities, the uncertainty involved in entity retrieval and the importance of web resources. Unlike traditional approaches, LG-ERM considers more rank influencing factors including the uncertainty of entity extraction, the style information of entities and the importance of Web sources, as well as the entity relationship. By combining local scoring and global aggregation in ranking, the query result can be more accurate and effective to meet users' needs. The experiments demonstrate the feasibility and effectiveness of the key techniques of LG-ERM.

FEW PROPOSED PROTOCOL FOR DEEP WEB CRAWLING PROCESS

Some examples of frameworks designed for extraction of deep web information are given below.

Search/Retrieval via URL

Search/Retrieval via URL (SRU) protocol is a standard XML-focused search protocol for internet search queries that uses contextual query language (CQL) for representing queries.

SRU is very flexible. It is XML-based and the most common implementation of SRU via URL, which uses the HTTP GET for message transfer. The SRU uses the representational state transfer (REST) protocol and introduces sophisticated technique for querying databases, by simply submitting URL-based queries For

example URL?version=1.1&operation=retrieve &query=dilip&maxRecords=12

This protocol is only considered useful when the information about the resource is predictable i.e. the query word is already planned from any source. With reference to our previous discussion the task of SRU is comes under values allotment and data extraction.

Z39.50

Z39.50 (ANSI/NISO Z39.50, Information Retrieval: Application Service Definition and Protocol Specification, 2003) is an ANSI/NISO standard that specifies a client/server-based protocol for searching and retrieving information from remote databases. Clients using the Z39.50 protocol can locate and access data in multiple databases. The data is not centralized in any one location. When a search command is initiated, the search is normally sent simultaneously in a broadcast mode to the multiple databases. The results received back are then combined into one common set. In a Z39.50 session, the Z39.50 client software that initiates a request for the user is known as the origin. The Z39.50 server software system that responds to the origin's request is called the target. This protocol is useful for extracting data from multiple sources simultaneously but here the search phrase must also defined using any other knowledge source. The task of the protocol can be considered in our values allotment and data extraction phase.

OPEN ARCHIVES INITIATIVE PROTOCOL FOR METADATA HARVESTING

The open archives initiative (OAI) (The Open Archives Initiative Protocol for Metadata Harvesting (Protocol Version 2.0), 2003) protocol for metadata harvesting (OAI-PMH) provides an interoperability framework based on the harvesting or retrieval of metadata from any number of widely distributed databases. Through the services of the OAI-PMH, the disparate databases are linked by a centralized index. The data provider agrees to have metadata harvested by the service provider. The metadata is then indexed by the harvesting service provider and linked via pointers to the actual data at the data provider address. This protocol has two major drawbacks. It does not make its resources accessible via dereferencable URIs, and it provides only limited means of selective access to metadata. This protocol not only provide the proper values allotment for query but also provide knowledge harvested from different source so it increases the accuracy by decreasing the unmatched query load.

PROLEARN QUERY LANGUAGE

The ProLearn Query Language (PLQL) developed (Campi, Ceri, Duvall, Guinea, Massart, & Ternier, 2008) by the PROLEARN "Network of Excellence", is query language, for repositories of learning objects. PLQL is primarily a query interchange format, used by source applications (or PLQL clients) for querying repositories or PLQL servers. PLQL has been designed with the aim of effectively supporting search over LOM, MPEG-7 and DC metadata. However, PLQL does not assume or require these metadata standards. PLQL is based on existing language paradigms like the contextual query language and aims to minimize the need for introducing new concepts.

For a given XML binding form, all relevant metadata standards for learning objects, it was decided to express exact search by using query paths on hierarchies by borrowing concepts from XPath. Thus, PLQL combines two of the most popular query paradigms, allowing its implementations to reuse existing technology from both fields i.e. approximate search (using information retrieval engines such as Lucene) and exact search (using XML-based query engines). This is simple

protocol and work as a mediator for transforming data extracted from on object to other. It is applicable in the phase of querying the resource with predefined query words.

HOST LIST PROTOCOL

The Host-List Protocol (HLP) model (Khattab, Fouad, & Rawash, 2009) is a periodical script designed to provide a way to inform web search engines about hidden hosts or unknown hosts. The virtual hosting feature, applied in apache web server allows one Apache installation to serve different actual websites. This virtual hosts feature will be the target during the design process of this model. The algorithm of the HLP model is such that it extracts hidden hosts, in the form of virtual hosts from apache web server using one of the PHP scripting language based open source technologies which is utilizing an open standard technology in the form of XML language, building a frontier of extracted hosts then sending such hosts frontier to the web search engines that support this protocol via HTTP request in an automatic fashion through a cron job. Hosts frontier is an XML file that lists virtual hosts extracted from the configuration file of the apache web server "httpd.conf" after verifying its configuration to take a decision about from where to extract virtual hosts from "httpd.conf". This protocol is designed to reduce the task or identifying virtual host on any server. It generate host list in XML format for crawler and provide the path for data extraction.

REALLY SIMPLE SYNDICATION

RSS stands for "Really Simple Syndication" "(Grossnickle et al., 2005). It is a technique to easily distribute a list of headlines, update notices, and sometimes content to a wide number of people. It is used by computer programs that organize those headlines and notices for easy reading. Most people are interested in many websites whose content changes on an unpredictable schedule. RSS is a better technique to notify the new and changed contents. Notifications of changes to multiple websites are handled easily, and the results are presented to user are well organized and distinct from email. RSS works through the website author to maintain a list of notifications on their website in a standard way. This list of notifications is called an "RSS Feed". Producing an RSS feed is very simple and lakhs of websites like the BBC, the New York Times and Reuters, including many weblogs now providing this feature. RSS provides very basic information to do its notification. It is made up of a list of items presented in newest to oldest order. Each item usually consists of a simple title describing the item along with a more complete description and a link to a web page with the actual information being described. This mechanism of extracting contents is very effective for site updating their contents daily but the loop whole is again generating proper feed from combination of database and generated page links. It comes into the category of values allotments and data extraction.

SITEMAP PROTOCOL

The sitemaps protocol (Sitemaps, 2009) allows a webmaster to inform search engines about URLs on a website that are available for crawling. A sitemap is an XML file that lists the URLs for a website. It allows webmasters to include additional information about each URL such as when it was last updated, how often it changes and how important it is in relation to other URLs in the website. This allows search engines to crawl the site more intelligently. Sitemaps are a URL inclusion proto-

Table 3. Comparison of different protocols in the context of deep web information extraction

Name	Techniques	Advantages	Limitations
SRU	Uses the REST protocol, send encoded query words through http get request.	Simple xml based request. Independent of underlining database.	Limitation of 256 character query. Responding server must be equipped for analyzing query.
Z39.50	ANSI/NISO client/server-based protocol.	The data can be extracted from multiple locations. Access data from multiple databases. Various results combined into one result set, regardless of their original format.	Complex technique and the searching is limited to the speed of the slowest server, No updates and support.
OAI-PMH	XML response over http.	Disparate databases are linked by a centralized index, Simple http based request. Much faster and independent of database.	Evaluation on xml. limitation in object access, exchange and transfer, No technique available for the user to know when the metadata was last harvested.
ProLearn Query Language	XML XQuery and Xpath based	Best suited for approximate search and exact search, supports hierarchical metadata structures.	Complex to implement and application development model is not mature.
Host List Protocol	Apache and xml	Retrieval of url from virtual host.	Extracting hidden hosts are limited to the root.
RSS	XML	Short bunch of information, real time update, generation of RSS and reading mechanism is easy to implement.	Dependent upon site administrator to generate these feed. Because auto feed generator will read only anchor tag.
Sitemap protocol	XML based, Site administrator's protocol	Easy to generate, Simple structure and node hierarchy.	Site administrator's involvement is needed, if the pages are hidden by Ajax or flash, explicit mentioning must required.

Table 4. Query words versus Results counts for surface web search engines

Query words	Surface Web Search Engine			
	Google.com	Yahoo.com	Bing/live.com	Ask.com
cloud computing	34,200,000	1,510,000	15,300,000	13,100,000
global warming	30,200,000	2,800,000	14,100,000	7,582,000
optical modulation	1,800,000	57,800	1,600,000	1,290,000
walmart	20,800,000	194,000,000	16,300,000	3,340,000
best buy	296,000,000	44,200,000	551,000,000	188,000,000
Dictionary	191,000,000	7,880,000	68,400,000	27,400,000
Astrology	25,300,000	40,400,000	22,600,000	7,790,000
Insurance	363,000,000	41,600,000	262,000,000	49,730,000

col and complement robots.txt, a URL exclusion protocol. Sitemaps are particularly advantageous on websites where some contents of web site is not linked with public pages and webmasters use rich Ajax or Flash contents that are not normally processed by search engines. Sitemaps helps to find out the hidden contents when submitted to

crawler and it do not replace the existing crawl-based mechanisms that search engines already use to discover URLs. Sitemap protocol will come under the taxonomical categorization of values allotment and data extraction. Table 3 depicts the comparison of different protocols for deep web information extraction.

Figure 4. Variation of results count versus query words for different search engines

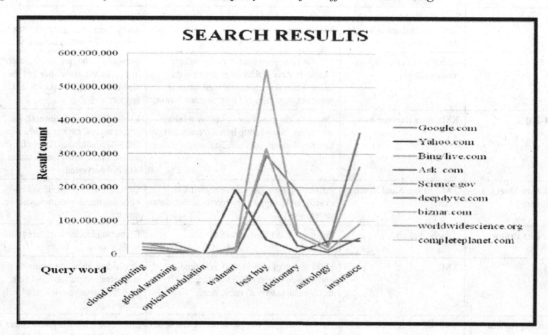

Figure 5. Snap shots showing results by different surface Web search engines in response to query words

Table 5. Query words versus results counts for deep web search engines

Query words	Deep Web Search Engine				
	science.gov	deepdyve.com	biznar.com	worldwide science.org	complete planet.com
cloud computing	32,318,709	1,229,954	108,609	52,763	5000
global warming	14,436,593	563,694	380,240	214,452	3702
optical modulation	1,774,195	1,513,665	116,407	442,284	621
Walmart	5,960,514	87,143	234,180	244	41
best buy	315,040,211	72,574	2,613,762	111,272	2208
dictionary	56,975,239	101,835	397,173	122,385	1087
astrology	14,101,342	87,964	132,384	3,404	290
insurance	92,056,043	170,433	2,795,986	291,304	5000

COMPARATIVE STUDY OF DIFFERENT SEARCH ENGINES

We have conducted search on different search engines some of them are surface web search engine which crawler frontier that retrieves only anchor tag and other are deep search engines which retrieves deep web information. The results are shown below in Table 4 and Table 5. From the analysis of these results, it is clear that surface search engines show more number of results compared to deep web search engine because still at present deep web search engine are not more efficient due to the lack of technological advancement and standards. Results count for query words related to technical literature are relatively more in number deep web search engines while results count of general query words are more in surface web search engines due to the numerous public posting and search engine optimization work. Figure 4 depicts the variation of results count versus query words for different search engines. Figure 5 depicts the snap shots showing results by different surface web search engines in response to query words. Figure 6 depicts the snap shots showing results by different deep web search engines in response to query words.

CONCLUDING REMARKS

After going through the exhaustive literature survey, a general framework is developed for understanding the entire web crawling mechanisms and to find out the limitations of general web crawling mechanism in searching the deep web. The entire

Figure 6. Snap shots showing results by different deep Web search engines in response to query words

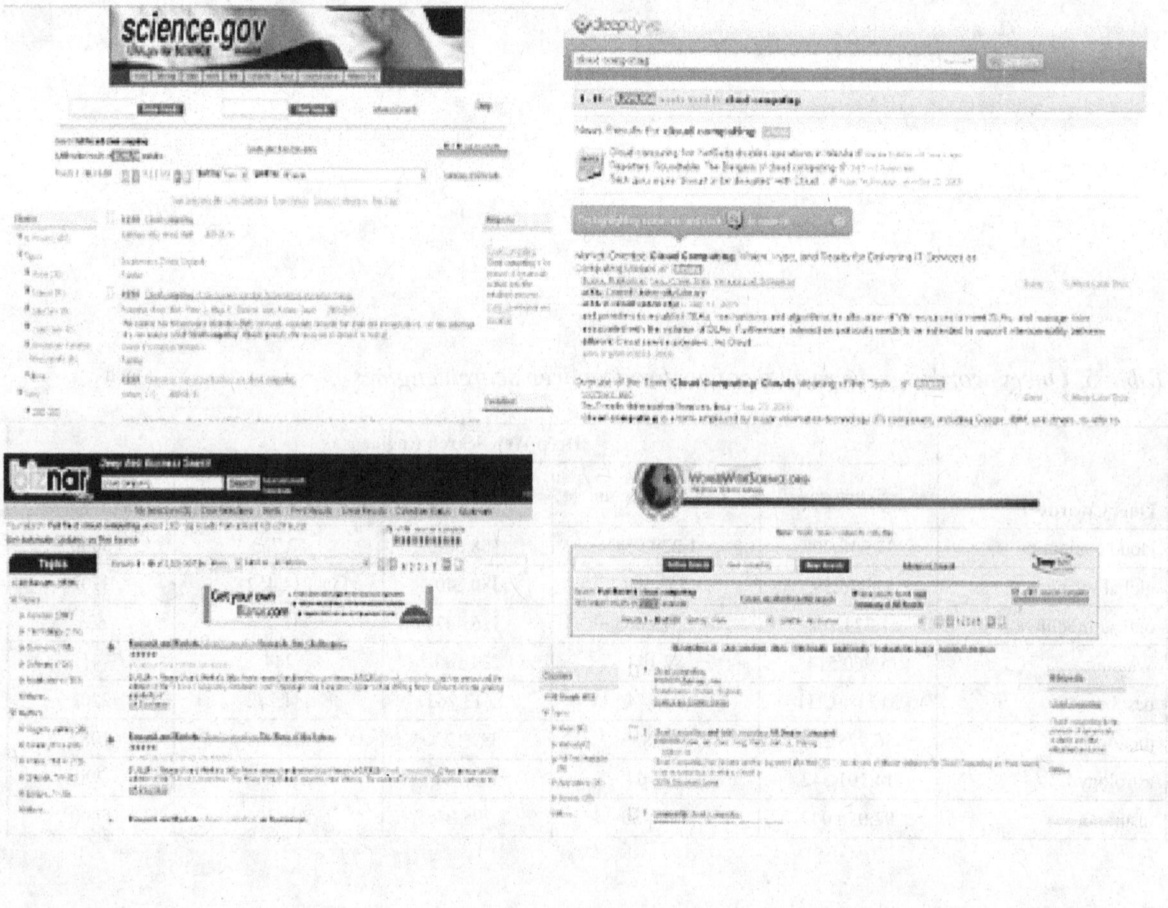

working of web crawling mechanisms in general is divided into four parts in the developed framework. The present web crawling mechanisms are analyzed and merged as per developed framework to understand their advantages and limitations in searching the deep web. Some existing deep web information extraction protocols are analyzed with their comparative study. We have also done a comparative study of surface web search engines and deep web search engines.

In query interface analysis, heuristic rules based techniques are the simplest one but does not guarantee even a solution. IRFA eliminates the problem of classification of high dimensional and sparse search interface data. Future work is to identify other techniques for feature waiting for generation of random forest. Various techniques of search form schema mapping are described and can be categorized as heuristic and formal techniques. In this regard an appropriate measure has to be defined to filter out false matching. In domain ontology identification the main area to be focused for research is concept relation learning to increase the efficiency of the system. In values allotment there is a scope for improvement by using hierarchy to detect more hypernymy relationship. In fuzzy comprehensive evaluation methods, the future research area include to find a model to detect the optimal configuration parameter to produce the results with high accuracy but MrCOM model has good fitness with highly accurate query strategy selection. In response analysis and navigation, most of the techniques do not work on real time domain and a large time is consumed in processing to find the desired result. In relevance ranking one of the latest techniques is LG-ERM which considers a larger number of ranking influencing factors such as the uncertainty of entity extraction, the style information of entity and the importance of web sources as well as entity relationship. By combining the techniques of local scoring and global aggregation in ranking, the query result can be made more accurate and effective. There exists an improper mapping

problem in the process of query interface analysis. Over traffic load problem arises from unmatched semantic query. Data integration suffers due to the lack of cost models at the global level in the process of proper values allotment. In distributed websites there is challenge for query optimization and proper identification of similar contents in the process of response analysis and navigation. The ranking of deep web contents are difficult due to the lake of reference to other sites.

On the basis of taxonomic classification and consequent analysis of different algorithms a conclusion is made that an open framework based deep web information extraction protocol is required to eliminate the limitations of present web crawling techniques in searching the deep web. Future work includes the design, analysis and implementation of open framework protocol for deep web information extraction, considering the fact that it must allow simple implementation without much modifying present architecture of web.

REFERENCES

Akilandeswari, J., & Gopalan, N. P. (2008). An Architectural Framework of a Crawler for Locating Deep Web Repositories Using Learning Multi-agent Systems. In *Proceedings of the 2008 Third International Conference on Internet and Web Applications and Services* (pp.558-562).

ANSI/NISO Z39. 50. (2003). *Information Retrieval: Application Service Definition and Protocol Specification*. Retrieved from http://www.niso.org/standards/standard_detail.cfm?std_id=465.

Arlotta, L., Crescenzi, V., Mecca, G., et al. (2003). Automatic annotation of data extracted from large Web sites. In *Proceedings of the 6th International Workshop on Web and Databases*, San Diego, CA (pp. 7-12).

Bhatia, K. K., & Sharma, A. K. (2008). A Framework for Domain Specific Interface Mapper (DSIM). *International Journal of Computer Science and Network Security*, 8, 12.

BrightPlanet.com LLC. (2000, July). *White Paper: The Deep Web: Surfacing Hidden Value.*

Campi, A., Ceri, S., Duvall, E., Guinea, S., Massart, D., & Ternier, S. (2008, January). Interoperability for searching Learning Object Repositories: The ProLearn Query Language. *D-LIB Magazine*.

Chen, S., Wen, L., Hu, J., & Li, S. (2008). Fuzzy Synthetic Evaluation on Form Mapping in Deep Web Integration. In *Proceedings of the International Conference on Computer Science and Software Engineering*. IEEE.

Cheng, T., Yan, X., & Chang, K. C. C. (2007). Entity Rank: searching entities directly and holistically. In *Proceedings of the VLDB*.

Cope, J., Craswell, N., & Hawking, D. (2003). Automated Discovery of Search Interfaces on the web. In *Proceedings of the Fourteenth Australasian Database Conference (ADC2003)*, Adelaide, Australia.

Crescenzi, V., Mecca, G., & Merialdo, P. (2001). RoadRunner: towards automatic data extraction from large Web sites. In *Proceedings of the 27th International Conference on Very Large Data Bases*, Rome, Italy (pp. 109-118).

Cui, Z., Zhao, P., Fang, W., & Lin, C. (2008). *From Wrapping to Knowledge: Domain Ontology Learning from Deep Web*. In *Proceedings of the International Symposiums on Information Processing*. IEEE.

Deitel, A., Faron, C., & Dieng, R. (2001). *Learning ontologies from rdf annotations*. Paper presented at the IJCAI Workshop in Ontology Learning.

Deng, X. B., Ye, Y. M., Li, H. B., & Huang, J. Z. (2008). An Improved Random Forest Approach For Detection Of Hidden Web Search Interfaces. In *Proceedings of the Seventh International Conference on Machine Learning and Cybernetics*, Kunming, China. IEEE.

Do, H. H., & Rahm, E. (2002, August). COMA-a System for Flexible Combination of Schema Matching Approaches. In *Proceedings of the 28th Intl. Conference on Very Large Databases (VLDB)*, Hong Kong.

Doan, A., Domingos, P., & Halevy, A. (2001). Reconciling schemas of disparate data sources: A machine-learning approach. In *Proceedings of the International Conference on Management of Data (SIGMOD)*, Santa Barbara, CA. New York: ACM Press.

Doan, A., Lu, Y., Lee, Y., & Han, J. (2003). Object matching for information integration: A profiler-based approach. *II Web*, 53-58.

Doan, A. H., Domingos, P., & Levy, A. (2000). Learning source descriptions for data integration. In *Proceedings of the WebDB Workshop* (pp. 81-92).

Galler, J., Chun, S. A., & An, Y. J. (2008, September). Toward the Semantic Deep Web. In *Proceedings of the IEEE Computer* (pp. 95-97).

Grossnickle, J., Board, T., Pickens, B., & Bellmont, M. (2005, October). *RSS—Crossing into the Mainstream*. Ipsos Insight, Yahoo.

He, B., & Chang, K. C. C. (2003). *Statistical schema matching across web query interfaces*. Paper presented at the SIGMOD Conference.

He, B., & Chang, K. C. C. (2006, March). Automatic complex schema matching across web query interfaces: A correlation mining approach. In *Proceedings of the ACM Transaction on Database Systems* (Vol. 31, pp. 1-45).

He, B., Zhang, Z., & Chang, K. C. C. (2005, June). *Meta Querier: Querying Structured Web Sources On the-fly*. Paper presented at SIGMOD, System Demonstration, Baltimore, MD.

He, H., Meng, W., Yu, C., & Wu, Z. (2005). Constructing interface schemas for search interfaces of Web databases. In *Proceedings of the 6th International Conference on Web Information Systems Engineering (WISE'05)* (pp. 29-42).

He, H., Meng, W., Yu, C. T., & Wu, Z. (2003). WISE-Integrator: An Automatic Integrator of Web search interfaces for e-commerce. In *Proceedings of the 29th International Conference on Very Large Data Bases (VLDB'03)* (pp. 357-368).

Hernandez, M., & Stolfo, S. (1995). *The merge/purge problem for large databases*. In *Proceedings of the SIGMOD Conference* (pp. 127-138).

Ibrahim, A., Fahmi, S. A., Hashmi, S. I., & Choi, H. J. (2008). Addressing Effective Hidden Web Search Using Iterative Deepening Search and Graph Theory. In *Proceedings of the 8th International Conference on Computer and Information Technology Workshops*. IEEE.

Jiang, F., Jia, L., Meng, W., Meng, X., & MrCoM. (2008). A Cost Model for Range Query Translation in Deep Web Data Integration. In *Proceedings of the Fourth International Conference on Semantics, Knowledge and Grid*. IEEE.

Kaljuvee, O., Buyukkokten, O., Molina, H. G., & Paepcke, A. (2001). Efficient Web form entry on PDAs. In *Proceedings of the 10th International Conference on World Wide Web (WWW'01)* (pp. 663- 672).

Khattab, M. A., Fouad, Y., & Rawash, O. A. (2009). Proposed Protocol to Solve Discovering Hidden Web Hosts Problem. *International Journal of Computer Science and Network Security, 9*(8).

Kou, Y., Shen, D., Yu, G., & Nie, T. (2008). *LG-ERM: An Entity-level Ranking Mechanism for Deep Web Query*. Washington, DC: IEEE.

Lage, P. B. G. J. P., Silva, D., & Laender, A. H. F. (2004). Automatic generation of agents for collecting hidden web pages for data extraction. *Data & Knowledge Engineering, 49*, 177–196. doi:10.1016/j.datak.2003.10.003

Liang, H., Zuo, W., Ren, F., & Sun, C. (2008). Accessing Deep Web Using Automatic Query Translation Technique. In *Proceedings of the Fifth International Conference on Fuzzy Systems and Knowledge Discovery*. IEEE.

Ling, Y.-Y., Liu, W., Wang, Z.-Y., Ai, J., & Meng, X.-F. (2006). Entity identification for deep web data integration. *Journal of Computer Research and Development*, 46-53.

Madhavan, J., Bernstein, P. A., & Rahm, E. (2001). *Generic Schema Matching with Cupid*. Paper presented at the 27th VLBB Conference, Rome.

Melnik, S., Garcia-Molina, H., & Rahm, E. (2002) Similarity Flooding: A Versatile Graph Matching Algorithm. In *Proceedings of the 18th International Conference on Data Engineering (ICDE)*, San Jose, CA.

Meng, X., Yin, S., & Xiao, Z. (2006). A Framework of Web Data Integrated LBS Middleware. *Wuhan University Journal of Natural Sciences, 11*(5), 1187–1191. doi:10.1007/BF02829234

Nie, Z., Ma, Y., Shi, S., Wen, J., & Ma, W. (2007). Web object retrieval. In *Proceedings of the WWW Conference*.

Niepert, M., Buckner, C., & Allen, C. (2007). A Dynamic Ontology for a Dynamic Reference Work. In *Proceedings of the (JCDL'07)*, Vancouver, British Columbia, Canada.

Qiang, B., Xi, J., Qiang, B., & Zhang, L. (2008). *An Effective Schema Extraction Algorithm on the Deep Web*. Washington, DC: IEEE.

Raghavan, S., & Garcia-Molina, H. (2001). Crawling the hidden Web. In *Proceedings of 27th International Conference on Very Large Data Bases (VLDB'01)* (pp. 129-138).

Sitemaps. (2009). *Sitemaps Protocol*. Retrieved from http://www.sitemaps.org

Song, H., Giri, S., & Ma, F. (2004). Data Extraction and Annotation for Dynamic Web Pages. In *Proceedings of EEE* (pp. 499-502).

Stojanovic, L., Stojanovic, N., & Volz, R. (2002). Migrating data intensive web sites into the semantic web. In *Proceedings of the 17th ACM Symposium on Applied Computing* (pp. 1100-1107).

Tejada, S., Knoblock, C. A., & Minton, S. (2002). Learning domain-independent string transformation weights for high accuracy object identification. In *Proceedings of the World Wide Web conference (WWW)* (pp. 350-359).

The Open Archives Initiative Protocol for Metadata Harvesting. (2003). *Protocol Version 2.0*. Retrieved from http://www.openarchives.org/OAI/2.0/openarchivesprotocol.htm

Tijerino, Y. A., Embley, D. W., Lonsdale, D. W., Ding, Y., & Nagy, G. (2005). Towards ontology generation from tables. *World Wide Web Journal*, 261-285.

Wu, W., Yu, C., Doan, A., & Meng, W. (2004). An interactive clustering-based approach to integrating source query interfaces on the Deep Web. In *Proceedings of the ACM SIGMOD International Conference on Management of Data (SIGMOD'04)* (pp. 95-106).

Yang, J., Shi, G., Zheng, Y., & Wang, Q. (2007). Data Extraction from Deep Web Pages. In *Proceedings of the International Conference on Computational Intelligence and Security*. IEEE.

Zhang, Z., He, B., & Chang, K. (2004). Understanding Web query interfaces: Best-effort parsing with hidden syntax. In *Proceedings of the ACMSIGMOD International Conference on Management of Data (SIGMOD'04)* (pp. 107-118).

Zhang, Z., He, B., & Chang, K. C. C. (2005). Light-weight Domain-based Form Assistant: Querying Web Databases On the Fly. In *Proceedings of the VLDB Conference*, Trondheim, Norway (pp. 97-108).

Zhao, P., & Cui, Z. (2007). Vision-based deep web query interfaces automatic extraction. *Journal of Computer Information Systems*, 1441–1448.

Zhao, P., Lin, C., Fang, W., & Cui, Z. (2007). A Hybrid Object Matching Method for Deep Web Information Integration. In *Proceedings of the International Conference on Convergence Information Technology*. IEEE.

Zhong, X., Fu, Y., Liu, Q., Lin, X., & Cui, Z. (2007). A Holistic Approach on Deep Web Schema Matching. In *Proceedings of the International Conference on Convergence Information Technology*. IEEE.

This work was previously published in International Journal of Information Technology and Web Engineering, Volume 5, Issue 1, edited by Ghazi I. Alkhatib, pp. 1-22, copyright 2010 by IGI Publishing (an imprint of IGI Global).

Chapter 6
Ripple Effect in Web Applications

Nashat Mansour
American University of Beirut, Lebanon

Nabil Baba
Lebanese American University, Lebanon

ABSTRACT

The number of internet web applications is rapidly increasing in a variety of fields and not much work has been done for ensuring their quality, especially after modification. Modifying any part of a web application may affect other parts. If the stability of a web application is poor, then the impact of modification will be costly in terms of maintenance and testing. Ripple effect is a measure of the structural stability of source code upon changing a part of the code, which provides an assessment of how much a local modification in the web application may affect other parts. Limited work has been published on computing the ripple effect for web application. In this paper, the authors propose, a technique for computing ripple effect in web applications. This technique is based on direct-change impact analysis and dependence analysis for web applications developed in the .Net environment. Also, a complexity metric is proposed to be included in computing the ripple effect in web applications.

1. INTRODUCTION

Web applications and regular software differ from each other in several ways. They differ in the inclusion of complex multi-tiered, heterogeneous architecture including web applications and database servers. The boundaries of a web application in web platforms are the boundaries of the World Wide Web itself. Unlike normal software environment in regular applications, web applications are subject to more challenges. One challenge is that of determining and controlling the propagation of change in the web application that arises from a local modification.

DOI: 10.4018/978-1-4666-0023-2.ch006

Ripple effect measure has been identified as an important measure of change propagation. Imagine a stone being thrown into a pond: it makes a sound as it enters the water and causes ripples to move outwards. Transferring this image into a source code is easy. The stone entering the water is now the change to the source code of a program, the effect of the change ripples across the source code via data and control flow. The ripple effect reflects how possible it is that a change to a particular form, method or class is going to adversely affect the rest of a program. It helps in determining the scope of the change and presents a measure of the program complexity. It can also be considered an indicator of the stability of a program. Ripple effect was one of the earliest metrics concerned with the structure of a system and how its modules interact (Shepperd, 1993). Ripple effect can show the maintainer what the effect of a local change will be on the rest of the web application. It can highlight the fragile parts of the web application to be restructured for better stable performance.

A method for computing ripple effect was developed early by Yau et al. (1978). Tools have been developed to produce ripple effect measures for procedural software using Yau et al.'s algorithm which is based on set theory. Black (2001) reformulated the ripple effect computation using matrix arithmetic and used this formulation to produce a software tool, REST. However, Black's algorithm computes the ripple effect only for procedural programs. Mansour and Salem (2006) proposed ripple effect techniques for object oriented (OO) program based on an analysis of object-oriented dependencies, relations, and propagations inside and outside classes. Elish and Rine (2003, 2005) studied structural and logical stability of object oriented design. They assessed the capability of OO design stability indicators to be used as predictors of measures of the metrics of a stability metrics suite. Mattsson and Bosch (2000) assessed stability of evolving OO frameworks. Grosser et al. (2002) used case-based reasoning to predict

software stability. Jazayeri (2002) applied retrospective analysis to successive software releases to evaluate its architectural stability with size and coupling measures. Further work on change impact analysis for object oriented programs can be found in: Rajlich (2001), Ryder and Tip (2001), Briand, Labiche, and Soccar (2002), Kung et al. (1994), and Lee, Offutt, and Alexander (2000).

In this paper we propose a technique for computing the ripple effect and logical stability for web applications focusing on .Net environment (ASP. Net and VB.Net) and using matrix arithmetic. Thus, the main contribution is in addressing the distinctive features of web application software. The computation of the ripple effect of a .Net web application is based on an analysis of ASP. Net and VB.Net (Bell, 2002) object-oriented dependencies, relations, and propagations locally and globally for ASP.Net (.aspx pages) and inside and outside classes for VB.Net and VBScripts. For brevity, this analysis is omitted herein and can be found in (Baba, 2007). Also, we propose a complexity metric for ASP.Net and VB.Net code to be included in computing the ripple effect. We compute the ripple effect for ASP.Net and VB.Net object-oriented at the code level. Both global and Form-Scope Propagation for .aspx pages (ASP. Net) and inter-class propagation and intra-class propagation for each class are considered. We also compute the architectural ripple effect at the system level. Each matrix used within the algorithm holds a particular type of information about the software under study.

This paper is organized as follows. Sections 2-7 describe our technique for computing the ripple effect and illustrates it with a running example. Section 8 concludes the paper.

2. WEB CODE COMPLEXITY METRIC

In this section, we propose a complexity metric for a .aspx page that contains no scripts, which is also applicable for .aspx.vb files. This metric is henceforth referred to as Web Code Complexity Metric (WCCM).

After analyzing the code, we calculate the WCCM, δ, as follows:

$$\delta = \sum \alpha + 1$$

Where α refers to an occurrence of a control that makes/triggers an event (Hyperlink, Submit Button, OnClick Buttons…etc). For VB.Net/VBscript code α represents each type of a control flow or decision point (If … then… else; while … end while… etc). Consider the example code shown in Figure 1. It corresponds to a page called "indexpage.aspx" where a user is asked to enter his/her Login and Password. If s/he forgot to write one of them, a signal "*" will be triggered to remind him/her to write in the field and not leave it blank. A validator summary is responsible to explain the type of error made. Applying the WCCM to this example code in Figure 1 yields:

1. The txt_pass_validator validates the presence of a text value in the login text box, it can be written as:

```
if (text_password.text = "\0") then
message (*)
```

This validator is acting as a decision point thus incrementing the value of

α by one so $\alpha =1$

2. The same goes for txt_login_validator; so $\alpha= 2$
3. The submit button is always considered as a decision point or control flow; so update $\alpha= 3$
4. The validation summary can not be considered because its value depends on the value

of the validators so it can be embedded under the same decision point:

```
if (text_password.text = "NULL") then
txt_pass: message (*)
VS: message(You forgot to type your
Password)
```

So the Complexity of Form1 in Figure 1 is δ =3+1= 4.

3. FORM-SCOPE CHANGE PROPAGATION

The computation of ripple effect in.aspx file is based on the effect of a change to a control on the functionality of another control, thus affecting the whole.aspx page.

The code in Figure 1 is the.aspx file called "indexpage.aspx" it shows the exact code that is written in ASP/Html only inheriting a code behind file called "indexpage.aspx.vb" without any script written in it. This code includes controls that affect the structure of the website. These are highlighted in grey. Other controls such as, labels, images, lines, and font do not affect the structure and are not considered.

Form-Scope Propagation due to a change in a control is based on the following five conditions derived from dependence analysis:

- An item is defined in a control scope.
- A control is set to validate another control. For example, in Figure 1, "txt_login_validator" is set to validate the "txt_login" web control.
- A control is set to use or be used by another control. For example, in Figure 1, the "VS" validation summary is set to use all the validators' values in the.aspx to write them in its summary.
- A control value (e.g., text box value) is used as an input value to a module in the

Figure 1. ASP/HTML code for indexpage.aspx presentation file

```
<%@ Page Language="vb" AutoEventWireup="false" Codebehind="indexpage.aspx.vb"
Inherits="TestLogin.indexpage"%>
<script runat="server">
   Sub Page_Load(Sender As Object, E As EventArgs)
          Welcome.Text = "Welcome to L.A.U!"
   End Sub
</script>
<HTML>          <HEAD><title>Lebanese American University Student Banner System</title>   </HEAD>
        <body MS_POSITIONING="GridLayout">
              <form id="Form1" method="post" runat="server">
              <asp:textbox id="txt_login" style="Z-INDEX: 101; LEFT: 312px; POSITION:
                     absolute; TOP: 176px" runat="server" Width="136px"
              Height="34px"></asp:textbox>
              <asp:label id="Label2" style="Z-INDEX: 108; LEFT: 160px; POSITION:
              absolute; TOP: 256px" runat="server" Width="144px" Height="32px"
              ForeColor="#004000" Font-Bold="True">Password</asp:label>
              <asp:requiredfieldvalidator id="txt_pass_validator" style="Z-INDEX: 106;
                     LEFT: 456px; POSITION: absolute; TOP: 248px" runat="server"
              ControlToValidate="txt_pass" ErrorMessage="*">
              </asp:requiredfieldvalidator><asp:textbox id="txt_pass" style="Z-INDEX: 102;
              LEFT: 312px; POSITION:          absolute; TOP: 248px" runat="server"
              Width="137px" Height="34px" TextMode="Password"></asp:textbox>
              <asp:label id="txt_label" style="Z-INDEX: 103; LEFT: 320px; POSITION:
                     absolute; TOP: 352px" runat="server" Width="208px" Height="24px"
              ForeColor="#C00000" Font-Bold="True" Font-Size="Smaller"
              Visible="False">Label</asp:label>
              <asp:button id="submit" style="Z-INDEX: 104; LEFT: 320px; POSITION:
              absolute; TOP: 304px" runat="server" Width="136px" Height="32px"
              ForeColor="#004000" Text="Submit"></asp:button>
              <asp:requiredfieldvalidator id="txt_login_validator" style="Z-INDEX: 105;
              LEFT: 456px; POSITION: absolute; TOP: 176px"runat="server"
              ControlToValidate="txt_login" ErrorMessage="*">
                     </asp:requiredfieldvalidator><asp:label id="Label1" style="Z-INDEX:
              107; LEFT: 160px; POSITION:absolute; TOP: 184px" runat="server"
              Width="144px" Height="32px" ForeColor="#004000" Font-
              Bold="True">Login</asp:label>
                            <asp:label id="Label3" style="Z-INDEX: 109; LEFT: 72px;
     POSITION:
              TOP: 16px" runat="server" Width="520px" Height="72px"
              Font-Size="XX-Large" Font-Names="Monotype Corsiva">Lebanese
              American University </asp:label>
              <asp:validationsummary id="VS" style="Z-INDEX: 110; LEFT:        72px;
              POSITION: absolute; TOP: 416px" runat="server" Width="184px"
              Height="40px"> </asp:validationsummary>
                     <p><asp:label id="Welcome" runat="server" /></p>
</form> </body></HTML>
```

script inside the.aspx file. For example, in Figure 1, the value of "txt_pass" is sent to a method in the code behind to check whether the password is correct.

- Page-Scope or inherited methods called by control events. For example, the "submit" button web control is calling a method called "button_click" inherited from code behind file sending it the values of the txt_login and txt_pass to check if they match a user's login and password or not.

Using these rules we can develop a vector V_{F1}, which is a 0-1 vector that represents the control definition in Form *F1*. Controls that satisfy any of the above conditions will be denoted by "1" and those which do not by "0". We shall use a short name for each control found in the Form (for example: txt_login will be written as "tlg"), note that labels and lines that are not interacting with the controls or has nothing to do with the functionality of the website will not be taken into consideration. So the dimension of the vector V_{F1} will be equal to the number of variables/controls found in the.net page that has direct relationship with the page's functionality. Referring to our example in Figure 1, the vector V_{F1} will be represented in Figure 2.Where: tlg represents txt_login; tpv represents txt_pass_validator; tp represents txt_pass; sub represents submit; tlv represents txt_login_validator; vs represents validationsummary "VS"; wl represents Welcome label

A 0-1 direct impact matrix S_F can be produced to show which control value may propagate to other controls within the form F. The rows and columns (r, c) of S_F represent each individual occurrence of a control. (r, c)=1 indicates a propagation from row *r* to column *c*, a change in *r* will directly affect *c* according to one or more of the above five mentioned conditions. Note that S_F is not bidirectional (symmetric) matrix. A brief description of this propagation will be shown after representing the matrix S_F in Figure 3.

Figure 2.

	tlg	tpv	tp	sub	tlv	vs	wl
$V_{F1} = ($	1	1	1	1	1	1	1 $)$

Figure 3.

$$S_{F1} = \begin{pmatrix}
 & tlg & tpv & tp & sub & tlv & vs & wl \\
tlg & 1 & 0 & 0 & 1 & 1 & 0 & 0 \\
tpv & 0 & 1 & 0 & 1 & 0 & 1 & 0 \\
tp & 0 & 1 & 1 & 1 & 0 & 0 & 0 \\
sub & 0 & 1 & 0 & 1 & 1 & 0 & 0 \\
tlv & 0 & 0 & 0 & 1 & 1 & 1 & 0 \\
vs & 0 & 0 & 0 & 0 & 0 & 1 & 0 \\
wl & 0 & 0 & 0 & 0 & 0 & 0 & 1
\end{pmatrix}$$

We observe in the above matrix that the value of *tlg* will propagate to itself of course and to *tlv* because *tlv* is a validator web control that checks if there is a value in *tlg* or not so it depends on its value to function. While the value of *tlg* will also propagate to the button *sub* because *sub* is sending the value in *tlg* to a method in code behind file to check whether that login value is available in the database or not.

4. PAGE-SCOPE CHANGE PROPAGATION

Propagation across forms and scripts in the.aspx presentation file is called *Page-Scope Change Propagation*. A change to a control can propagate to other method in a script or to another control in another form if it falls under the following three conditions (derived from dependencies analysis):

- The control value works as an input parameter to a scripted method in the.aspx file. For example,
    <INPUT style="Z-INDEX: 101; LEFT: 144px; POSITION: absolute; TOP: 176px"type="submit"value="Submit" id="Submit1" name="Submit1" onClick="check(login.text)">

- The control's value is a returned value from a method in a script under a certain event. For example,

```
a = "Hello"
TextBox1.Text = a
```

The value of the webform TextBox1 will be the value of variable "*a*" defined in the script.

- A control is setting a value or working with another control in another form.

The matrix X_{F1} will be made from number of columns equal to the number of forms and methods within the page and the number of rows is equal to the number of controls used. Thus, X_{F1} for our example is presented in Figure 4.

That is, there is no propagation from any control in Form *F1*, because the Page-Scope Change Propagation involves no flow of program change across forms and scripts. Whereas the label "Welcome" takes its value from the page_load (PL) script. The Page-Scope Change Propagation of all controls in *F1* can be found by finding the Boolean product of S_{F1} and X_{F1} (Figure 5).

We used the Boolean matrix product of S_{F1} and X_{F1} to maintain consistency with Form-Scope Change Propagation computation which is also Boolean. In order to indicate the amount of propagation from web controls in a Form, assume Form1 to another Form or method in a script we should find the standard matrix product of V_{F1} and $S_{F1}X_{F1}$ in Figure 6.

This result states that there are zero propagation from Form1 to Form1 and one propagation from Form1 to Page_Load (PL) script. Note that the only label "Welcome" is propagating to Page_Load. Of course if the form had more forms or scripts then the result will be different.

Figure 4.

$$X_{F1} = \begin{array}{c} \\ tlg \\ tpv \\ tp \\ sub \\ tlv \\ vs \\ wl \end{array} \begin{array}{cc} F1 & PL \\ \begin{pmatrix} 0 & 0 \\ 0 & 0 \\ 0 & 0 \\ 0 & 0 \\ 0 & 0 \\ 0 & 0 \\ 0 & 1 \end{pmatrix} \end{array}$$

The complexity measurement for the ind-expage.aspx (Form Scope) is $\alpha_{F1} = 4$ and the complexity measurement of the script page_load α_{PL} is 1. Finding the product of $V_{F1}\ S_{F1} X_{F1}$ and Complexity matrix C (see Figure 7) will represent the complexity-weighted total control-change propagation for Form1.

The maximum value that $V_{F1}.\ S_{F1}.\ X_{F1}.$ C can have is $\sum_{i=1}^{\Omega} \theta_i \delta_i$. We normalize the number obtained by the product $V_{F1}.\ S_{F1}.\ X_{F1}.$C by dividing it with the maximum value that can be obtained which is $\sum_{i=1}^{\Omega} \theta_i \delta_i$ to give the mean complexity-weighted control-change propagation per control in Form1.

In our code the $\sum_{i=1}^{\Omega} \theta_i \delta_i = 7 \times 4 + 1 \times 1 = 29$. Therefore, ripple effect of Form1 is defined as:

$$RE_{F1} = (V_{F1}.\ S_{F1}.\ X_{F1}.\ C) / \sum_{i=1}^{\Omega} \theta_i \delta_i = 1/29 = 0.034$$

From the above-mentioned results, we conclude that the ripple effect of any.aspx file that contains one form and no scripts (i.e., following one way in creating.aspx file) is very small and, thus, the logical stability is high.

Figure 5.

$$S_{F1}X_{F1} = \begin{pmatrix} 1 & 0 & 0 & 1 & 1 & 0 & 0 \\ 0 & 1 & 0 & 1 & 0 & 1 & 0 \\ 0 & 1 & 1 & 1 & 0 & 0 & 0 \\ 0 & 1 & 0 & 1 & 1 & 0 & 0 \\ 0 & 0 & 0 & 1 & 1 & 1 & 0 \\ 0 & 0 & 0 & 0 & 0 & 1 & 0 \\ 0 & 0 & 0 & 0 & 0 & 0 & 1 \end{pmatrix} \begin{pmatrix} 0 & 0 \\ 0 & 0 \\ 0 & 0 \\ 0 & 0 \\ 0 & 0 \\ 0 & 0 \\ 0 & 1 \end{pmatrix} = \begin{pmatrix} 0 & 0 \\ 0 & 0 \\ 0 & 0 \\ 0 & 0 \\ 0 & 0 \\ 0 & 0 \\ 0 & 1 \end{pmatrix}$$

Figure 6.

$$V_{F1} S_{F1} X_{F1} = \begin{pmatrix} 1 & 1 & 1 & 1 & 1 & 1 & 1 \end{pmatrix} \begin{pmatrix} 0 & 0 \\ 0 & 0 \\ 0 & 0 \\ 0 & 0 \\ 0 & 0 \\ 0 & 0 \\ 0 & 1 \end{pmatrix} = \begin{matrix} F1 & PL \\ (0 & 1) \end{matrix}$$

Figure 7.

$$V_{F1} \cdot S_{F1} \cdot X_{F1} \cdot C = (\ 0\ \ 1\ ) \begin{bmatrix} 4 \\ 1 \end{bmatrix} = (\ 1\ )$$

5. INTER-CLASS CHANGE PROPAGATION

Calculating the inter-class change propagation means that the webpage contains a code behind file or a script written in Visual Basic (in the.Net environment). That is the developer used another way of creating web pages. The computation of ripple effect is based on the effect where a change to a variable will affect the rest of the webpage.

Figure 8 gives the code-behind file of index-page.aspx presentation file, mainly it is made from one class called "indexpage" and two methods one called "page_load" that is called once the page is loaded, and another called "submit_click" that is called once the submit button in index.aspx is clicked. Simply, this code creates a database connection and opens it to be accessed through a query called "cmd" in line 11 in page_load method. Then, when the user enters his login and password and clicks on the submit button the submit_click method will be called to execute the query in line 14 and checks whether a user is found (in line 16). If the user is found, the page will be redirected to another one called personal.aspx in line 19. If not, the user will have a note saying that the user_login and password are incorrect and should be entered again. Clearly in the page_load method in line 9 we are defining the connection and we are using it in line 10 where we are opening the connection. The connection is also used in line 11 to create a query called "cmd" so any change that might happen in line 9 will affect line 10 thus affecting the query "cmd" in line 11. Clearly propagation takes place from definitions to uses of variables and via assignments.

Based on the work of Mansour and Salem (2006), inter-class ripple computation due to a change in a variable will mostly be based on the following six conditions:

1. The variable is defined in an assignment statement. For example 'cmd' in:

```
11 cmd = New OleDbCommand("select *
from Login where UserLogin
= '" & txt_login.Text.Trim & "' and
UserPassword ='"
& txt_pass.Text.Trim & "'", cn)
```

2. The variable is assigned a value which is read as an input. For example 'dr' in line 16:

```
16 dr.Read()
```

3. The variable is calling a function or a method that is essential to the rest of the program to work. Example 'cn' in

```
10 cn.Open()
```

4. The variable is an input parameter to a method. An example not found in the above figure can be 'ccounter' in:

```
Public double calculate (double
ccounter)
```

5. The variable takes a returned value from a called method. An example can be 'dr' in:

```
12 dr = cmd.ExecuteReader
```

6. The variable is in Class-Scope, shared or inherited variable. The same example can be taken from line 6 because cmd is a global variable defined globally for all the methods in class indexpage (see Table 1).

Figure 8. Code written in VB.Net for code behind indexpage.aspx.vb

```
1    Imports System.Data.OleDb

2    Public Class indexpage
3       Inherits System.Web.UI.Page
4       Public Shared user_ID As Long
5       Dim cn As OleDbConnection
6       Dim cmd As OleDbCommand
7       Dim dr As OleDbDataReader
8       Private Sub Page_Load(ByVal sender As System.Object, ByVal e As
            System.EventArgs) Handles MyBase.Load
9          cn = New OleDbConnection("Provider=Microsoft.Jet.OLEDB.4.0;
             Data Source=C:\Inetpub\wwwroot\TestLogin\test.mdb;")
10         cn.Open()
11         cmd = New OleDbCommand("select * from Login where UserLogin
             = '" & txt_login.Text.Trim & "' and UserPassword ='" &
             txt_pass.Text.Trim & "'", cn)
12         dr = cmd.ExecuteReader

13      End Sub

14      Private Sub submit_Click(ByVal sender As System.Object, ByVal e
15         As System.EventArgs) Handles submit.Click
16         dr.Read()
17         If dr.HasRows Then
18            user_ID = dr(0)
19            Response.Redirect("Personal.aspx")
20         Else
21            txt_login.Text = ""
22            txt_pass.Text = ""
23            txt_label.Visible = True
24            txt_label.Text = "Check your Login/Password"
25         End If
26         dr.Close()
27         cn.Close()
        End Sub
28
     End Class
```

In VB.Net if a variable is defined in a block scope then it is not visible outside that scope. When creating the matrix *Vm* we will be working under the VB.Net visibility rules. Intuitively any global or shared values on the right hand side of assignments should count. Based on the conditions above we will develop a vector V_{load} and move on in calculating the page_load ripple effect. There-

fore, V_{load} will be as presented in Figure 9. Where cn_{d9} represents cn = New OleDbConnection("Provider...) defined in line 9, cn.o represents cn.Open() in line 10, cmd represents cmd = New OleDbCommand("select * from ...) in line 11, txt_l represents the use of text_login in line 11, txt_p represents the use of text_pass in line 11, cn_{u11} represents the use of cn

Figure 9.

$$V_{load} = \begin{array}{ccccccc} cn_{d9} & cn.o & cmd & txt\_l & txt\_p & cn_{u11} & dr \\ (1 & 1 & 1 & 1 & 1 & 1 & 1) \end{array}$$

in line 11 as a parameter, and dr represents dr executing cmd query "*dr = cmd.ExecuteReader*" in line 12.

The variables are taken in the order of appearance in the page_load method, if two variables appeared in the matrix having the same name then they will be differentiated by a "*u*" or a "*d*" meaning either used or defined followed by the line number. A 0-1 direct impact matrix S_{laod} can be produced to show which variable may propagate to other variables within the method page_load. The rows and columns of S_{load} will represent each individual occurrence of a variable. Propagation will be shown from row i to column j. A brief description of this propagation will be shown after representing the matrix S_{load} in Figure 10.

We observe in the above matrix that the value of *txt_l* will propagate to itself of course and to *cmd*, while the value of *cmd* will propagate to *dr* where *dr* will be used to execute and read the value returned from the query. So the *txt_l* will also propagate to *dr*. We can notice that S_F is both reflexive and transitive; that is every variable occurrence is assumed to propagate to itself and if a variable *v* propagates to variable *s* and variable *s* propagates to variable *n* then variable *v* will also propagate to variable *n*. As mentioned earlier, S_F shows how a variable's value can reach or affect another variable in the code.

6. INTRA-CLASS CHANGE PROPAGATION

Propagation across classes and methods in a script (in.aspx) or code behind file.aspx.vb is called *intra-class change propagation*. A change

to a variable can propagate to other method in a script or to another other method in a class found in code behind if it falls under one of the following conditions:

- The variable is an inherited, shared or a global variable. For example cmd in Table 2.

The *cmd* query command is global to all the methods in class indexpage, a change in *cmd* in any method might affect *dr*.

- The variable is an input parameter in an Intra-class message. An example can be counter1 in:

```
Average = Grades.calculate (counter1)
```

- The variable or object is returned by a method to another method in a different class. An example can be *age* in: Return (age)

In the example indexpage.aspx.vb we can analyze the method *page_load* according to the above conditions. For example, the definition of a

Table 1.

```
2 Public Class indexpage
3 Inherits System.Web.UI.Page
4 Public Shared user_ID As Long
5 Dim cn As OleDbConnection
6 Dim cmd As OleDbCommand
..
..
..
12 dr = cmd.ExecuteReader
```

Figure 10.

$$S_{load} = \begin{array}{c} \\ cn_{d9} \\ cn.o \\ cmd \\ txt\_l \\ txt\_p \\ cn_{u11} \\ dr \end{array} \begin{array}{ccccccc} cn_{d9} & cn.o & cmd & txt\_l & txt\_p & cn_{u11} & dr \\ \left(\begin{array}{ccccccc} 1 & 1 & 1 & 0 & 0 & 1 & 0 \\ 0 & 1 & 1 & 0 & 0 & 1 & 0 \\ 0 & 0 & 1 & 0 & 0 & 0 & 1 \\ 0 & 0 & 1 & 1 & 0 & 0 & 0 \\ 0 & 0 & 1 & 0 & 1 & 0 & 0 \\ 0 & 0 & 1 & 1 & 1 & 1 & 0 \\ 0 & 0 & 0 & 0 & 0 & 0 & 1 \end{array}\right) \end{array}$$

new connection to a database *cn* is a global variable, so any change in that definition will affect all the methods in the class. The same thing goes for the query command *cmd*. While *dr* is defined globally then its execution in page_load will affect the values read in submit_click method. Because the values returned in dr.Read() in line 16 depends directly on the execution of *cmd* in line 12 "*dr = cmd.ExecuteReader*". We can represent the propagation of these variable through the class using a 0-1 matrix X_{load} as shown in Figure 11.

Intra-class change propagation involves the flow of program/variable change across classes, that is why the column *P_load* is all zeros. By finding the Boolean product of S_{load} and X_{laod} we can find the Intra-class change propagation of all variables in *page_load* (see Figure 12). Therefore the $S_{load}X_{load}$ will be as pictured in Figure 12.

Now in order to indicate the amount of propagation from variables in the load method, we

should find the standard matrix product of V_{load} and $S_{load}X_{load}$ as seen in Figure 13.

A complexity measure should also be taken into account in any.aspx.vb. A matrix C can be presented, representing complexity measurement. We use our WCCM measure. Hence, for our running example, the matrix C is picture in Figure 14.

The product of $V_{load}S_{load}X_{laod}$ and C is picture in Figure 15.

We normalize this number by dividing it by $\sum_{i=1}^{\Omega}\theta_i\delta_i$ to give the mean complexity-weighted control-change propagation per variable in

Table 2.

```
2 Public Class indexpage
3 Inherits System.Web.UI.Page
4 Public Shared user_ID As Long
5 Dim cn As OleDbConnection
6 Dim cmd As OleDbCommand
 ..
 ..
 ..
 12 dr = cmd.ExecuteReader
```

Figure 11.

$$X_{load} = \begin{array}{c} \\ cn_{d9} \\ cn.o \\ cmd \\ txt\_l \\ txt\_p \\ cn_{u11} \\ dr \end{array} \begin{array}{cc} P\_load & sub\_click \\ \left(\begin{array}{cc} 0 & 1 \\ 0 & 1 \\ 0 & 1 \\ 0 & 0 \\ 0 & 0 \\ 0 & 0 \\ 0 & 1 \end{array}\right) \end{array}$$

Figure 12.

$$S_{load}\, X_{laod} \; = \; \begin{pmatrix} 1 & 1 & 1 & 0 & 0 & 1 & 0 \\ 0 & 1 & 1 & 0 & 0 & 1 & 0 \\ 0 & 0 & 1 & 0 & 0 & 0 & 1 \\ 0 & 0 & 1 & 1 & 0 & 0 & 0 \\ 0 & 0 & 1 & 0 & 1 & 0 & 0 \\ 0 & 0 & 1 & 1 & 1 & 1 & 0 \\ 0 & 0 & 0 & 0 & 0 & 0 & 1 \end{pmatrix} \begin{pmatrix} 0 & 1 \\ 0 & 1 \\ 0 & 1 \\ 0 & 0 \\ 0 & 0 \\ 0 & 0 \\ 0 & 1 \end{pmatrix} = \begin{pmatrix} 0 & 3 \\ 0 & 2 \\ 0 & 2 \\ 0 & 1 \\ 0 & 1 \\ 0 & 1 \\ 0 & 1 \end{pmatrix}$$

Figure 13.

$$V_{load}\, S_{load}\, X_{laod} = \begin{pmatrix} 1 & 1 & 1 & 1 & 1 & 1 & 1 \end{pmatrix} \begin{pmatrix} 0 & 3 \\ 0 & 2 \\ 0 & 2 \\ 0 & 1 \\ 0 & 1 \\ 0 & 1 \\ 0 & 1 \end{pmatrix} = \begin{pmatrix} 0 & 11 \end{pmatrix}$$

Page_load. In our code $\sum_{i=1}^{\Omega} \theta_i \delta_i$ is equal to 37. Therefore, ripple effect of page_load is defined as:

$$RE_{load} = (V_{load} \cdot S_{load} \cdot X_{load} \cdot C) / \sum_{i=1}^{\Omega} \theta_i \delta_i = 33/37$$
$$\doteq 0.89$$

7. AGGREGATING THE RIPPLE EFFECT CALCULATION

The calculation of the total ripple effect of a website does not cover only the intra- and inter-class ripple effects. A website is usually composed of a number of pages, each can be developed differently as illustrated in Figure 16 where a website is made up of 3 pages. Next, we show how to compute an aggregate ripple effect for a whole web application.

Referring to the example in Figure 16, we note that page1 is made from two files one is the.aspx

Figure 14.

$$C = \begin{matrix} P\_load \\ Sub\_click \end{matrix} \begin{pmatrix} 1 \\ 3 \end{pmatrix}$$

Figure 15.

$$V_{load} . S_{load} . X_{laod} . C \quad = \quad (\quad 0 \quad 11 \quad) \begin{bmatrix} 1 \\ 3 \end{bmatrix} \quad = 33$$

Figure 16. A 3-page website, each developed in a different way

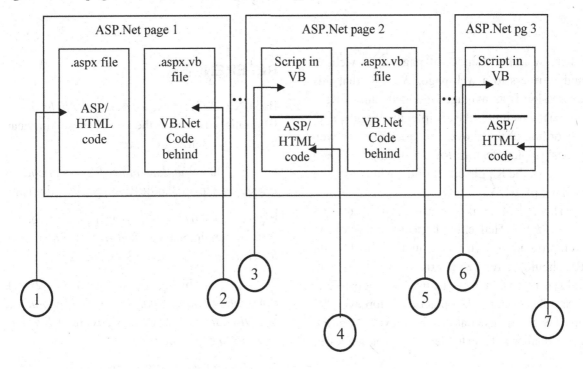

file and the second is the code behind file.aspx.vb. In order to calculate the ripple effect of page 1, we have to calculate the *RE* of the.aspx file. Following arrow number 2 in Figure 16, we find that it is pointing to the code behind that might contain more than one method. Hence, the aggregate *RE* of the whole.aspx.vb file is given by

$$1/n \sum_{i=1}^{n} \frac{(Vmi . Smi . Xmi . C)}{\sum_{i=1}^{\Omega} \theta_i \delta_i}$$

where m refers to methods and n is the number of methods in the code behind. The same goes for arrows number 3, 5 and 6. The total ripple effect of the whole.aspx page is given by

$$1/l \sum_{i=1}^{l} \frac{(Vki . Ski . Xki . C)}{\sum_{i=1}^{\Omega} \theta_i \delta_i}$$

Where *k* refers to methods or forms and *l* is the number of all methods and forms in the.aspx file. The.aspx and.aspx.vb are strongly connected; one cannot work without the other. This is why we consider the total ripple effect of the.net page as a whole will be the sum of the ripple effect of.aspx and.aspx.vb.

After finding the ripple effect of each page, the aggregate ripple effect of the whole web program is given by

$$RE website = 1/w \sum_{i=1}^{w} REpi$$

where w is the number of pages in a website and *pi* refers to a web page. We note that this expression is an average weighted sum of REs, where the weight is the complexity of each page. According to the above rule, assuming the web pages' ripple effect as: *REp1* = 4, *REp2* = 6, and *REp3* = 5, the Ripple effect of the whole website will be: (4+6+5)/3 = 5.

The logical stability measure is defined as *LS = 1/ (RE+1)*. Hence, the logical stability of the web program is 0.16. These numbers of LS and RE should be read as relative and not absolute values. That is, they can be used to compare different design and coding practices. Moreover, RE and LS can be used at different levels from the individual page level to the whole program level.

8. CONCLUSION

We have presented a technique for computing the ripple effect in web programs, in the.Net environment, addressing the features of these web programs. That is, this technique is based on generating change propagation matrices from dependence relations detected in the various parts of a web program and on matrix arithmetic. It also employs a web complexity metric that provides weight coefficients for propagation information.

Our technique has been demonstrated using a case study and has shown to be useful for quantifying the ripple effect and measuring the logical stability. Further work can be done to empirically

show that this technique for computing the ripple effect is useful for reducing maintenance effort and time. However, this technique suffers from the limitation presented by the computation cost of code parsing and generating of propagation matrices and vectors. This is particularly serious for large-scale web programs. Also, this technique does not include semantic and run-time dependences.

REFERENCES

Baba, N. (2007). *Ripple Effect in Web Applications*. Unpublished master's thesis, Lebanese American University, Beirut, Lebanon.

Bell, E. (2002). *Fundamentals of Web applications using. Net*. Upper Saddle River, NJ: Prentice Hall.

Black, S. (2001). Computing ripple effect for software maintenance. *Software Maintenance. Research and Practice*, 13(4), 263–278.

Briand, L., Labiche, Y., & Soccar, G. (2002). *Automating Impact Analysis and Regression test Selection Based on UML Designs (Tech. Rep. No. SCE- 02-04)*. Carleton.

Elish, M., & Rine, D. (2003). Investigation of Metrics for Object-Oriented Design Logical Stability. In *Proceedings of the 7th European Conference On Software Maintenance And Reengineering*. Washington, DC: IEEE Computer Society.

Elish, M., & Rine, D. (2005). Indicators of Structural Stability of Object Oriented Designs: A Case Study. In *Proceedings of the 29th Annual NASA/IEEE Software Engineering Workshop* (pp. 183-192).

Grosser, D., Sahraoui, H., & Valtchev, P. (2002). Predicting Software Stability Using Case-Based Reasoning. In *Proceedings of the 17th IEEE International Conference on Automated Software Engineering* (pp. 295-298).

Jazayeri, M. (2002). On architectural stability and evolution. In *Proceedings of the 7th Ada-Europe International Conference on Reliable Software Technologies* (pp. 13-23). London: Springer Verlag.

Kung, D., Gao, J., Hsia, P., Wen, F., Toyoshima, Y., & Chen, C. (1994). Change Identification in Object Oriented Software Maintenance. In *Proceedings of the International Conference on Software Maintenance* (pp. 201-211).

Lee, M., Offutt, J., & Alexander, R. (2000). Algorithmic Analysis of the Impact of Changes to Object-Oriented Software. In. *Proceedings of TOOLS, 2000,* 61–70.

Mansour, N., & Salem, H. (2006). Ripple Effect in Object Oriented Programs. *Journal of Computational Methods in Sciences and Engineering, 6*(5-6), 23–32.

Mattsson, M., & Bosch, J. (2000). Stability assessment of evolving industrial object oriented Frameworks. *Journal of Software Maintenance: Research and Practice, 12,* 79–102..doi:10.1002/(SICI)1096-908X(200003/04)12:2<79::AID-SMR204>3.0.CO;2-A

Rajlich, V. (2001). *Propagation of changes in Object Oriented Programs (Tech. Rep.).* Detroit, MI: Wane State University.

Ryder, B., & Tip, F. (2001). Change Impact Analysis for Object-Oriented programs. In *Proceedings of the ACM Workshop on Program Analysis for Software Tools and Engineering* (Vol. 10, pp. 46-53).

Shepperd, M. (1993). Software Engineering Metrics: *Vol. I. Measures and Validations.* London: McGraw Hill.

Yau, S., Collofello, J., & McGregor, T. (1978). Ripple effect analysis of software maintenance. In. *Proceedings of COMPSAC, 78,* 60–65.

This work was previously published in International Journal of Information Technology and Web Engineering, Volume 5, Issue 2, edited by Ghazi I. Alkhatib, pp. 1-15, copyright 2010 by IGI Publishing (an imprint of IGI Global).

Chapter 7
Finer Garbage Collection in LINDACAP

Nur Izura Udzir
Universiti Putra Malaysia, Malaysia

Hamidah Ibrahim
Universiti Putra Malaysia, Malaysia

Sileshi Demesie
Bahir Dar University, Ethiopia

ABSTRACT

As open systems persist, garbage collection (GC) can be a vital aspect in managing system resources. Although garbage collection has been proposed for the standard LINDA, it was a rather course-grained mechanism. This finer-grained method is offered in LINDACAP, a capability-based coordination system for open distributed systems. Multicapabilities in LINDACAP enable tuples to be uniquely referenced, thus providing sufficient information on the usability of tuples (data) within the tuple-space. This paper describes the garbage collection mechanism deployed in LINDACAP, which involves selectively garbage collecting tuples within tuple-spaces. The authors present the approach using reference counting, followed by the tracing (mark-and-sweep) algorithm to garbage collect cyclic structures. A time-to-idle (TTI) technique is also proposed, which allows for garbage collection of multicapability regions that are being referred to by agents but are not used in a specified length of time. The performance results indicate that the incorporation of garbage collection techniques adds little overhead to the overall performance of the system. The difference between the average overhead caused by the mark-and-sweep and reference counting is small, and can be considered insignificant if the benefits brought by the mark-and-sweep is taken into account.

DOI: 10.4018/978-1-4666-0023-2.ch007

INTRODUCTION

The LINDA coordination model (Gelernter, 1985; Gelernter, 1989) offers an alternative to the conventional point-to-point communication framework with regard to coordinating and synchronising agents' activities. The shared data space known as tuple-spaces (TSs) provides a medium for communication and facilitates the coordination among the interacting agents—agents communicate with each other via the tuple-space where they write and retrieve data (known as tuples). The clear separation between the coordination and the computation concerns relieves the agents of the messy aspects of communication, leaving them free to concentrate their time and space for other more crucial aspects of computation. This paradigm allows for anonymous interaction between agents separated in time and space: communicating agents need not know each other's identity, and also the data can be retrieved any time after it has been placed in the tuple-space. In the original LINDA model, there are three primitive operations which enable agents to manipulate the tuple-space: **out** to write a tuple, **rd** to perform read and **in** to read and remove the data. The tuples are retrieved associatively in a non-deterministic fashion: the retrieval operation may return any matching tuple; and, if a number of agents are waiting for a tuple of the same template, a matching tuple, when available, may be given to any one of them. Interacting via the TS where there is no direct communication between them, the communicating agents are decoupled in name, space and time—they need not know each other's identity, nor co-exist at the same time in order to communicate with each other—providing a flexible coordination mechanism suitable for open, heterogeneous systems. LINDA's popularity is shown in its commercial variants such as Sun's JavaSpaces (Freeman, 1999) and IBM's TSpaces (Wyckoff, 1998).

In open implementations of LINDA, agents or active entities can join and leave the system at anytime: they do not need to be defined at compile time. Agents which are not executing at the same time, can communicate via tuple-spaces. An agent can also store a tuple in a tuple-space to be used by some other agents at any time in the future. In general, agents running on different machines and operating systems can be compiled separately, and written in different languages. On the other hand, in closed implementations, all agents that wish to interact must be known at compile time. The information derived at compile time can be used to control the system and optimize the application in the best possible way, including when to remove which object when it becomes unnecessary. Applications for open systems are, in general, intended to have a longer running time than those for closed systems. In LINDA's open system, tuples created are never removed from the system as it is difficult to decide if the tuples are still in-use or otherwise. As memory is a finite resource, this may lead to system failure due to memory exhaustion—the system will eventually run out of memory space if unusable objects are not reclaimed. Therefore, the implementations of open LINDA systems need to address the problem of memory management in order to be of practical use. One way of managing memory is by means of garbage collection (GC), which attempts to reclaim memory that is no longer used by an application.

The garbage collection mechanism has been implemented in Ligia (Menezes, 2000) to garbage collect unnecessary tuple-spaces using a graph, where reference information is maintained. However, tuples within the tuple-space cannot be garbage collected selectively: in order to perform garbage collection on tuples, information about their usefulness should be maintained for the garbage collector. This is not possible in the standard LINDA as tuples cannot be individually referenced—for the purpose of maintaining the information related to their usefulness—as they are nameless. In LINDACAP (Udzir, 2006; Udzir, 2007), with the introduction of *multicapabilities*, i.e. capabilities for classes of objects, tuples within a tuple-space can now be referenced. Therefore, the information about the accessibility of tuples can be maintained by adopting the graph structure used in Ligia. As a result, garbage collection on tuples has been implemented in order to remove unnecessary tuples with specified template and subsequently reclaim memory space.

This paper describes the deployment of garbage collection in LINDACAP. The implementation is based on reference counting. Unfortunately, this technique does not apply to cyclic garbage, i.e., two or more multicapability regions referring to one another. In order to overcome this, the implementation of garbage collection in LINDACAP is extended using the mark-and-sweep algorithm to allow cyclic garbage collection.

There may also be the case where some agents may still have references to certain groups of tuples (represented by multicapabilities regions) but they no longer use the tuples/groups: the agents may have forgotten about these tuples which are consuming memory space. Based on the graph, the regions will not be considered as garbage as they are agents referring to them. A mechanism should be provided to allow these regions to be garbage collected. Therefore, in this paper, we also describe a rather simple, but useful, additional technique, i.e., the incorporation of time-to-idle (TTI) for the purpose of reclaiming memory allocated for groups of tuples that have been 'idle' for a certain length of time.

RELATED WORK AND MOTIVATION

Resource management, for example, is vital in distributed systems that involve ubiquitous and persistent computing. Poor presource management, at best leads to poor performance. One resource that needs to be managed is memory, which is limited but can be reclaimed through garbage collection. Garbage collection is the process of searching and automatically reclaim unused memory cells to avoid memory exhaustion. The algorithms are based on traversing a tracing

graph representing the memory to analyse and determine the cell's usefulness (Menezes, 1997; Menezes, 2000). In order to do so, the kernel must have knowledge of which objects are being referenced by which agents.

The reference counting algorithm has been around since 1960 when it was developed by Collins (Collins, 1960)—each reference to a memory cell increases the cells's counter, while a de-referencing decreases the counter; memory cells are reclaimed when their reference counter reach zero. This method, however, fails to reclaim cyclic structures, where the counters never reach zero (McBeth, 1963). A solution to this problem was introduced by combining tracing (mark-and-sweep) and reference counting (Friedman, 1979; Lins, 2008).

Agent registration in Ligia helps the kernel to identify any agent in the system. Tuple-spaces are also identified by their unique handle. As a result of this, a graph with reference information for each tuple-space has been proposed to maintain information about the usefulness of tuple-spaces for the garbage collector. Thus, a garbage collection mechanism has been implemented in Ligia, to garbage collect unused tuple-spaces. However, tuples within a tuple-space that are no longer used by agents cannot be removed by the garbage collector unless the whole tuple-space becomes garbage. This is particularly problematic in the case of the UTS as the default tuple-space can never be garbage collected in Ligia. In order to perform garbage collection on certain tuples, information about their accessibility should be maintained, and this can be done using multicapabilities in LINDACAP.

Garbage collection of tuples in JavaSpaces (Freeman, 1999) uses leasing in which tuples within the space are associated with a limited time frame (i.e., leases). As a result, the garbage collector agent can remove these tuples after a certain period of time elapses (when the lease expires). The drawback of this approach is that it is difficult to estimate in advance how long a particular object will be of use. They vary according to applications, for instance, we might want to reclaim all objects generated by a particular agent, or to clean up any object left over a particular protocol.

Object (tuple) retrieval in SecOS (Bryce, 1999) is different from Lᴉɴᴅᴀ. Matching does not require the number of fields of the template and the target tuple to be equal. Thus, an agent can use an empty template to retrieve any tuple without gaining access to its contents. As a result, tuples can be deleted periodically by the garbage collector agent. Since tuples are deleted randomly without considering any tuple's usage information, the garbage collector agent cannot selectively delete unnecessary tuples. In a later version of SecOS (Vitek, 2003), each tuple is tagged with some information (similar to a lease) before being inserted into the space. The garbage collector agent can use this information and removes the tuple with particular tags. However, tagging of tuples is performed by agents voluntarily but not automatically by the system.

In Scope (Merrick, 2001), a streaming mechanism is used to perform an input operation upon all tuples in a given scope without affecting the other agents which are using the same tuples. This streaming mechanism can be used to access fresh tuples in the relevant scope and these tuples can be tagged with a scope representing the time-slices in which their leases expire. A sibling agent can use these tags to search for tuples whose lease has expired, and remove them.

Garbage collection in shared memory systems has also been discussed in XMem (Wegiel, 2008), where unused shared objects are identified and reclaimed, using conventional tracing approach. Their more fine-grained tracing garbage collection is an improvement as compared to the course-grained block-based reference counting garbage collection of Microsoft's Singularity (Fähndrich, 2006).

In Lᴉɴᴅᴀᴄᴀᴘ, tuples within a tuple-space are grouped together based on a specified template and can be referred to using a multicapability

with a unique identity. This allows the kernel to keep track of the tuples usability for the garbage collector. Therefore, garbage collection of tuples has been implemented in Lᴉɴᴅᴀᴄᴀᴘ to remove selective tuples (a multicapability region) rather than removing the whole tuple-space (Udzir, 2006; Udzir, 2007; Udzir, 2008). The implementation is based on the reference counting technique, in which the links that reference each multicapability region is maintained, and the storage for these regions can be reclaimed when the reference counter is decreased to zero. Unfortunately this approach has two problems. First, the reference counts for multicapability regions that are part of a cycle (i.e. two or more multicapability regions that refer to one another) will be non-zero. Thus, this cyclic garbage cannot be garbage collected. In order to overcome these problems, a tracing based technique called the mark-and-sweep method, which is based on Dijkstra's on-the-fly proposal (Dijkstra, 1978) is also implemented.

Lᴉɴᴅᴀᴄᴀᴘ

Lᴉɴᴅᴀᴄᴀᴘ extends Lᴉɴᴅᴀ with a capability-based control mechanism to provide a more refined control for open distributed systems without losing the flexibility of Lᴉɴᴅᴀ. Objects visibility to agents in Lᴉɴᴅᴀᴄᴀᴘ are defined by the capabilities (similar to tickets) they hold. Capabilities can provide information not only on a particular object, but also on which methods of the object an agent is permitted to invoke. Specific information about an agent's 'knowledge' is potentially very useful and can be manipulated in a variety of ways.

However, unlike access control lists (ACLs), capabilities can only be applied to named objects, such as tuple-spaces, but not the nameless tuples. To overcome this, we have introduced multicapabilities (Udzir, 2006; p. 25), i.e., capabilities for a class of objects: whereas a permission in a uni-capability allows an action on the *object* it refers to, a multicapability allows the action to be performed on an *element* of the class. Throughout this paper, we shall use the term 'capability' to refer to capabilities in general, and the terms 'uni-capability' or 'multicapability' accordingly when referring to a specific class.

Extending uni-capabilities which are pairs of object identifier *obj*, and rights (**[obj,{i,r,o}]**), each multicapability is a triple of an identifier, the template of tuples it refers to, and a set of rights, i.e., [*u,t,p*], where *u* will be used as a unique (unforgeable) identifier to a collection of objects of the specified template *t*, and *p* is a set of operations allowed on an element of the group. For example, a multicapability for a template of two integers is [α, ⟨?int,?int⟩ ,{i,r,o}], where i, r and o are permissions to perform destructive and non-destructive read, and write, respectively.

The capability data (unique identifier, reference/template, and permissions) are assumed to be securely encapsulated in the capability and only interpretable by the kernel when the capability is presented for verification.

Consequently, there are two kinds of capability in LINDACAP: a uni-capability (TS-capability) that holds the set of operations permitted on the tuple-space it refers to, and a multicapability which holds the set of operations to be performed on a group of tuples (multicapability region) within the tuple-space. Every tuple-space and tuple operation requires two-level capabilities: a TS-capability for the target tuple-space, and a multicapability for a specified template (a formal for a tuple pattern).

Whenever an agent requests for a new capability, the kernel returns a unique capability with full rights. Therefore, tuples of the same pattern may be referred to by different multicapabilities. For example, if an agent makes two separate requests for a multicapability for the template ⟨?str,?int,?int⟩ , it will get *two* multicapabilities, each different from the other (i.e., identified by different tags), although they correspond to the same template. If one of the multicapabilties is given to another agent, the second agent can

only 'see' tuples in the multicapability group it holds, but not those in the other group. Hence, multicapabilities can provide a partitioning of a tuple-space—enabling certain operations to be performed on a tuple of a specific group, but not on one of another group, even though both groups have the same template.

In LINDACAP, communications between two or more agents can be performed by passing capabilities among them. Requesting a capability even for the same template will not allow an agent to retrieve data produced by another agent: they need to use the same capability to access the same (set of) data. Hence the capability needs to be passed from the producer to the other agent via a tuple-space. Capabilities can be elements of tuples and stored within tuple-spaces. These tuples can be retrieved by another agent that has access to the tuple-spaces. The agent will then be able to access the objects referred to by these capabilities. The universal primordial capability,

$$cc = [\aleph, \ \langle ?cap \rangle \ ,\{r,o\}],$$

for a capability type is used to pass capabilities (encapsulated within tuples) via tuple-spaces, or at least via the universal tuple-space (UTS). For this purpose, a TS-capability (with full rights) for the UTS, and *cc* are given to each agent whenever it connects to the kernel.

Passing capabilities through tuple-spaces is just like passing any ordinary data, using the standard input/output primitives. For example, passing an arbitrary capability **cap1** entails writing the capability into a tuple-space,

```
ts.out(cc<cap1>);
```

With the universal primordial capability, *cc*, this tuple may be retrieved by another agent,

```
ts.rd(cc<?cap>);
```

Agents may duplicate, or make restricted copies, or even combine capabilities (in sum or subtraction operations) in their possession. There are two types of operations: unary and binary. Unary operations involve a single capability, where a copy of a capability, while having the same unique identifier as the original, may have a restricted template and/or rights. However, it is not possible to add rights, nor to generalize the template in a multicapability. For example, a copy of multicapability $[\alpha, \langle ?int, 3 \rangle, \{i,r,o\}]$ may have one or more rights removed; and its template restricted to $\langle \mathbf{56,3} \rangle$ —or any integer value as its first element—but not generalized to $\langle ?int, ?int \rangle$ as the second element has been specialized to integer 3 and cannot be altered.

Binary operations 'combine' two (or more) capabilities in an agent's possession. We present in this paper two simple binary operations: sum and subtraction.

The sum operation (+) produces a multicapability referring to the template of either multicapability, and represents permissions if any one of the multicapabilities grants that permission. A tuple produced using the sum of two multicapabilities can be retrieved using either multicapability, provided the action is permitted by the multicapability used in the attempted retrieval. To further elaborate on this operation, let us consider the following multicapabilities:

$c1 = [\alpha, \langle ?int, ?char \rangle, \{i,r,o\}]$

$c2 = [\beta, \langle 3, ?char \rangle, \{r,o\}]$

Writing a 'sum' tuple of these multicapabilities, e.g.,

ts1.out((c1+c2) <1,'a'>);

produces (into tuple-space *ts1*) a tuple that can be accessed using either *c1* or *c2*, or both multicapabilities. However, as the multicapabilities grant different rights, then any agent who has *c1* (or both multicapabilities) can read or remove the tuple, while those with only *c2* can only read it.

Reading using the sum operation,

ts1.rd((c1+c2)<3,'a'>);

enables the reader to search both groups 'simultaneously'. Without sum, the reader must perform two separate reads, e.g.,

ts1.rd(c1<3,'a'>);

ts1.rd(c2<3,'a'>);

with the risk of being blocked on the first group if no matching tuple is available, before it has the chance to search the second group (where the tuple might exist). Therefore, using sum in a read operation reduces the probability of being blocked by half. Indeed, what is more, possible deadlocks can be avoided a sequence of two input operations may deadlock, whereas the sum (which is equivalent to a parallel combination of two input operations) will only block until a tuple becomes available in either group.

The subtract operation (−) restricts a capability relative to another. For instance, if we have a third multicapability,

$c3 = [\gamma, \langle 3, ?char \rangle, \{i\}]$

an output operation using $(c1 - c3)$ will write a tuple of template $\langle ?int, ?char \rangle$, excluding any tuple whose first element is integer 3. This tuple will be written into group *c1*, and can only be accessed by agents holding *c1*, but not accessible with *c3*.

An input operation using $(c1 - c3)$ would yield a tuple matching the template of *c1*, except those matching the template of *c3*. Since the permission(s) in *c3* is also subtracted from *c1*, non-destructive reading is disallowed.

The combinatorial (binary) operations offer richer possibilities for capabilities to be manipulated to provide a finer control on objects' visibility to agents in open systems.

Multicapabilities in LINDACAP can be used to implement some applications which are not feasible in the standard LINDA model, such as garbage collection of tuples. Garbage collection requires the knowledge of the references to objects, meaning that the objects must be uniquely identifiable, which tuples are not—they are referred to using associative pattern matching, not by names. Multicapabilities enable selective groups of tuples to be referenced and consequently garbage collected, rather than garbage collecting the whole tuple-space as in Ligia.

GARBAGE COLLECTION OF TUPLES

Garbage collection is the process of deallocating memory spaces previously allocated to objects but the objects are no longer in use. Most garbage collection algorithms are implemented using directed graphs (Dijkstra, 1978) which are traversed to locate unused nodes (representing memory cells allocated to objects). Although there are a number of garbage collection algorithms for distributed systems, we focus on two approaches: reference counting and tracing.

In reference counting, each memory cell (node) allocated to an object is associated with a counter of the number of references to itself. The counter is increased every time a reference is made to it from a 'new' subject, and decreased whenever a reference to it is severed/deleted. The object is considered garbage when its reference count reaches nil, and the memory can be reclaimed.

The tracing approach involves searching the graph and marking all nodes that are reachable from the root (or roots of the subgraphs) (Dijkstra, 1978), hence all unmarked nodes are considered

garbage: nodes that are not reachable from the roots are not accessible by any 'living' node, and therefore are unnecessary.

A tracing algorithm is usually described based on colours: all nodes are initialized as garbage (white), and those accessible from the root is marked black (not garbage). The basic tracing algorithm is called "mark and sweep", where the garbage collector traverses the graph of references and marks each object it encounters, and then sweeps the graph to remove unmarked objects.

The implementation in Ligia, however, was restricted to tuple-spaces: we can either garbage collect, or keep, the whole tuple-space, but not selectively garbage collect certain tuples within the tuple-space. The main problem in introducing garbage collection for tuples is the lack of sufficient information about their 'usage'. This information can be maintained if we can reference a particular tuple, or a group of tuples of a certain type—something that is not possible in LINDA. While tuple-spaces have unique identities, tuples (and templates) are anonymous: they are referred to by values instead of names. Thus it is difficult to employ garbage collection on tuples without modifying the model: giving unique names to tuples will certainly break one of the fundamental characteristics of LINDA, i.e., associative retrieval. However, with the introduction of multicapabilities, this can be avoided as multicapabilities enable the system to reference a collection of nameless tuples to perform garbage collection on them, by garbage collecting the multicapability regions themselves. As multicapabilities provide a means to refer to tuples, we can perform garbage collection on the tuples (Figure 1).

In fact, this mechanism is better than Ligia in the sense that it provides a finer control over the system as we can now selectively garbage collect only a certain region specified by a given multicapability, rather than the whole tuple-space. In Menezes's Ligia, the UTS (including its contents) can never be garbage collected, therefore any

Figure 1. Garbage collection in Ligia and LINDACAP

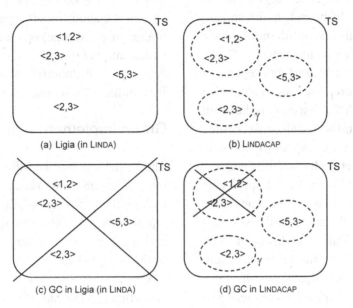

(a) Ligia (in LINDA)   (b) LINDACAP

(c) GC in Ligia (in LINDA)   (d) GC in LINDACAP

tuple put into it will persist in the system forever until explicitly removed, or the system terminates (Menezes, 2000). The number of tuples in the UTS may grow, and some may become unusable over time, thus consuming valuable memory space. With multicapabilities, it becomes possible to garbage collect some of these tuples as we can specify region(s) in the UTS to be garbage collected—without having to remove the whole UTS—thus providing a finer control over the system.

Garbage Collection in LINDACAP

Our earlier papers on LINDACAP (Udzir, 2007; Udzir, 2006; Udzir, 2005) discuss why capabilities are suitable for open systems, mainly due to their 'distributivity'. Multicapabilities have been implemented in a distributed manner in the sense that the kernel does not need to maintain any information pertaining to the capabilities. However, in order to implement garbage collection, the kernel must keep track of the capabilities (reference and permission on objects, including groups of objects, i.e., multicapability regions) against the agents holding them. For example, to determine if an object is garbage or not, the kernel must know which agent has the reference (capability) to the object: if no agent has the reference to the object, then it is garbage and can be reclaimed.

Having a (centralized) list maintained by the kernel undermines the distributed properties of capabilities. This would seem like a retrograde step—contradicting the ideal notion of capabilities need not be maintained by the kernel—but some carefully weighted decisions need to be made: do we want a system with distributed capabilities, or do we want a system that provides garbage collection and a deadlock detection mechanism? It is obvious that garbage collection, at least, is vital for persistent systems: if systems are meant to exist for a long period of time, there has to be some kind of scheme to ensure that the system does not run out of resources, such as memory. Of course, there is no guarantee that memory exhaustion can be fully prevented—as memory is a finite resource—but a system without some kind of mechanism to reclaim unused memory spaces would be disastrous, as memory exhaustion would be unavoidable. In the end, the advantages of these applications far outweigh the need to have a distributed capability management.

In LINDACAP, each multicapability region (or rather the data structure representing it) holds a list of references to it, with the first entry is the agent that created the region in the first place. Every time a multicapability region is referenced (by an agent, or an object, or another multicapability region), a new reference is added to the list specifying the identity of the agent/object/multicapability. Conversely, an entry in the reference list is deleted when a reference to the multicapability is removed. Eventually, if there are no more references left, the multicapability region is garbage collected.

In this implementation phase, the kernel keeps a list of all capabilities that have been passed through it under the following circumstances:

- Every time an agent requests and gets a new multicapability, or creates (and gets the capability for) a new TS, or
- Every time a tuple containing a capability is retrieved by an agent.

The *tuple monitoring* mechanism reported in (Menezes, 1998) helps the kernel keeps track of the (TS) references being passed within the system. LINDACAP also incorporates tuple monitoring, extended to capabilities being passed as tuple elements, which will be essential for garbage collection. To perform garbage collection of tuples, the graph structure used in Ligia (for tuple-spaces) has been adapted in LINDACAP (as described in (Udzir, 2008)) to keep track of the usability of tuples (multicapability regions) in the system. The nodes in the graph are made of agents (depicted as rectangles) and multicapability regions (ellipses). The root of the graph is the default multicapability *cc* where agent nodes linked to when they connect to the system. Each multicapability node is linked to the agent nodes holding a reference to it, and each agent must be linked to all multicapabilities

it holds. Whenever a multicapability (in a tuple) is written into another multicapability region, then a bridge (directed edge) is created between the two nodes, and the weight of the bridge is a counter representing the number of tuples containing the first multicapability that exist in another region.

Graph Implementation

The kernel can use the information supplied by the multicapabilities to keep track of the usage of tuples in the system, eventually removing unusable tuples to reclaim memory space. Although the garbage collection mechanism for TSs (adopted from Ligia) has been incorporated in LINDACAP, the discussion in this section focuses on the garbage collection mechanism with regard to tuples (and multicapability regions).

As capabilities are first-class values, it is possible to remove unused multicapability groups. It is more practical and efficient to re-create a new multicapability later when needed, rather than retaining the old unused ones which have the probability of not being used ever again.

The graph data structure used in Ligia can be adapted for this purpose. In the graph to implement tuple-garbage collection in LINDACAP, the nodes represent the agents and multicapability groups, connected by edges. Each multicapability node is tagged with a counter field to represent the number of references to them. The basic structure of the graph is as follows:

1. There is a node representing the primordial capability *cc*, which is the root of the graph. This node is used as the starting point of the tracing phase of mark-and-sweep to decide if a given node is garbage based on its accessibility from this starting point.

As mentioned above, the universal multicapability *cc* is the default multicapability for a capability type tuple/template—when an agent connects to the kernel (or a LINDACAP server), it is given a copy of the universal multicapability *cc* to enable them to share information among them. With regard to the graph creation, a node representing the *cc* multicapability is created upon the start of a LINDACAP server. As the root of the graph, and therefore the starting point for the tracing phase, this node is never garbage collected as its counter field is set to infinity. *cc* will continue to exist as long as the LINDACAP server lives.

2. There is a unique node representing each agent, which is created whenever an agent connects to a LINDACAP server. Since all agents automatically gets access to the *cc* multicapability, all agent nodes are linked to the *cc* node.

In order to keep track of tuples usability to perform garbage collection in LINDACAP, agents have to be registered upon their connection to the LINDACAP kernel. Thus, LINDACAP is able to identify the agents whenever they execute a primitive that could alter the graph. During agent registration, an agent requests for, and receives a unique identifier from the kernel. From there on, the agent uses this identifier when executing the LINDA primitives. The use of this unique identifier is transparent to the agents themselves, as the identifier is assembled in the messages sent to the kernel.

The graph structure is updated: the link is created between the node representing the agent and *cc* node. Figure 2a shows the graph situation when two agents connect and register themselves to the kernel.

3. There is a unique node representing each multicapability.

When an agent made a request to the LINDACAP kernel for a new multicapability, the kernel will return a unique multicapability for the specified template, different from any other multicapability, even those for templates of the same pattern. LINDACAP provides a primitive to create a multi-

capability object. The **newcap** primitive returns a multicapability object with a unique identifier which has the default full rights for **in**, **rd**, and **out** operations. At this point, the graph is updated accordingly. The kernel can identify the creator of multicapability as the message requesting the creation will contain the agent identifier. Using tuple monitoring (Menezes, 1998) the kernel can extract this identifier, create a node to represent the new multicapability in the graph, and create an arc linking the agent to the multicapability just created (Figure 2b).

4. The links between agents and multicapabilities are always via direct edges.
    ◦   Whenever an agent connects to the kernel, an edge is created between it and the *cc* node (Figure 2a).
    ◦   Whenever an agent creates a new multicapability, an edge is created between the agent and the node representing the multicapability (Figure 2b).

(It is important to see the difference between creating a node representing a multicapability in the *reference graph* here, and the *actual physical* creation of the multicapability region (representing the collection of tuples) referred to by that multicapability, which is created when the first tuple is outed using the multicapability.)

5. A multicapability node must be linked to its creator (agent) node.
6. An agent node must be linked to all multicapabilities it knows (see Section 'Keeping track of capabilities' for more details on how an agent can acquire knowledge of a multicapability).
7. The links between multicapability regions are done using labelled directed edges, called bridges. When a tuple containing a multicapability, say *mc2* is deposited into a multicapability region, say *mc1*, then a bridge

Figure 2. A graph representation of the references/capabilities in LINDACAP

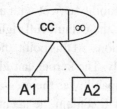

```
A1: lindacap.connect();
A2: lindacap.connect();
```

(a) Agents connecting to the kernel.
Each has reference to cc.

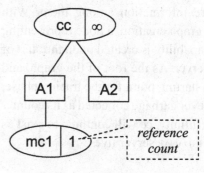

```
A1: mc1 = newcap(<...>);
```

(b) Agent A1 creates new multicapability.
mc1's reference count is set to 1: it is
currently known to A1 only.

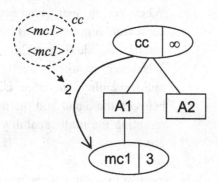

```
A1: out( cc<mc1> );
    out( cc<mc1> );
```

(c) A1 puts tuple <mc1> into cc, twice.
The reference count is now 3:
2 tuples <mc1> in region cc,
and 1 held by A1.

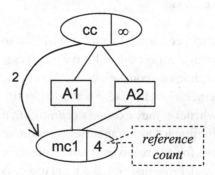

```
A2: rd( cc<?cap> );
```
(assuming no other capability in cc)

(d) A2 reads tuple <mc1> from cc.
The reference count is now 4:
2 tuples still in region cc,
and 1 each held by A1 and A2.

is created between them in the reference graph, directed from *mc1* to *mc2*. (Again, the bridge created is between the nodes in the graph, not between the physical regions, as the *mc2* region might not be created yet—no tuple **out**ed into it yet). The weight of the bridge is a counter representing the number of multicapabilities *mc2* (in tuples) that exist in *mc1*. If there are two capabilities *mc2* within *mc1*, the weight is 2 (Figure 2c).

The counter is decreased every time the tuple containing multicapability *mc2* is removed (via **in** or **inp**) from *mc1*, but remains the same if the multicapability is only read (**rd** or **rdp**). (Note that, except in certain cases where a new multicapability for a capability type is explicitly created, a capability tuple can only be seen using the primordial capability *cc* which does not allow destructive read, i.e., **in** or **inp**. However, with the sum operation (see Udzir, 2006; Udzir, 2007, such a tuple can be removed).

To explain the graph structure in Figure 2:

Figure 2a: Agents *A1* and *A2* connects to the kernel, and gets the default primordial capability *cc*. In the graph, an edge is created between the *cc* node and each node representing each agent.

Figure 2b: Each multicapability node is associated with its reference counter: the counter is set to 1 when the multicapability is created, implying that it is held by one agent—its creator. Therefore, in the diagram, when agent *A1* creates a new multicapability *mc1*, an edge is created between node *A1* and node *mc1*, with the reference count 1 indicating there is 1 reference to *mc1*, i.e., by *A1*.

Figure 2c: When the multicapability is written (in a tuple) using another multicapability, e.g. *cc* (i.e., written into region *cc*), a bridge is created from node *cc* to node *mc1*, indicating there is a reference to *mc1* from within *cc*. The reference counter associated with *mc1* is increased to indicate two references to it (from *A1* and now from *cc*). When a second tuple containing *mc1* is written into *cc*, as in Figure 2c, this reference counter is increased to 3 (meaning two references (in tuples) in *cc* and one reference from *A1*). The label on the bridge from node *cc* to node *mc1* is also updated to 2, indicating two references (tuples) to *mc1* in region *cc*.

Figure 2d: When agent *A2* reads a tuple containing *mc1*, a new edge is created between *A2* and *mc1*, indicating that *A2* now has a reference to *mc1*. The reference counter for *mc1* is increased to 4: one reference from *A1*, one reference from *A2*, and two references (in tuples) from within *cc*. Note that, the references/tuples in *cc* remains the same because *A2* only *reads* the tuple containing *mc1* (as *cc* does not allow **in**). If the capability tuple were removed—assuming the same diagram

as in Figure 2d except that *cc* is replaced with a different capability for capability type (e.g., *c2*) which allows **in**—then the bridge label will be decreased to 1 indicating there is only one tuple left containing *mc1* in the region *c2*. The reference counter on the other hand will remain the same, i.e., 3, as the reference has only been moved from the region *c2* to the agent *A2*.

Graph Algorithms

The algorithm for the graph construction, i.e when an agent performs an **out** operation is presented in Algorithm 1.

The pseudocodes for updating the graph when agents retrieve tuples from multicapability regions are given in Algorithm 2. When an agent executes an **rd** operation, the process is as presented in Algorithm 2.

When an agent executes an **in** operation, the pseudocode is as presented in Algorithm 3.

Keeping Track of Capabilities

We have established that the kernel needs to maintain some kind of information regarding the references—which agent knows about which object in order to perform garbage collection. There are three ways for an agent to acquire a capability (either for a tuple-space or a group of tuples), and how the kernel may keep the information it needs:

Algorithm 1.

```
IF agent A1 outs a tuple into mc1 THEN
IF NOT Linked(A1, mc1) THEN
Link(A1,mc1);
IF the tuple contains multicapabilty mc THEN
IF the agent outs to cc THEN
Link(cc, mc);
IF NOT Linked(A1, mc) THEN
Link(A1, mc);
 out(tuple);
```

1. The agent creates a tuple-space (therefore obtaining the TS-capability in return), or requests a new multicapability for some template of tuples. The kernel can simply record the agent's identity against the newly created capability.

2. The agent has retrieved a tuple containing a capability for the object. To obtain this information, it is necessary for the kernel to implement tuple monitoring (Menezes, 1998)—which has also been extended to monitor tuples containing multicapabilities—to enable the kernel to keep track of capabilities being passed as tuple elements.

3. The agent has been spawned by a parent agent, and the capability has been passed from the parent. As discussed in (Jacob, 2000), this is a rather complicated case, and the solution relies on process registration and 'deregistration' (Menezes, 1998), which have been incorporated in the LINDACAP implementation—all newly spawned agents must register themselves and the capabilities they hold, and must 'unregister' with the kernel before terminating.

Termination ordering assures that the termination message is the last message (from a given agent) to arrive in the kernel (Menezes, 2000). Agent termination is an operation that can generate garbage, as capabilities may be deleted, which would result in the loss of references to some objects. Therefore, termination ordering should also be observed to avoid race conditions.

Garbage Collection using Reference Counting

In the graph, each multicapability node is attached with a counter field indicating the number of references to the multicapability region, either references held by agents, or in tuples within another multicapability. As illustrated in Figure 2 this counter is increased every time a new reference is made to the multicapability region, i.e., when:

- A copy of the tuple containing the multicapability is written into a TS, or
- A capability tuple is read by an agent.

Algorithm 2.

```
IF agent A1 reads a tuple in mc1 using a specified template THEN
IF NOT Linked(A1, mc1) THEN
Link(A1, mc1);
IF the tuple contains multicapabilty mc THEN
IF NOT Linked(A1,mc) THEN
Link(A1, mc);
 rd(template);
```

Algorithm 3.

```
IF agent A1 removes a tuple in mc1 using a specified template THEN
IF NOT Linked(A1, mc1) THEN
Link(A1, mc1);
IF the tuple contains multicapabilty mc THEN
IF NOT Linked(A1,mc) THEN
Link(A1, mc);
IF an agent is withdrawing from cc THEN
Unlink(cc, mc);
 in(template);
```

A **rd** operation does not 'move' a tuple to the retrieving agent, but rather it returns a duplicate copy of the tuple—hence increasing the reference counter. **in** on the other hand, removes a tuple to be given to the retrieving agent. Therefore, an **in** operation does not increase the counter value.

Naturally, a node (representing a multicapability region) is deleted (i.e., garbage collected) if and when the reference (counter) to it becomes nil. The counter is decreased every time a reference to the multicapability is deleted, i.e., the edge(s) connected to the node is/are deleted. This transpires when:

- A multicapability is revoked, or
- A tuple containing a multicapability referring to the region is deleted, or
- An agent holding the multicapability dies.

Any one of these circumstances decreases the reference counter in the multicapability node. The multicapability region is considered garbage when the counter reaches zero. Being the default multicapability, the *cc* (root) node is never garbage collected: its reference counter is $m = \infty$.

Garbage Collection for Cyclic Garbage

Garbage collection based on reference counting does not garbage collect multicapability regions that are part of a cyclic structure. Passing of multicapabilities in LINDACAP allows the creation of cyclic structures between multicapability regions. In this paper, garbage collection based on the mark-and-sweep algorithm (Dijkstra, 1978) has been implemented to remove cyclic garbage using the graph structure in LINDACAP (as described in (Udzir, 2008)) to keep track of the usability of tuples in the system. Mark-and-sweep has two separate phases:

1. **Marking phase:** After the graph has been initialized by marking all objects (multicapability regions) as garbage, traversal is initiated from the root; all objects it encounters are marked as non garbage.
2. **Sweeping phase:** Unmarked objects are freed, and the resulting memory is made available to the executing program.

Figure 3a depicts a possible graph situation at some point during the execution of a LINDACAP system. A cyclic reference exists between *mc1* and *mc2* when a tuple containing *mc1* is written in *mc2*, and vice versa. When agents *A1* and *A2* terminate, the reference counters for *mc1*, *mc2*, *mc3*, and *mc4* are decreased by 1. The reference counting takes effect and identifies and collects the objects (shown in dotted lines in Figure 3b), i.e., all links to/from both *A1* and *A2*, and node *mc4*, whose counter is now zero.

The regions *mc1* and *mc2* are now garbage as they are no longer referenced by any agent but they are not garbage collected as each of their reference counter is 1: they are cyclic garbage. Therefore, in this case, the garbage collection using reference counting is incomplete as it does not collect cyclic garbage.

This problem can be solved using the mark-and-sweep technique. Firstly, the graph is initialized by marking all nodes (regions) as garbage (white), then the graph is traversed starting from the root (*cc* node) and every node in the path is marked non-garbage (black) to identify this node as being in-use. At the end of the first step, all unmarked (white) nodes are considered as garbage—as they are not accessible from the root—and are 'swept'. Figure 3c depicts the graph after the tracing: all the nodes left unpainted are considered garbage and are collected, including regions *mc1* and *mc2*.

Figure 3. Tuples (multicapability regions) being garbage collected (b) reference counting and (c) tracing (mark-and-sweep)

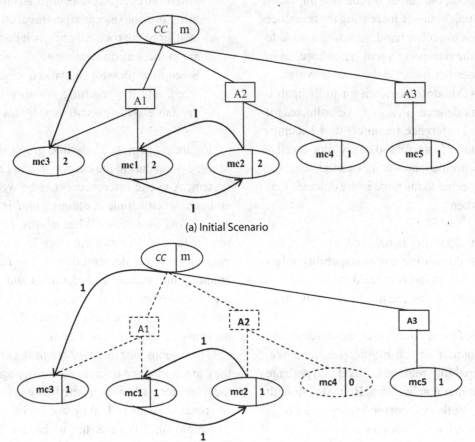

(a) Initial Scenario

(b) GC by reference counting (after A1 and A2 died)

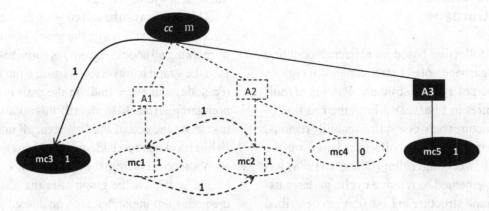

(c) GC by tracing (mark-and-sweep)

Time-To-Idle (TTI) Technique

Both reference counting and mark-and-sweep techniques cannot garbage collect multicapability regions as long as the agent that holds the reference to these regions exist in the system. If these regions are no longer in use by the agent, they should be removed from the system but the agent may forget to delete them. These regions are consuming memory resources, which may cause memory exhaustion. To overcome this problem, a time-to-idle (TTI) technique is proposed.

TTI is the period for which the object is guaranteed to stay idle (not in use). When a link is made to a multicapability object, TTI is added to the current time and the resulting expiration date is attached to the object. To guarantee that the object will continue to exist, it should be refreshed before it expires. This can be done when an object is referenced. When the object is about to expire, the kernel can send a notification message to all agents possessing a reference (capability) to this object at a specified time before the object is about to be garbage collected, and any agent wanting to use the object must refresh it. The TTI for a multicapability region is chosen automatically by the system when the object is created. Every object is treated equally: the same TTI value is chosen for every object in the system.

Implementation

We implemented the LINDACAP system with garbage collection using reference counting and mark-and-sweep separately, together with the time-to-idle technique. The system was implemented using Java (JDK 1.5.0). The particular choice of Java was based on the idea that Java provides the support to run the system in heterogeneous environment.

The implementation of LINDACAP uses distributed central server as their tuple-space distribution strategy (Carriero, 1986), in which the kernel is distributed among several machines (Figure 4),

which would result in the distribution of workload among the servers. The communication among servers occurs via sockets. LINDACAP implementation has the following characteristics:

- There is a public universal tuple-space (UTS) that is accessible by all.
- All agent communications are done solely in terms of tuple-space operations.
- Every tuple-space, with the exception of the UTS, is explicitly created by agents in the system.
- Tuple-spaces in the kernel are flat structured: every tuple-space is created at the same level, that is, tuple-spaces cannot be created inside others.
- Capabilities are first class objects.
- There is a single type of capabilities, referred to as *?cap* in templates.

The architecture is divided into two basic systems: the server and agent systems (Figure 4).

Server System

Each server is multithreaded: a new thread is created for each request made by an agent. The server is composed of the following basic components:

1. Agent Connection

This component is responsible for dealing with all communications with agents. It is implemented by the **Connection** Java class which keeps reading the socket for requests. When an agent requests a connection, the server assigns a name to the agent (**AgentId**): this name is guaranteed to be unique and will be used in all ensuing communications. Baasically, a thread is normally created when the message is received from an agent that has never communicated with this server before.

The message sent from the agent system is attached with the **AgentId**, which identifies the agent sending the message: this is very important

Figure 4. Architecture of the LINDACAP system

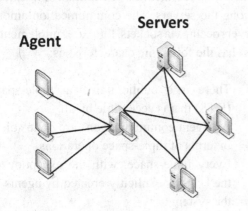

Agent

Servers

for the garbage collector which needs to know which agents are executing each operation. The **TupleMonitoring** method which is implemented in the agent connection module extracts information for the garbage collector from the message sent by the agents. If the multicapabilities are involved in the operation, the garbage collection structure is updated accordingly. When an agent disconnects, the corresponding connection is closed and the **Socket** object is collected by the Java garbage collector.

2.   Garbage Collector

In the server system classes **GCReference-Counting**, **GCMarkAndSweep**, and **GCIdleCollector** are implemented, which extend the class **thread**. When the first object of the class **LindacapServer** is created, a **GCIdleCollector** object is also created. The method **sleep** in the class **thread** is used to stop this thread for a determined period of time to reduce the resource consumption by the garbage collector. The execution of this thread removes multicapability regions with expired TTI and reclaims the memory space. Prior to the expiration of a TTI of a multicapability region, a notification message will be sent to all agents that has access to this region, as a reminder for them to refresh the TTI if the region is still needed.

GCReferenceCounting is executed when an agent holding the reference terminates. The execution of the **GCReferenceCounting** invokes relevant methods and these methods linearly search for nodes (multicapability regions) whose **Reference Counter** is zero, remove these multicapability regions, and reclaim the memory space. The other garbage collection thread is **GCMarkAndSweep**, which is executed when there is a possible existence of a cyclic garbage: two or more multicapability regions that refer to one another. The execution of this object involves several separate steps. First, the **color** variable in all the multicapability objects set to **0** (marked **white**). After this initialization, the marking phases starts. Starting from **cc**, the color of all multicapability regions which are reachable from the root **cc**, is set to **1** (marked **black**). Reaching the end of the graph, it proceeds to remove all nodes whose **color** variable is still 0 (i.e. **white**). The **thread** method **sleep** is used to avoid both these garbage collection objects from consuming resources. Therefore, the **sleep** method is used to control the execution of the object, which would help to reduce resource consumption.

Agent System

The agent system is composed basically of one component, i.e., the **Communication System**, which is responsible for providing all the agents with coordination capabilities. An agent has to request a connection to the **LindacapServer** and only then can it use the primitives. When the request is accepted, a unique name (**AgentId**) is assigned to the agents and a communication channel using a socket object is established.

Algorithms

The pseudocodes for the mark-and-sweep and the time-to-idle techniques are given in Algorithm 4. For mark-and-sweep, the pseudocode is as presented in Algorithm 4.

Algorithm 4.

```
IF agent A1 terminates THEN
FOR(all mci, Is_Linked(A1,mci)) DO
Un_Link(A1,mci);
// Mark-and-Sweep Algorithm
// Initialization step
FOR (all mci in the system) DO
Mark(white);
// mark phase starts
FOR(all mci, referenced by cc) DO
//all multicapability regions that are
//directly or indirectly accessible by
//the cc will be marked black (non-garbage).
Mark(black);
//sweep phase starts
FOR (all mci, is_Mark(white)) DO
Sweep();
```

The psuedocode for Time-To-Idle (TTI) is given in Algorithm 5.

A simple and naive strategy for implementing garbage collection is to run the garbage detection algorithm every time the reference to an object has been decreased to zero/nil; and one reason for a reference to be deleted is when the agent holding the reference dies. It is also known that performing garbage collection can be an expensive operation (in terms of kernel load). Therefore, it is more efficient to garbage collect only when needed, i.e., when there is insufficient memory space available. Even though this strategy involves a larger amount of work to be carried out at one time compared to the former, garbage collection is likely to be performed less often.

Algorithm 5.

```
IF an Agent get access to mc1 THEN
add(TTI,mc1);
FOR every interval of time CHECK
FOR(each mci in the system) DO
IF (TTI of mci is about to expired &&
IsNotNotifiedYet(mci)) THEN
FOR(all Aj, Is_referred(mci,Aj) DO
Send(Notification Message);
ChangeStatus(mci,notified);
ELSE IF (TTI value is expired and
isNotified (mci)) THEN
 GarbageCollect(mc1);
```

Some mechanisms may be adopted to increase the efficiency of the algorithms, e.g. the Jump_stack data structure by Lins (Lins, 2000); or the 'critical link' concept (pointer operation that creates a cycle) (Lins, 2008) where the mark-scan algorithm will be run whenever an attempt is made to delete a pointer to the cell pointed by the critical link. These algorithms have been adapted in shared memory architectures (de Araújo Formiga, 2007) as well as in distributed environments (Lins, 2006; Lins, 2003). Baker et al. (2009) propose lazy pointer stacks, which performs accurate garbage collection in such uncooperative environments, with the implementation of a real-time concurrent garbage collection algorithm.

EXPERIMENTAL RESULTS

The reason for garbage collection is to avoid memory exhaustion. As discussed in earlier sections, Ligia (Menezes, 2000) only performs garbage collection on TSs, and therefore does not garbage collect tuples in the UTS. This can lead to disastrous consequences. Multicapabilities enable some of these tuples to be garbage collected, as has been discussed in the previous section. Experiments have been carried out to demonstrate this.

These experiments compared two capability systems for memory exhaustion: one incorporates the garbage collection mechanism for *tuples*, whereas the other does not. The characteristics for the systems are:

1. Both systems deployed garbage collection of *TSs*, based on Ligia.
2. All interactions are via the UTS, which is not being garbage collected—the TSs garbage collection mechanism (Menezes, 2000) cannot be performed on UTS. Therefore, we can be certain that these experiments only concern garbage collection on tuples, and not on TSs.

3. Each agent requests their own multicapability with no capabilities being passed among the agents, which implies that all the agents used different multicapability regions. Thus, these regions cannot be referenced by any other agents, and are considered garbage when the agents creating them die.

Results for GC using Reference Counting

The experiments involved running a group of agents in limited memory space, where each agent requests a new multicapability, **outs** a number of tuples into the UTS using the newly acquired multicapability, and then dies after explicitly deleting the multicapability. The agents' code snippet is given in Algorithm 6.

As expected, the server with no tuple-garbage collection eventually ran out of memory, whereas the server with garbage collection did not encounter the same problem, in fact was able run indefinitely.

Figure 5 depicts the results of the experiments: with (a) 5000 tuples **out**ed per iteration, (b) 7000 tuples, and (c) 10000 tuples. Each graph shows that without garbage collection on tuples, the system crashes (runs out of memory) sooner than its GC-enabled counterpart, i.e., after approximately 67000 tuples have been deposited in the UTS, whereas there is no such concern with the system with garbage collection on tuples.

When garbage collection on tuples is not present, tuples written by previous agents are left accumulating in the UTS (in effect, stored in the kernel); this adds considerable overhead to the system. Although none of the tuples would be

Algorithm 6.

```
cap = newcap(<int,int>); f
or (t = 0; t < 10000; t++)
UTS.out(cap<1,t>);
del cap;
```

used again, they are consuming the memory resources—causing the system to crash if the memory is not reclaimed (by the garbage collector). With tuple-garbage collection incorporated, we have a cleaner structure, hence better performances for agents—thus, tuple-GC provides the system with a major advantage in terms of reliability.

Of course, no garbage collection mechanism can completely eliminate the memory exhaustion problem, as memory is a finite resource—there is always a possibility of completely consuming the resource without creating any garbage.

Incorporating the garbage collection mechanism incurs some overhead. Again, the two LINDACAP systems are compared: both incorporate capabilities, but only one has the garbage collection mechanism on tuples. Figure 6 illustrates that no significant increase in overhead (measured by completion time for each agent) is imposed by the garbage collection algorithm.

As can be seen from the graph, the larger the number of tuples **out**ed, the less overhead (proportionally) incurred. Comparing the time taken by each system (with, and without tuple-GC) to write a certain number of tuples, in Figure 6a, the time difference in **out**ing 100 tuples is 0.001 second (only 1.5% increase in overhead), whereas writing 1000 tuples produces a difference of 0.023 second (i.e., 6%, which is the same time difference as in Figure 6b). In Figure 6b, where tuples are written in increments 1000, the system with garbage collection on tuples took 0.24 second longer (7%) to write 10000 tuples, compared to the one without tuple-GC.

Results for GC using Mark-and-Sweep

The main reason for garbage collection is to overcome memory exhaustion. In order to show the occurrence of memory exhaustion, an experiment was performed with limited memory size (the heap size is limited to 12517376 kb) where agents

Figure 5. Memory exhaustion problem

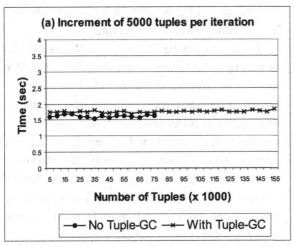

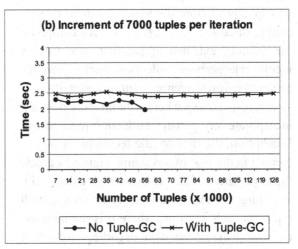

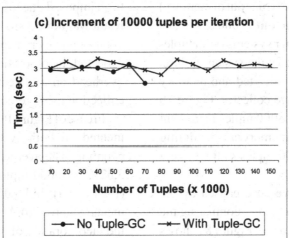

Figure 6. Overhead added by tuple garbage collection scheme

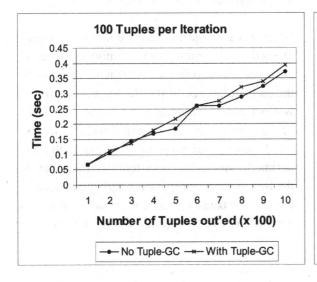

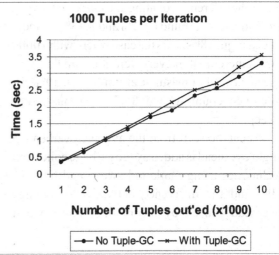

repeatedly write tuples until the server exhausts its capacity to store the tuples. In these experiments, three LINDACAP systems were compared for memory exhaustion: the first two systems incorporate garbage collection mechanisms for tuples based on reference counting and mark-and-sweep, respectively, whereas the third one does not incorporate any garbage collection mechanism. In addition, the time-to-idle technique is implemented in the first two systems. Figure 7 depicts the experiments that cause memory exhaustion. In Figure 7a an agent repeatedly writes n x 3000 tuples in each iteration, where n is the iteration number. In the system without garbage collection on tuples, the server crashed (runs out of memory) when an agent was about to **out** 15000 tuples at a time, whereas both the other systems (with tuple-garbage collection) tolerates the system failure, because unused tuples are deleted and memory is reclaimed by the garbage collector. Figure 7b depicts similar results for 5000 tuples increment in each iteration, when the memory crashed during the attempt to write 20000 tuples in the system without garbage collection.

Figure 8 shows the overhead caused by the tuple garbage collection schemes, comparing the three systems described above. Figure 8a shows the case where the agent stores tuples with the number of tuples are increased by 5000 in every iteration, and the average overheads measured due to the reference counting and mark-and-sweep techniques are around 6.92% and 8.33%, respectively. Figure 8b shows the case where with 10000 tuples increment in every iteration, and the average overheads measured are around 6.59% for reference counting and 8.67% for mark-and-sweep.

As can be seen from the average overhead, a slight overhead is added by the garbage collection techniques using mark-and-sweep and time-to-idle, and the difference between the average overhead caused by the mark-and-sweep and

reference counting is small and can be considered insignificant if the benefits brought by the mark-and-sweep is taken into account.

Incorporating the TTI garbage collection technique in a system imposes some overhead. In order to show the overhead due to TTI technique, two LINDACAP systems are compared. Both systems have garbage collection mechanism using reference counting or mark-and-sweep algorithm, but only one has TTI garbage collection mechanism. Figure 9 depicts the experimental results where reference counting is incorporated on both systems but only one has TTI. In Figure 9a, where agents store tuples in 5000 increments, the average overhead imposed due to TTI was around 3.35%. The result was almost similar for the experiment with 10000 tuples increment, i.e., 4.28% (Figure 9b). As expected, incorporating TTI imposes significant overhead on a system that already has garbage collection using reference counting algorithm.

In case of Figure 10, mark-and-sweep is implemented on both systems instead of reference counting. The average overhead due to TTI was 3.62% for 5000 tuples increments in each loop (Figure 10a), and 5.05% for 10000 tuples increments (Figure 10b). According to the above results, the incorporation of TTI technique adds a slight overhead on the performance of the system.

Garbage Collection Memory Usage

When any garbage collection algorithm is implemented in any system, some memory overhead is expected. In order to show the memory usage by the garbage collection algorithms, an experiment was performed where an agent creates a number of multicapability regions and therefore increasing the number of nodes in the graph. In this experiment, two LINDACAP systems were compared: the first system incorporate a garbage collection mechanism based on reference counting whereas a mark-and-sweep algorithm is implemented in

Figure 7. Memory exhaustion problem

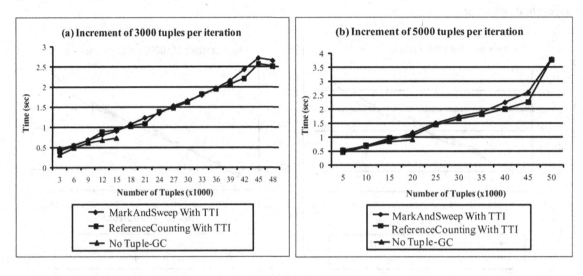

Figure 8. Overhead added by the tuple garbage collection schemes

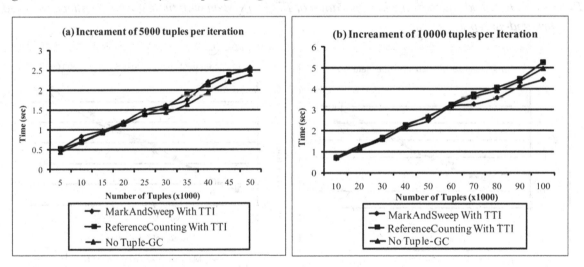

the second system. In addition, the time to idle technique is implemented on both systems. The memory usage due to each garbage collection algorithm was collected.

In Figure 11a an agent loops and **out**s $n \times 100$ multicapability regions in each iteration, where n is the iteration number. The system with mark-and-sweep garbage collection algorithm uses more memory resource compared to the one with reference counting. This shows the fact that in tracing-based algorithms such mark-and-sweep, unreachable (garbage) objects are not reclaimed immediately when they become unreachable. This dramatically increases the memory overhead and degrades system performance. This can be a problem in a system that interacts with a human user or that must satisfy real-time execution constraints. Figure 11b depicts the same case but where the increment was 500 regions per iteration.

Figure 9. Overhead added by the incorporation of TTI in a system with tuple garbage collection based on reference counting

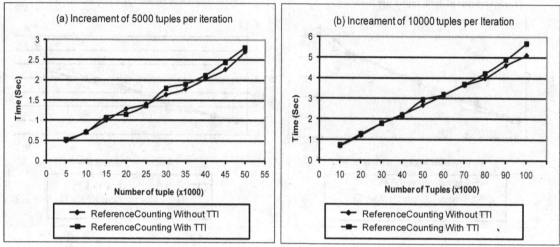

Figure 10. Overhead added by the incorporation of TTI in a system with tuple garbage collection based on mark-and-sweep algorithm

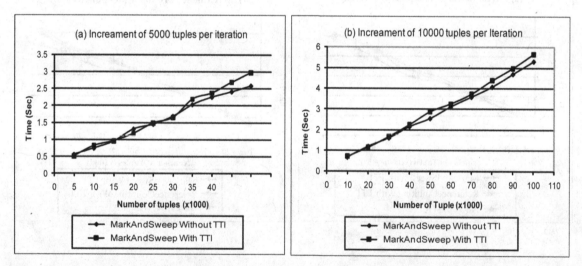

FINER GARBAGE COLLECTION

With capabilities, a finer garbage collection mechanism can be achieved. Ligia only incorporates object/TS references, but L<small>INDACAP</small> can manipulate the reference *and* the associated permissions (in the capability) for finer control: rights can be encoded in the edges. Consider an example as illustrated in Figure 12. A tuple containing a capability for

ts2 is emitted into *ts1*, forming a bridge from *ts1* to *ts2*. *ts2* itself contains a capability tuple for *ts3*, creating another bridge from *ts2* to *ts3*.

In a non-capability system, if there exists *at least one* agent, A1 that has the reference for *ts1*, then as long as A1 is alive, none of the three TSs can be garbage, as there is always a possibility that A1 might retrieve the tuple ⟨ts2⟩ in *ts1*, and subsequently gaining access to *ts3* via *ts2*.

Figure 11. Memory usages by garbage collection schemes

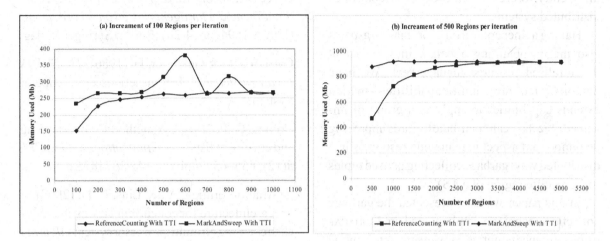

Figure 12. An example for a finer garbage collection

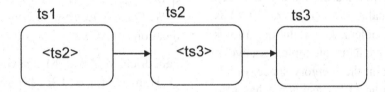

(Here it is assumed that no intervention occurs resulting in the tuples being removed except by A1.) In LINDACAP on the other hand, the rights in the capability for *ts1* held by A1 is relevant: if the capability only grants **out** permission, then A1 cannot read the tuple ⟨ts2⟩ from *ts1*, and therefore cannot access *ts2*, and subsequently *ts3*, too. Hence, *ts2* and *ts3* can be garbage collected. In the case of *ts1* itself, although *ts1* is not garbage—since A1 still has the capability for it and may write into it—any reference/capability within it is garbage. In fact any *tuple* within it is garbage. The tuple space *ts1* can be regarded as a "null", where anything can be written into it, but they will not be retained—much like a dumping ground—and therefore can be flagged as such, for optimization purposes, to ensure that tuples stored in it will not increase memory usage.

A *null* capability—i.e., a capability that does not grant anything, and holding such a 'null' capability is useless for the agent's *access* to the object—can be useful for garbage collection. For instance, if the kernel knows that all tuples containing the capability for an object (a TS or a group of tuples) are null, and no other agent holds a copy of the capability, then it knows that the object is garbage, as no operation can be performed on it, therefore the object can be reclaimed.

CONCLUSION

Capabilities represent the information on 'who knows about what operation on a certain object'. The more the kernel knows of the system's behaviour, the better, more optimised coordination can be achieved, thus increasing the system's efficiency. The extra information, supplied by the capabilities given to the agents, can provide the facility to create a finer level of control in distributed systems.

Having a finer control on coordination aspects also means obtaining a better, more efficient decentralised resource management. We have demonstrated how multicapabilities—which extends capabilities to apply to a group of unnamed objects—can contribute towards improving one important aspect in managing resources in a distributed way: garbage collecting unused tuples (or a specific region of a TS).

In this paper, we have presented the garbage collection technique implemented in LINDACAP using reference counting. However, this technique alone does not garbage collect unused tuples which are part of a cyclic structure. Therefore, we have also incorporated the mark-and-sweep garbage collection algorithm (Dijkstra, 1978) as a solution to this problem. This technique allows the system to remove unusable tuples in a cyclic structure and reclaim the memory space. In addition, a time-to-idle (TTI) technique has also been proposed to reclaim the memory storage for multicapability regions that are being referred to by agents but no longer in use.

The performance results indicate that the incorporation of garbage collection techniques adds a slight (almost insignificant) overhead to the overall performance of the system. And also the difference between the average overhead caused by the mark-and-sweep and reference counting is small. This is not such a big loss compared to a possibly more disastrous consequence, such as a memory crash, especially in open, persistent systems.

REFERENCES

Baker, J., Cunei, A., Kalibera, T., Pizlo, F., & Vitek, J. (2009). Accurate garbage collection in uncooperative environments revisited. *Concurrency and Computation*, *21*(12), 1572–1606.. doi:10.1002/cpe.1391

Bryce, C., Oriol, M., & Vitek, J. (1999). *A coordination model for agents based on secure spaces* (LNCS 1594, pp. 4-20). Berlin: Springer Verlag.

Carriero, N., & Gelernter, D. (1986). *The S/Net's Linda kernel*. ACM Transactions on Computer Systems.

Collins, G. E. (1960). A method for overlapping and erasure of lists. *Communications of the ACM*, *3*(12), 655–657. doi:10.1145/367487.367501

de Araújo Formiga, A., & Lins, R. D. (2007). A new architecture for concurrent lazy cyclic reference counting on multi-processor systems. *Journal of Universal Computer Science*, *13*(6), 817–829.

Dijkstra, E. W., Lamport, A. J. L., Scholten, C. S., & Steffens, E. F. M. (1978). *On-the-fly garbage collection: An exercise in cooperation*. Communications of ACM.

Fähndrich, M., Aiken, M., Hawblitzel, C., Hodson, O., Hunt, G., Larus, J. R., & Levi, S. (2006). Language support for fast and reliable message-based communication in singularity OS. In *Proceedings of the EuroSys 2006 Conference* (pp. 177-190). New York: ACM.

Freeman, E., Hupfer, S., & Arnold, K. (1999). *JavaSpaces: Principles, Patterns, and Practice. The Jini Technology Series*. Reading, MA: Addison-Wesley.

Friedman, D. P., & Wise, D. S. (1979). Reference counting can manage the circular environments of mutual recursion. *Information Processing Letters*, *8*(1), 41–45. doi:10.1016/0020-0190(79)90091-7

Gelernter, D. (1985). Generative communication in LINDA. *ACM Transactions on Programming Languages and Systems*, *7*(1), 80–112.. doi:10.1145/2363.2433

Gelernter, D. (1989). *Multiple tuple spaces in* LINDA (LNCS 366, pp. 20-27). Berlin: Springer Verlag.

Jacob, J. L., & Wood, A. (2000). *A principled semantics for inp* (LNCS 1906, pp. 51-66). Berlin: Springer Verlag.

Lins, R. D. (2002). An efficient algorithm for cyclic reference counting. *Information Processing Letters, 83*(3), 145–150. doi:10.1016/S0020-0190(01)00328-3

Lins, R. D. (2003). *An efficient multi-processor architecture for parallel cyclic reference counting* (LNCS 2565, pp. 111-139). Berlin: Springer.

Lins, R. D. (2006). *New algorithms and applications of cyclic reference counting* (LNCS 4178, pp. 15-29). Berlin: Springer.

Lins, R. D. (2008). Cyclic reference counting. *Information Processing Letters, 109*(1), 71–78. doi:10.1016/j.ipl.2008.09.009

McBeth, J. H. (1963). On the reference counter method. *Communications of the ACM, 6*(9), 575. doi:10.1145/367593.367649

Menezes, R. (2000). *Resource Management in Open Tuple Space Systems.* Unpublished doctoral dissertation, University of York, York, UK.

Menezes, R., & Wood, A. (1997). Garbage collection in open distributed tuple space systems. In *Proceedings of the 15th Brazilian Computer Networks Symposium (SBRC'97)* (pp. 525-543).

Menezes, R., & Wood, A. (1998). Using tuple monitoring and process registration on the implementation of garbage collection in open LINDA-like systems. In *Proceedings of the 10th IASTED International Conference Parallel and Distributed Computing and Systems*, Las Vegas, NV (pp. 490-495).

Merrick, I. (2001). *Scope-Based Coordination for Open Systems.* Unpublished doctoral dissertation, University of York, York, UK.

Udzir, N. I. (2006). *Capability-Based Coordination for Open Distributed Systems.* Unpublished doctoral dissertation, University of York, York, UK.

Udzir, N. I., Muda, Z., Sulaiman, M. N., Zulzalil, H., & Abdullah, R. (2008). Refined garbage collection for open distributed systems with multicapabilities. In *Proceedings of the 3rd International Symposium on Information Technology 2008 (ITSim'08).*

Udzir, N. I., Wood, A. M., & Jacob, J. L. (2005). *Coordination with multicapabilities* (LNCS 3454, pp. 79-93). Berlin: Springer Verlag.

Udzir, N. I., Wood, A. M., & Jacob, J. L. (2007). Coordination with multicapabilities. *Science of Computer Programming: Special Issue on Coordination Models and Languages, 64*(2), 205–222.

Vitek, J., Bryce, C., & Oriol, M. (2003). Coordinating processes with secure spaces. *Science of Computer Programming, 46*(1-2), 163–193. doi:10.1016/S0167-6423(02)00090-4

Wegiel, M., & Krintz, C. (2008). XMem: type-safe, transparent, shared memory for cross-runtime communication and coordination. In *Proceedings of the 2008 ACM SIGPLAN conference on Programming language design and implementation.*

Wyckoff, P., McLaughry, S., Lehman, T., & Ford, D. (1998). TSpaces. *IBM Systems Journal, 37*(3), 454–474. doi:10.1147/sj.373.0454

This work was previously published in International Journal of Information Technology and Web Engineering, Volume 5, Issue 3, edited by Ghazi I. Alkhatib, pp. 1-26, copyright 2010 by IGI Publishing (an imprint of IGI Global).

Chapter 8
A Model for Ranking and Selecting Integrity Tests in a Distributed Database

Ali A. Alwan
Universiti Putra Malaysia, Malaysia

Hamidah Ibrahim
Universiti Putra Malaysia, Malaysia

Nur Izura Udzir
Universiti Putra Malaysia, Malaysia

ABSTRACT

Checking the consistency of a database state generally involves the execution of integrity tests on the database, which verify whether the database is satisfying its constraints or not. This paper presents the various types of integrity tests as reported in previous works and discusses how these tests can significantly improve the performance of the constraint checking mechanisms without limiting to a certain type of test. Having these test alternatives and selecting the most suitable test is an issue that needs to be tackled. In this regard, the authors propose a model to rank and select the suitable test to be evaluated given several alternative tests. The model uses the amount of data transferred across the network, the number of sites involved, and the amount of data accessed as the parameters in deciding the suitable test. Several analyses have been performed to evaluate the proposed model, and results show that the model achieves a higher percentage of local processing as compared to the previous selected strategies.

INTRODUCTION

A database state is said to be consistent if and only if it satisfies the set of integrity constraints. A database state may change into a new state when it is updated either by a single update opera-tion (insert, delete or modify) or by a sequence of updates (transaction). If a constraint is false in the new state, the new state is inconsistent, the enforcement mechanism can either perform compensatory actions to produce a new consistent state, or restore the initial state by undoing the

DOI: 10.4018/978-1-4666-0023-2.ch008

update operation. The steps, generate integrity tests, which are queries composed from the integrity constraints and the update operations and run these queries against the database, which check whether all the integrity constraints of the database are satisfied, are referred to as *integrity checking* (Ali, Hamidah, & Nur Izura, 2009; Ibrahim, Gray, & Fiddian, 2001; Ibrahim, 2006) is the main focus of this paper.

The growing complexity of modern database applications plus the need to support multiple users has further increased the need for a powerful integrity subsystem to be incorporated into these systems. Therefore, a complete integrity subsystem is considered to be an important part of any modern DBMS. The crucial problem in designing this subsystem is the difficulty of devising an efficient algorithm for enforcing database integrity against updates (Ibrahim, Gray, & Fiddian, 2001). Thus, it is not surprising that much attention has been paid to the maintenance of integrity in centralized databases. A naïve approach is to perform the update and then check whether the integrity constraints are satisfied in the new database state. This method, termed *brute force checking*, is very expensive, impractical, and can lead to prohibitive processing costs. Enforcement is costly because the evaluation of integrity constraints requires accessing large amounts of data, which are not involved in the database update transition. Hence, improvements to this approach have been reported in many research papers (Martinenghi, 2005; Mc-Cune & Henschen, 1989; Nam, 1998; Nicolas, 1982; Qian, 1989; Simon & Valduriez, 1989). The problem of devising an efficient enforcement is more crucial in a distributed environment.

The brute force strategy of checking constraints is worse in the distributed context since the checking would typically require data transfer as well as computation leading to complex algorithms to determine the most efficient approach. Allowing an update to execute with the intension of aborting it at commit time in the event of constraints viola-

tion is also inefficient since rollback and recovery must occur at all sites which participated in the update. Moreover, devising an efficient algorithm for enforcing database integrity against update is extremely difficult to implement and can lead to prohibitive processing costs in a distributed environment (Grefen, 1993; Ibrahim, Gray, & Fiddian, 2001). A comprehensive survey on the issues of constraint checking in centralized, distributed and parallel databases is provided in (Feras, 2006; Ibrahim, 2006). Works in the area of constraint checking for distributed databases concentrate on improving the performance of the checking mechanism by executing the complete and sufficient tests when necessary. None of the work has look at the potential of support test in enhancing the performance of the checking mechanism. Also, the previous works claimed that the sufficient test is cheaper than the complete test and its initial integrity constraint. They depend solely on the assumption that the update operation is submitted at the site where the relations to be updated is located, which is not necessary the case. Thus, the aim of this paper is to analyze the performance of the checking process when various types of integrity tests are considered without concentrating on certain type of test as suggested by previous works. The most suitable test is selected from the various alternative tests in determining the consistency of the distributed databases. Here, suitable means the test that minimizes the amount of data transferred across the network, the amount of data accessed, and the number of sites involved during the process of checking the constraints.

In this paper we provide the previous works related to this research, the basic definitions, notations, and examples which are used in the rest of the paper. In addition, discussion on the importance of considering different types of integrity tests is given. Next we introduce our proposed model to rank and to select integrity tests. Finally, the results of the proposed model are extensively discussed.

RELATED WORKS

For distributed databases, a number of researchers have looked at the problem of semantic integrity checking. Although many researchers have been conducted concerning the issues of integrity constraint checking and maintaining in distributed databases but they failed to exploit the available information at the target site and explore various types of integrity tests to ensure local checking can always be achieved. This is briefly shown in Table 1, where columns labeled 1, 2, 3, 4, 5, 6, 7, 8, and 9 represent the work by Simon and Valduriez (1989), Qian (1989), Mazumdar (1993), Gupta (1994), Nam (1998), Ibrahim, Gray, and Fiddian (2001), Ibrahim (2002), Madiraju, Sunderraman, and Haibin (2006), and Soumya, Madiraju, and Ibrahim (2008) respectively.

Simon and Valduriez (1989) constructed a simplification method for integrity constraints expressed in terms of assertions for central databases and extended it to distributed databases. This method produces at assertion definition time, differential pre-tests called compiled assertions, which can be used to prevent the introduction of inconsistencies in the database. The cost of integrity checking is reduced because only the data subject to update is checked in this approach.

Qian (1989) argued that most approaches derive simplified forms of integrity constraints from the syntactic structure of the constraints and the update operation without exploiting knowledge about the application domain and the actual implementation of the database. Qian (1989) shows that distributed constraints can be translated into constraints on the fragments of a distributed database, given the definition of the fragmentation, and offers a framework for constraint reformulation. The constraint reformulation algorithm used to derive sufficient conditions can potentially be very inefficient because it searches through the entire space of eligible reformulation for the optimal one. Using heuristic rules to restrict the reformulation step may miss some optimal reformulation.

Mazumdar (1993) works aim at minimizing the number of sites involved in evaluating the integrity constraints in a distributed environment. In his approach the intention is to reduce the non-locality of constraints by deriving sufficient conditions not only for the distributed integrity constraints given, but also for those arising as tests for particular transactions. His method relies on a standard backchaining approach to finding the sufficient conditions.

Gupta (1994) presented an algorithm to generate parameterized local tests that check whether an update operation violates a constraint. This algorithm uses the initial consistency assumption, an integrity constraint assertion that is expressed in a subset of first order logic, and the target relation to produce the local test. This optimization

Table 1. Summary of the previous studies

Criteria		1	2	3	4	5	6	7	8	9
Types of integrity constraints	Domain	√	√	√			√	√		
	Key	√	√	√		√	√	√		
	Referential	√	√	√	√	√	√	√		
	Semantic	√	√	√		√	√	√	√	√
	Transition							√		
Types of tests	Complete	√	√			√				
	Sufficient		√	√	√		√	√	√	√
	Support									

technique allows a global constraint to be verified by accessing data locally in a single database where the modification is made. However, this approach is only useful in situations where each site of a distributed DBMS contains one or more intact relations since it does not consider any fragmentation rules.

Nam (1998) proposed an approach to find efficient complete local tests for the integrity constraints that are expressible as conjunctive queries with negated subgoals CQC:'s. He considered the problem of determining whether an update to a distributed database preserves global consistency, when not all the underlying relations are accessible and when the update is made to the relations that are accessible. For CQC:'s where the remote predicates do not occur more than once, consistency of local insertions and local deletions can be tested completely in time polynomial in the size of the local relations and updates. Further, these tests can be generated at compile time in the form of safe, nonrecursive Datalog (or SQL) queries, whose size is linear in the size of the constraint in the best case and exponential in the worst case. However, he assumed single occurrences of the remote predicates in the constraint query.

Ibrahim, Gray, and Fiddian (2001) contributed to the solution of constraint checking in a distributed database by demonstrating when it is possible to derive from global constraints localized constraints. They have proved that examining the semantics of both the tests and the relevant update operations reduces the amount of data transferred across the network. The simplified tests have reduced the amount of data that needed to be accessed and the number of sites that might be involved. Ibrahim (2002) extended the work in Ibrahim, Gray, and Fiddian (2001) by considering the transition constraints.

The work proposed by Madiraju, Sunderraman, and Haibin (2006) focuses on checking global constraints involving aggregates in the presence of updates. The algorithm takes as input an update statement, a list of global constraints involving

aggregates and granules. The sub-constraint granules are executed locally at remote sites and the algorithm decides if a constraint is violated based on these sub-constraint executions. The algorithm performs constraints checking before the updates and thus saves time and resources on rollback. This approach is limited as they only consider semantic integrity constraints involving both arithmetic and aggregate predicates. Other types of integrity constraints that are important and are frequently used in database applications are not being considered.

Soumya, Madiraju, and Ibrahim (2008) proposed a technique to achieve optimization of constraint checking process in distributed databases by exploiting technique of parallelism, compile time constraint checking, localized constraint checking, and history of constraint violations. The architecture mainly consists of two modules: constraint analyzer and constraint ranker for analyzing the constraints and for ranking the constraints, respectively, for systems with relational databases. They achieved optimization in terms of time by executing the constraints in parallel with mobile agents.

From these works, it can be observed that most of the previous studies proposed an approach to deriving simplified form of the initial integrity constraint with the sufficiency property, since the sufficient test is known to be cheaper than the complete test and its initial integrity constraint as it involves less data to be transferred across the network and can always be evaluated at the target site, i.e., only one site will be involved during the checking process. The previous approaches assume that an update operation will be executed at a site where the relation specified in the update operation is located, which is not always true. For example, consider a relation R that is located at site 1. An insert operation into R is assumed to be submitted by a user at site 1 and a sufficient test is used to validate the consistency of the database with respect to this update operation, which can be performed locally at site 1. But if the same update operation is submitted at a different site, say 2, the sufficient test is no longer appropriate as it will definitely access information from site 1 which is now remote to site 2. Therefore, an approach is needed so that local checking can be performed regardless the location of the submitted update operation. This means given an integrity constraint several integrity tests should be generated. Having these alternatives of tests, then selecting the most suitable test to be evaluated becomes an issue that needs to be tackled. Also, the approach must be able to cater the important and frequently used integrity constraint types.

PRELIMINARIES

Our approach has been developed in the context of relational databases. A database is described by a database schema, D, which consists of a finite set of relation schemas, $<R_1, R_2, ..., R_m>$. A relation schema is denoted by $R(A_1, A_2, ..., A_n)$ where R is the name of the relation (predicate) with n-arity and A_i's are the attributes of R. A relational distributed database schema is described as (D, IC, AS) where IC is a finite set of integrity constraints and AS is a finite set of allocation schemas.

Database integrity constraints are expressed in prenex conjunctive normal form with the range restricted property. A conjunct (literal) is an atomic formula of the form $R(u_1, u_2, ..., u_k)$ where R is a k-ary relation name and each u_i is either a variable or a constant. A positive atomic formula (positive literal) is denoted by $R(u_1, u_2, ..., u_k)$ whilst a negative atomic formula (negative literal) is prefixed by $\neg$. An (in)equality is a formula of the form $u_1 OP u_2$ (prefixed with $\neg$ for inequality) where both u_1 and u_2 can be constants or variables and $OP \in \{<, \leq, >, \geq, \diamond, =\}$.

Throughout this paper the *company* database is used, as given in Figure 1. This example has been used in most previous works related to the area of constraint checking (Feras, 2006; Gupta, 1994; Ibrahim, Gray, & Fiddian, 2001; Ibrahim, 2006).

Table 2 summarizes the integrity tests generated for the integrity constraints listed in Figure 1 using the simplification methods proposed by Nicolas (1982), Ibrahim (1998), and Ali, Hamidah, and Nur Izura (2009). It is not the intention of this paper to present the simplification methods as readers may refer to Nicolas (1982), Ibrahim (1998), and Ali, Hamidah, and Nur Izura (2009). Based on Table 2, integrity tests 1, 2, 3, 6, 7, 10, 13, 15, 18, 20, 23, 25, 26, 28, 29, and 31 are the complete tests. A complete test is a test that has both the sufficiency and the necessity properties and is used to verify if an update operation leads a consistent database state to either a new consistent state or an inconsistent state. Meanwhile integrity tests 11, 16, 21, 27, 30, and 32 are the sufficient tests, i.e., tests that verify if an update operation leads a consistent database state to a new consistent database state (McCarroll, 1995). Both tests are derived using the update operation and the integrity constraint to be checked as the input. Thus, they are the non-support tests. Support tests are 4, 5, 8, 9, 12, 14, 17, 19, 22, and 24, where these tests are generated based on the update operation and other integrity constraint as the support. For example test 4 and test 12 are generated using *IC*-5 as the support. In this example complete tests are derived using the algorithm proposed by Nicolas (1982); complete/sufficient tests are generated using the algorithm proposed by Ibrahim (1998); while support tests are produced using the algorithm which is proposed by us Ali, Hamidah, and Nur Izura (2009). These algorithms adopt the substitution techniques and absorption rules to generate integrity tests (Nicolas, 1982; Ibrahim, 1998).

THE SUPPORT TEST

The integrity tests of integrity constraints are used to prove the safety of changes made by an update operation. The execution of the integrity tests results in an easier checking process compared to

Figure 1. The Company static integrity constraints

Schema:

emp(eno, dno, ejob, esal); dept(dno, dname, mgrno, mgrsal); proj(eno, dno, pno)

Integrity Constraints:

Domain Constraint

'A specification of valid salary'

IC-1: $(\forall w \forall x \forall y \forall z)(emp(w, x, y, z) \rightarrow (z > 0))$

Key Constraints

'Every employee has a unique eno'

IC-2: $(\forall w \forall x1 \forall x2 \forall y1 \forall y2 \forall z1 \forall z2)(emp(w, x1, y1, z1) \wedge emp(w, x2, y2, z2) \rightarrow (x1 = x2) \wedge (y1 = y2) \wedge (z1 = z2))$

'Every department has a unique *dno*'

IC-3: $(\forall w \forall x1 \forall x2 \forall y1 \forall y2 \forall z1 \forall z2)(dept(w, x1, y1, z1) \wedge dept(w, x2, y2, z2) \rightarrow (x1 = x2) \wedge (y1 = y2) \wedge (z1 = z2))$

Referential Integrity Contraints

'The *dno* of every tuple in the *emp* relation exists in the *dept* relation'

IC-4: $(\forall t \forall u \forall v \forall w \exists x \exists y \exists z)(emp(t, u, v, w) \rightarrow dept(u, x, y, z))$

'The *eno* of every tuple in the *proj* relation exists in the *emp* relation'

IC-5: $(\forall u \forall v \forall w \exists x \exists y \exists z)(proj(u, v, w) \rightarrow emp(u, x, y, z))$

'The *dno* of every tuple in the *proj* relation exists in the *dept* relation'

IC-6: $(\forall u \forall v \forall w \exists x \exists y \exists z)(proj(u, v, w) \rightarrow dept(v, x, y, z))$

General Semantic Integrity Contraints

'Every manager in *dept* 'D1' earns > £4000'

IC-7: $(\forall w \forall x \forall y \forall z)(dept(w, x, y, z) \wedge (w = 'D1') \rightarrow (z > 4000))$

'Every employee must earn ≤ to the manager in the same department'

IC-8: $(\forall t \forall u \forall v \forall w \forall x \forall y \forall z)(emp(t, u, v, w) \wedge dept(u, x, y, z) \rightarrow (w \leq z))$

'Any department that is working on a project P_1 is also working on project P_2'

IC-9: $(\forall x \forall y \exists z)(proj(x, y, P_1) \rightarrow proj(z, y, P_2))$

---

the execution of the initial integrity constraints. For each integrity constraint in this case called target integrity constraint and its appropriate update template(s), integrity test(s) is derived. This test is either complete or sufficient. Support test is also derived for the target integrity constraint and its appropriate update operation using the other integrity constraints (called non-target integrity constraints) as support.

For the following analysis consider the worst case scenario where the *emp*, *dept*, and *proj* relations are located at three different sites, *S1*, *S2*, and *S3*, respectively. We assume that there are 500 employees (500 tuples), 10 departments (10 tuples), and 100 projects (100 tuples). We also assume that the following update operations are submitted by three different users at three different sites.

Table 2. Integrity tests of the integrity constraints of Figure 1

IC-i	Update template	Integrity test
IC-1	insert $emp(a, b, c, d)$	1. $d > 0$
IC-2	insert $emp(a, b, c, d)$	2. $(\forall x2 \forall y2 \forall z2)(\neg emp(a, x2, y2, z2) \vee [(b = x2) \wedge (c = y2) \wedge (d = z2)])$
		3. $(\forall x1 \forall y1 \forall z1)(\neg emp(a, x1, y1, z1))$
		4. $(\exists v \exists w)(proj(a, v, w))$
		5. $(\exists t \exists u \exists w)(dept(t, u, a, w))$
IC-3	insert $dept(a, b, c, d)$	6. $(\forall x2 \forall y2 \forall z2)(\neg dept(a, x2, y2, z2) \vee [(b = x2) \wedge (c = y2) \wedge (d = z2)])$
		7. $(\forall x1 \forall y1 \forall z1)(\neg dept(a, x1, y1, z1))$
		8. $(\exists t \exists v \exists w)(emp(t, a, v, w))$
		9. $(\exists u \exists w)(proj(u, a, w))$
IC-4	insert $emp(a, b, c, d)$	10. $(\exists x \exists y \exists z)(dept(b, x, y, z))$
		11. $(\exists t \exists v \exists w)(emp(t, b, v, w))$
		12. $(\exists u \exists w)(proj(u, b, w))$
	delete $dept(a, b, c, d)$	13. $(\forall t \forall v \forall w)(\neg emp(t, a, v, w))$
		14. $(\forall u \forall w)(\neg proj(u, a, w))$
IC-5	insert $proj(a, b, c)$	15. $(\exists x \exists y \exists z)(emp(a, x, y, z))$
		16. $(\exists v \exists w)(proj(a, v, w))$
		17. $(\exists t \exists u \exists w)(dept(t, u, a, w))$
	delete $emp(a, b, c, d)$	18. $(\forall v \forall w)(\neg proj(a, v, w))$
		19. $(\forall t \forall u \forall w)(\neg dept(t, u, a, w))$
IC-6	insert $proj(a, b, c)$	20. $(\exists x \exists y \exists z)(dept(b, x, y, z))$
		21. $(\exists u \exists w)(proj(u, b, w))$
		22. $(\exists t \exists v \exists w)(emp(t, b, v, w))$
	delete $dept(a, b, c, d)$	23. $(\forall u \forall w)(\neg proj(u, a, w))$
		24. $(\forall t \forall v \forall w)(\neg emp(t, a, v, w))$
IC-7	insert $dept(a, b, c, d)$	25. $(a <> \text{'D1'}) \vee (d > 4000)$
IC-8	insert $emp(a, b, c, d)$	26. $(\forall x \forall y \forall z)(\neg dept(b, x, y, z) \vee (d \leq z))$
		27. $(\exists t \exists v \exists w)(emp(t, b, v, w) \wedge (w \geq d))$
	insert $dept(a, b, c, d)$	28. $(\forall t \forall v \forall w)(\neg emp(t, a, v, w) \vee (w \leq d))$
IC-9	insert $proj(a, b, P1)$	29. $(\exists z)(proj(z, b, P2))$
		30. $(\exists x)(proj(x, b, P1))$
	delete $proj(a, b, P2)$	31. $(\forall x)(\neg proj(x, b, P1))$
		32. $(\exists z)(proj(z, b, P2) \wedge (z <> a))$

a. A user A submits an insert operation, insert $emp(E1, D4, J5, 4500)$ at $S1$.

b. A user B submits an insert operation, insert $emp(E2, D4, J6, 3000)$ at $S2$.

c. A user C submits an insert operation, insert $emp(E3, D3, J5, 4500)$ at $S3$.

This scenario is chosen as it represents all possibilities of cases, which includes the possibility that the cost of checking the constraints locally is more expensive than the cost of checking the constraints globally (involving remote sites) and vice versa. The integrity constraints that might be violated by these update operations are *IC*-1, *IC*-2, *IC*-4, and *IC*-8. This is based on the well-known update theorem (Nicolas, 1982).

Table 3 compares the initial integrity constraints and the various types of integrity tests with regards to the amount of data transferred across the network, the amount of data accessed, and the number of sites involved for the simple update operations (a), (b), and (c), and allocation as given above. The parameters and the measurement used in this comparison are taken from Ibrahim, Gray, and Fiddian (2001). This is discussed in more detail in following section. The symbol δR denotes the size of the relation R, i.e., $\delta emp = 500$, $\delta dept = 10$, and $\delta proj = 100$.

Update operation (a) represents the assumption made by the previous researches where the update operation is submitted at the site where the relation is located. For this update, integrity tests 1, 2, 3, 11, and 27 are the local tests. For the update operation (b), integrity tests 1, 5, 10, and 26, while for the update operation (c), integrity tests 1, 4, and 12 are the local tests. From the analysis, it is clear that local checking can be achieved by not only selecting sufficient test as suggested by previous works (Gupta, 1994; Ibrahim, Gray, & Fiddian, 2001; Ibrahim, 2002; Ibrahim, 2006; Madiraju, Sunderraman, & Haibin, 2006; Mazumdar, 1993; Qian, 1989; Simon & Valduriez, 1989) but also by evaluating the sufficient, complete or support tests depending on the information that is available at the target site. We have also showed that necessary, complete, sufficient, and support tests can be local or global tests depending on where the update operation is submitted and the location of the relation(s) specified in the integrity tests (refer to Table 3). Most importantly, we have proved that in some cases, support tests can benefit the distributed database, as shown by update operation (b) with test 5 (*IC*-2) and update operation (c) with tests 4 (*IC*-2) and 12 (*IC*-4), where local constraint checking can be achieved. This simple analysis shows that applying different type of integrity tests gives different impacts to the performance of the constraint checking. Integrating these various types of integrity tests during constraint checking and not concentrating on certain types of integrity tests (as suggested by previous works) can enhance the performance of the constraint mechanisms. Thus, developing an approach that can minimize the number of sites involved and the amount of data transferred across the network during the process of checking the integrity of the distributed databases is important.

INTEGRITY TESTS RANKING AND CLASSIFICATION

We introduce a model to rank and select the suitable test to be evaluated given several alternative tests. The model uses the properties of the tests, the amount of data transferred across the network, the number of sites participated, and the amount of data accessed as the criteria in deciding the suitable test. In addition, classifying the integrity tests that have been ranked and selected into global tests and local tests is performed to identify the number of local processing and global processing that may have to be performed. This approach consists of four phases, namely: analyzing the update operation, selecting the integrity tests, ranking and selecting the suitable integrity test, and classifying the integrity tests. In the following the phases of the approach for ranking and classifying the integrity tests are elaborated.

Analyzing the Update Operation

This phase analyzes the syntax of the actual update operations that are submitted by users. It checks that the name of relation and the number of attributes/columns which are specified in the real update operation are the same as the name of relation and the number of attributes/columns that are specified in the database schema.

Table 3. Estimation of the amount of data transferred (T), the amount of data accessed (A), and the number of sites involved (σ)

Update operation, UO	Integrity constraint, IC	The amount of data transferred, T	The amount of data accessed, A	The number of sites, σ	Integrity test, IT	The amount of data transferred, T	The number of sites, σ	The amount of data accessed, A	Type of integrity test
(a)	IC-1	0	δemp	1	1	0	1	0	Local complete test
	IC-2	0	$2\delta emp$	1	2, 3	0	1	δemp	Local complete test
					4	$\delta proj$	2	$\delta proj$	Global support necessary test
					5	$\delta dept$	2	$\delta dept$	Global support necessary test
	IC-4	$\delta dept$	$\delta emp + \delta dept$	2	10	$\delta dept$	2	$\delta dept$	Global complete test
					11	0	1	δemp	Local sufficient test
					12	$\delta proj$	2	$\delta proj$	Global support sufficient test
	IC-8	$\delta dept$	$\delta emp + \delta dept$	2	26	$\delta dept$	2	$\delta dept$	Global complete test
					27	0	1	δemp	Local sufficient test
(b)	IC-1	δemp	δemp	1	1	0	1	0	Local complete test
	IC-2	δemp	$2\delta emp$	2	2, 3	δemp	2	δemp	Global complete test
					4	$\delta proj$	2	$\delta proj$	Global support necessary test
					5	0	1	$\delta dept$	Local support necessary test
	IC-4	δemp	$\delta emp + \delta dept$	2	10	0	1	$\delta dept$	Local complete test
					11	δemp	2	δemp	Global sufficient test
					12	$\delta proj$	2	$\delta proj$	Global support sufficient test
	IC-8	δemp	$\delta emp + \delta dept$	2	26	0	1	$\delta dept$	Local complete test
					27	δemp	2	δemp	Global sufficient test
(c)	IC-1	δemp	δemp	1	1	0	1	0	Local complete test
	IC-2	δemp	$2\delta emp$	2	2, 3	δemp	2	δemp	Global complete test
					4	0	1	$\delta proj$	Local support necessary test
					5	$\delta dept$	2	$\delta dept$	Global support necessary test
	IC-4	$\delta emp + \delta dept$	$\delta emp + \delta dept$	3	10	$\delta dept$	2	$\delta dept$	Global complete test
					11	δemp	2	δemp	Global sufficient test
					12	0	1	$\delta proj$	Local support sufficient test
	IC-8	$\delta emp + \delta dept$	$\delta emp + \delta dept$	3	26	$\delta dept$	2	$\delta dept$	Global complete test
					27	δemp	2	δemp	Global sufficient test

Figure 2. The algorithm for selecting integrity tests

```
Algorithm-SIT (Select the integrity tests for a given update
operation)
Input: Update Operation, U
        Update Templates, UT = {UT1, UT2, …, UTn}
        List of Integrity Tests, IT = {IT1, IT2, …, ITm}
Output: A set of selected integrity tests, ST
BEGIN
    ST = { }
    For each UTi do
      IF the relation of U = the relation of UTi and the type of
        update operation U = the type of update operation of UTi
      THEN ST = ST ∪ ITj of UTi
END
```

Selecting the Integrity Tests

The main process of this phase is to identify and select the integrity tests which might be violated given an update operation submitted by a user. This phase is achieved by comparing the actual update operation with the update templates that have been generated. This comparison includes checking the name of relation and type of update operation. If both the actual update operation and update template have the same relation name and type of update operation, then the integrity tests of the update template are selected. Figure 2 illustrates the algorithm for selecting the integrity tests.

We now present detailed example for selecting integrity tests to clarify the processing in this phase. We assume that the update operation submitted by a user is an insert operation into the *emp* relation. The integrity constraints that might be violated given this insert operation are *IC*-1, *IC*-2, *IC*-4, and *IC*-8.

Update operation: insert *emp*(*E*1, *D*2, *J*5, 5000)

The integrity constraints and the integrity tests which need to be checked are:

IC-1: $(\forall w \forall x \forall y \forall z)(emp(w, x, y, z) \rightarrow (z > 0))$
1. $d > 0$

IC-2: $(\forall w \forall x1 \forall x2 \forall y1 \forall y2 \forall z1 \forall z2)(emp(w, x1, y1, z1) \wedge emp(w, x2, y2, z2) \rightarrow (x1 = x2) \wedge (y1 = y2) \wedge (z1 = z2))$
2. $(\forall x2 \forall y2 \forall z2)(\neg emp(a, x2, y2, z2) \vee [(b = x2) \wedge (c = y2) \wedge (d = z2)])$
3. $(\forall x1 \forall y1 \forall z1)(\neg emp(a, x1, y1, z1))$
4. $(\exists v \exists w)(project(a, v, w))$
5. $(\exists t \exists u \exists w)(dept(t, u, a, w))$

IC-4: $(\forall t \forall u \forall v \forall w \exists x \exists y \exists z)(emp(t, u, v, w) \rightarrow dept(u, x, y, z))$
10. $(\exists x \exists y \exists z)(dept(b, x, y, z))$
11. $(\exists t \exists v \exists w)(emp(t, b, v, w))$
12. $(\exists u \exists w)(proj(u, b, w))$

IC-8: $(\forall t \forall u \forall v \forall w \forall x \forall y \forall z)(emp(t, u, v, w) \wedge dept(u, x, y, z) \rightarrow (w \leq z))$
26. $(\forall x \forall y \forall z)(\neg dept(b, x, y, z) \vee (d \leq z))$
27. $(\exists t \exists v \exists w)(emp(t, b, v, w) \wedge (w \geq d))$

Ranking and Selecting the Suitable Integrity Test

Most of the works in integrity constraints checking focused on techniques to simplify integrity constraints with the assumption that the simplified forms of the constraints are cheaper than the initial constraints. Thus, the simplified form is evaluated (instead of the initial constraint) to verify the consistency of the database.

Most of the efficiency measurements consider a single cost component and are more applicable for measuring the cost of evaluating integrity constraints in a centralized environment rather than a distributed environment. Hsu and Imielinski (1985) for example measured the simplicity of an integrity constraint in terms of the notion of its *checking space*. The checking space of a constraint $IC\text{-}j(v_1, v_2, ...,v_n)$ is defined as $v_1 \times v_2 \times ... \times v_n$ where × is the Cartesian product operator, and the v_i's are the range variables in the *IC-j*. A constraint *IC-i* is said to be simpler than a constraint *IC-j* if the checking space of *IC-i* is smaller than the checking space of *IC-j*. The checking space of a constraint is a rough measure of the complexity of its evaluation and the number of variables it has to access. This measurement was later used by other researchers. Mazumdar (1993) proposed a metric called scatter, σ, to capture the amount of non-local access necessary to evaluate a constraint.

Ibrahim, Gray, and Fiddian (2001) use three parameters, namely: the amount of data transferred, T, the number of sites involved, σ (which is the scatter metric proposed by Mazumdar (1993), and the amount of data accessed, A.

a. *T* provides an estimate of the amount of data transferred across the network. It is measured based on the following formula, $T = \Sigma^n_{i=1} dt_i$, where dt_i is the amount of data transferred from site *i* to the target site.

b. A provides an estimate of the amount of data accessed, which is related to the number and the size of relations specified in a given constraint or integrity tests, *IC*. This measurement indirectly indicates the size of the checking space. It is based on the following formula, $A_{(R1, R2, ..., Rn)} = \delta R1 + \delta R2 + ... + \delta Rn$ where the R*i*'s are the relations specified in *IC* and δRi is the size of R*i*.

c. σ gives a rough measurement of the amount of non-local access necessary to evaluate a constraint or simplified form. This is measured by analyzing the number of sites that might be involved in validating the constraint.

A set of heuristic rules is also introduced in Ibrahim, Gray, and Fiddian (2001). Since, they only consider two types of integrity tests (complete and sufficient) and depend solely on the assumption that the update operation is always submitted at the site where the relation specified in the update is located, thus these rules are not general enough. These rules do not consider support tests and there is no ranking between the types of tests, i.e., all tests are considered the same.

We argue that tests should be ranked as they have different probability of being true or false in a given database state. Thus, we suggest complete test should have the highest property, followed by sufficient test, and lastly by support test. This is because complete test has both the sufficiency and the necessity properties, sufficient test can only verify for valid database state, and most of the support test is either sufficient test or necessary test. As mentioned in the literature, the amount of data transferred across the network is the most critical factor; therefore we suggest the amount of data transferred to have the highest priority in the ranking model, followed by the number of sites involved, and the amount of data accessed. Based on these arguments, we have proposed a ranking model as shown in Figure 3. Each value in the box, i.e., 1, 2, 3, ..., *P*, is the *rank value* where *P* is the maximum rank value. The *rank value* of a test, Test$_i$, with respect to *T* is denoted by Rank$_T$. Similar notation is used for indicating the rank value of a test with respect to σ and A. Thus, we can calculate the total rank value for a given test by simply adding the rank values for each of the parameter, i.e., Rank-*Test*$_i$ = Rank$_T$ + Rank$_\sigma$ + Rank$_A$. A test with the lowest total rank value is said to be the most suitable test. The ranking model is designed as follows:

Figure 3. The proposed ranking model

Parameter / Type of Test	Complete, C	Sufficient, S	Support, Sup	Remarks
$T = 0$	1	2	3	If $T \neq 0$, the tests are rank accordingly based on the amount of data transferred. Rank value begins with 4, 5, 6, ..., P_T
$\sigma = 1$	4	5	6	If $\sigma \neq 1$, the tests are rank accordingly based on the number of sites involved. Rank value begins with 7, 8, 9, ..., P_σ
$\mathscr{A}_C = \mathscr{A}_S = \mathscr{A}_{Sup}$	7	8	9	If $\mathscr{A}_C \neq \mathscr{A}_S \neq \mathscr{A}_{Sup}$, the tests are rank accordingly based on the amount of data accessed. Rank value begins with 10, 11, 12, ..., $P_\mathscr{A}$.

Note: P_T (P_σ and $P_\mathscr{A}$, respectively) is the maximum rank value assigned to a test based on T (σ and $\mathscr{A}$, respectively).

1. If the amount of data transferred of a given test is 0 (i.e., the test is a local test), then depending on its property, a value of 1, 2, and 3 is assigned to the $Rank_T$ if the test is a complete, sufficient, and support, respectively. Otherwise for each non local test ($T \neq 0$), the tests are ordered according to the value of T and the test with the lowest T, a value of 4 is assigned to its $Rank_T$. The next lowest, a value of 5 is assigned to its $Rank_T$ and so on.

2. If the number of sites involved in checking a given test is 1, then depending on its property, a value of 4, 5, and 6 is assigned to the $Rank_\sigma$ if the test is a complete, sufficient, and support, respectively. The rank value begins with 4 (and not 1, 2, or 3) to show that the number of sites has lower priority than the amount of data transferred. Otherwise, for each test with $\sigma \neq 1$, the tests are ordered according to the number of sites involved and the test with the lowest σ, a value of 7

is assigned to its Rank_σ. The next lowest, a value of 8 is assigned to its Rank_σ and so on. Also, note that a test with $\text{Rank}_\sigma = 4$ (5 and 6, respectively) will definitely not be assigned a $\text{Rank}_T = 4$ (5 and 6, respectively) since $\text{Rank}_T = 4$ (5 and 6, respectively) indicates that the test is a nonlocal test while $\text{Rank}_\sigma = 4$ (5 and 6, respectively) denotes that the test is a local test. Although they have the same rank value, i.e. 4, but after adding the rank value for both T and σ, the local test will definitely have lower total rank value compared to the nonlocal test.

3. If the amount of data accessed for each of the test is the same, then depending on its property, a value of 7, 8, and 9 is assigned to the Rank_A if the test is a complete, sufficient, and support, respectively. The rank value begins with 7 (and not 1, 2, …, 6) to show that the amount of data accessed has the lowest priority compared to the amount of data transferred and the number of sites involved. Otherwise, for each test with different amount of data accessed, these tests are ordered according to the amount of data accessed and the test with the lowest A, a value of 10 is assigned to its Rank_A. The next lowest, a value of 11 is assigned to its Rank_A and so on. Also, note that a test with $\text{Rank}_A = 7$ (8 and 9, respectively) can be assigned a $\text{Rank}_\sigma = 7 (8, 9, …, P_\sigma)$ which indicate that the test is a nonlocal complete test (nonlocal sufficient test and nonlocal support test, respectively) and the amount of data accessed is the same for all the alternative tests.

Figure 4 presents the algorithm, *Algorithm-RIT*, for the ranking technique. To illustrate the technique for ranking integrity tests, three cases are considered: (i) centralized database (all relations are located at the same site), (ii) average case, and (iii) worst case, the same scenario as given in the previous section. The same sizes of relations and update operations as given in the previous section

are used. *IC*-4 is used to demonstrate the model, i.e. tests 10 (complete), 11 (sufficient), and 12 (support) are compared.

Based on the result shown in Table 4, complete test, C, is selected, as it is the most suitable test for centralized database. Since all tests have similar characteristics with regards to T (=0) and σ (=1), the only different are the properties of the tests and A. Mazumdar scatter metric alone is not able to select the suitable test as these tests have the same scatter metric, $\sigma = 1$, while Ibrahim, Gray, and Fiddian, (2001) will select the test with the lowest A. If the tests have the same amount of data accessed, then no solution is given in (Ibrahim, Gray, & Fiddian, 2001). In our model, complete test will be selected.

Table 5 presents an average case with several different scenarios. Here, we assume that two of the relations are located at the same site while the other relation is located at a different site, and update operation is submitted at any of these sites. From the results, we observed that local test is always selected regardless the type of the tests. In cases where more than one local test is available, then the tests are rank according to the type and the amount of data accessed (this scenario is similar to the case (i) centralized database discussed earlier).

Table 6 presents the worst case scenario. In this case each relation is located at different sites. The test that is selected is the local test ($T = 0$ and $\sigma = 1$). As mentioned, most of the previous works assumed that the update operation is submitted at the site where the relation specifies in the update is located, and thus the sufficient test is always selected. In our model, the suitable test (with the lowest total rank value) is selected and this test can be complete, sufficient or support.

In addition, we analyze the behaviour of the ranking and selecting model when the number of tuples in each relation is equivalent. This is shown by Table 7 and Table 8 which demonstrate the results of an average case (iv) and the worst case (v), respectively. Here, we assume that there are

Figure 4. Algorithm for ranking integrity tests

```
Algorithm-RIT
Input: Integrity constraint, IC-i, update operation, U, Integrity tests, {T1,
T2, …, Tm}
Output: List of tests which has been ranked
BEGIN
K = { }, L = { }, M = { }, Rank_T = 0, Rank_σ = 0, Rank_√ = 0
Get the list of integrity tests based on IC-i and U, Ti = {T1, T2, …, Tm}
For each Ti do
  BEGIN
    IF Ti is C and T = 0 THEN Rank_T = 1 ELSE K = K ∪ {Ti}
    IF Ti is S and T = 0 THEN Rank_T = 2 ELSE K = K ∪ {Ti}
    IF Ti is Sup and T = 0 THEN Rank_T = 3 ELSE K = K ∪ {Ti}
    IF Ti is C and σ = 1 THEN Rank_σ = 4 ELSE L = L ∪ {Ti}
    IF Ti is S and σ = 1 THEN Rank_σ = 5 ELSE L = L ∪ {Ti}
    IF Ti is Sup and σ = 1 THEN Rank_σ = 6 ELSE L = L ∪ {Ti}
    IF Ti is C and √Ti = √Tj for j = {1, 2, …, m} and i <> j THEN Rank_√ = 7
    ELSE M = M ∪ {Ti}
    IF Ti is S and √Ti = √Tj for j = {1, 2, …, m} and i <> j THEN
    Rank_√ = 8 ELSE M = M ∪ {Ti}
    IF Ti is Sup and √Ti = √Tj for j = {1, 2, …, m} and i <> j THEN  Rank_√
    = 9 ELSE M = M ∪ {Ti}
    Rank-Test_Ti = Rank_T + Rank_σ + Rank_√
  END
Sort the elements of K according to T
Rank_T = 4
For each element j in K do
  BEGIN
    IF j = 1 THEN
        Rank-Test_Test-j = Rank-Test_Test-j + Rank_T
        Temp-T = T of element j
    ELSE
      IF T of element j = Temp-T THEN
          Rank-Test_Test-j  = Rank Test_Test-j + Rank_T
      ELSE
          Rank-Test_Test-j = Rank-Test_Test-j + Rank_T + 1
          Temp-T = T of element j
          Rank_T = Rank_T + 1
    END
Sort the elements of L according to σ
Rank_σ = 7
For each element j in L do
  BEGIN
    IF j = 1 THEN
        Rank-Test_Test-j = Rank-Test_Test-j + Rank_σ
        Temp-σ = σ of element j
    ELSE
      IF σ of element j = Temp-σ THEN
          Rank-Test_Test-j = Rank- Test_Test-j + Rank_σ
      ELSE
          Rank-Test_Test-j = Rank-Test_Test-j + Rank_σ + 1
          Temp-σ = σ of element j
          Rank_σ = Rank_σ + 1
    END
Sort the elements of M according to √
Rank_√ = 10
For each element j in M do
  BEGIN
    IF j = 1 THEN
        Rank-Test_Test-j = Rank-Test_Test-j + Rank_√
        Temp-√ = √ of element j
    ELSE
      IF √ of element j = Temp-√ THEN
          Rank-Test_Test-j = Rank-Test_Test-j + Rank_√
      ELSE
          Rank-Test_Test-j = Rank-Test_Test-j + Rank_√ + 1
          Temp-√ = √ of element j
          Rank_√ = Rank_√ + 1
    END
END
```

100 employees (100 tuples), 100 departments (100 tuples), and 100 projects (100 tuples).

We believed that the ranking model will select the most appropriate integrity tests even when the number of tuples becomes huge. Table 9 presents the results of the worse case (vi) where there are 1500 *employees* (1500 tuples), 10 *departments* (10 tuples), and 200 *projects* (200 tuples). We have evaluated the proposed model with several other cases and several different types of integrity constraints, and in all cases the technique is able to select the suitable test as expected.

Classifying the Integrity Tests

This section presents the last phase in the approach for ranking and classifying integrity tests. This phase classifies the integrity tests that have been ranked and selected. Complete, sufficient, and support tests can be classified as either local or global tests depending on where the actual update operation is submitted and the location of the relation(s) specified in the integrity tests. If the test can be performed locally, then the test is local. In contrary, if data need to be transferred across the network then the test is global.

For the example given in *Selecting the Integrity Tests* and based on the algorithm for ranking the integrity tests in *Ranking and Selecting the Suitable Integrity Test*, we clarify the process of classifying the integrity tests. Assume that the update operation is submitted at site $S1$ where the *emp* relation is located.

Update operation: insert *emp*($E1$ $D2$, $J5$, 5000)

The integrity tests that are selected based on the given insert operation which are classified as local tests are listed below.

1. $d > 0$
2. $(\forall x2 \forall y2 \forall z2)(\neg emp(a, x2, y2, z2) \vee [(b = x2) \wedge (c = y2) \wedge (d = z2)])$
11. $(\exists t \exists v \exists w)(emp(t, b, v, w))$

Table 4. Case (i) centralized database

Update is submitted at site:	Location of Relations		Rank-$Test_i$	Test Selected
S1	S1	emp, dept, proj	$C = 1 + 4 + 10 = 15$ $S = 2 + 5 + 12 = 19$ $Sup = 3 + 6 + 11 = 20$	C

Table 5. Case (ii) average case

Update is submitted at site:	Location of Relations		Rank-$Test_i$	Test Selected
S1	S1	emp, dept	$C = 1 + 4 + 10 = 15$ $S = 2 + 5 + 12 = 19$ $Sup = 4 + 7 + 11 = 22$	C
	S2	proj		
S2	S1	emp, dept	$C = 4 + 7 + 10 = 21$ $S = 5 + 7 + 12 = 24$ $Sup = 3 + 6 + 11 = 20$	Sup
	S2	proj		
S1	S1	emp, proj	$C = 4 + 7 + 10 = 21$ $S = 2 + 5 + 12 = 19$ $Sup = 3 + 6 + 11 = 20$	S
	S2	dept		
S2	S1	emp, proj	$C = 1 + 4 + 10 = 15$ $S = 5 + 7 + 12 = 24$ $Sup = 4 + 7 + 11 = 22$	C
	S2	dept		
S1	S1	dept, proj	$C = 1 + 4 + 10 = 15$ $S = 4 + 7 + 12 = 23$ $Sup = 3 + 6 + 11 = 20$	C
	S2	emp		
S2	S1	dept, proj	$C = 4 + 7 + 10 = 21$ $S = 2 + 5 + 12 = 19$ $Sup = 5 + 7 + 11 = 23$	S
	S2	emp		

27. $(\exists t \exists v \exists w)(emp(t, b, v, w) \land (w \geq d))$

Another example, assume that the same update operation is submitted at site S3 where the *proj* relation is located. The integrity tests that are selected based on the given insert operation is classified as follows:

Local test:

1. $d > 0$
4. $(\exists v \exists w)(proj(a, v, w))$
12. $(\exists u \exists w)(proj(u, b, w))$

Global test:

26. $(\forall x \forall y \forall z)(\neg dept(b, x, y, z) \lor (d \leq z))$

RESULT AND DISCUSSION

Most of the previous researchers argued that the amount of data transferred is the most critical factor that influences the performance of integrity constraint checking in distributed database environment (Gupta, 1994; Ibrahim, 1998; Ibrahim, Gray, & Fiddian, 2001; Ibrahim, 2002; Madiraju, Sunderraman, & Haibin, 2006; Simon & Valduriez, 1989; Soumya, Madiraju, & Ibrahim, 2008). Strategies which are capable of verifying and maintaining the consistency of a database state at a local site without involving any remote site are more effective. Most of the previous works focus on maximizing the percentage of local integrity constraint checking ($T \approx 0$). That means, minimizes the percentage of global integrity constraint check-

Table 6. Case (iii) worst case

Update is submitted at site:	Location of Relations		Rank-*Test$_i$*	Test Selected
$S1$	$S1$	emp	$C = 4 + 7 + 10 = 21$	S
	$S2$	dept	$S = 2 + 5 + 12 = 19$	
	$S3$	proj	$Sup = 5 + 7 + 11 = 23$	
$S2$	$S1$	emp	$C = 1 + 4 + 10 = 15$	C
	$S2$	dept	$S = 5 + 7 + 12 = 24$	
	$S3$	proj	$Sup = 4 + 7 + 11 = 22$	
$S3$	$S1$	emp	$C = 4 + 7 + 10 = 21$	Sup
	$S2$	dept	$S = 5 + 7 + 12 = 24$	
	$S3$	proj	$Sup = 3 + 6 + 11 = 20$	

Table 7. Case (iv) average case

Update is submitted at site:	Location of Relations		Rank-*Test$_i$*	Test Selected
$S1$	$S1$	emp dept	$C = 1 + 4 + 7 = 12$	C
	$S2$	proj	$S = 2 + 5 + 8 = 15$	
			$Sup = 4 + 7 + 9 = 20$	
$S2$	$S1$	emp, dept	$C = 4 + 7 + 7 = 18$	Sup or C
	$S2$	proj	$S = 4 + 7 + 8 = 19$	
			$Sup = 3 + 6 + 9 = 18$	
$S1$	$S1$	emp, proj	$C = 4 + 7 + 7 = 18$	S
	$S2$	dept	$S = 2 + 5 + 8 = 15$	
			$Sup = 3 + 6 + 9 = 18$	
$S2$	$S1$	emp, proj	$C = 1 + 4 + 7 = 12$	C
	$S2$	dept	$S = 4 + 7 + 8 = 19$	
			$Sup = 4 + 7 + 9 = 20$	
$S1$	$S1$	dept, proj	$C = 1 + 4 + 7 = 12$	C
	$S2$	emp	$S = 4 + 7 + 8 = 19$	
			$Sup = 3 + 7 + 9 = 19$	
$S2$	$S1$	dept, proj	$C = 4 + 7 + 7 = 18$	S
	$S2$	emp	$S = 2 + 5 + 8 = 15$	
			$Sup = 4 + 7 + 9 = 20$	

Table 8. Case (v) worst case

Update is submitted at site:	Location of Relations		Rank-*Test$_i$*	Test Selected
$S1$	$S1$	emp	$C = 4 + 7 + 7 = 18$	S
	$S2$	dept	$S = 2 + 5 + 8 = 15$	
	$S3$	proj	$Sup = 4 + 7 + 9 = 20$	
$S2$	$S1$	emp	$C = 1 + 4 + 7 = 12$	C
	$S2$	dept	$S = 4 + 7 + 8 = 19$	
	$S3$	proj	$Sup = 4 + 7 + 9 = 20$	
$S3$	$S1$	emp	$C = 4 + 7 + 7 = 18$	Sup or C
	$S2$	dept	$S = 4 + 7 + 8 = 19$	
	$S3$	proj	$Sup = 3 + 6 + 9 = 18$	

Table 9. Case (vi) worst case

Update is submitted at site:	Location of Relations		Rank-Test$_i$	Test Selected
S1	S1	emp	$C = 4 + 7 + 10 = 21$ $S = 2 + 5 + 12 = 19$ $Sup = 5 + 7 + 11 = 23$	S
	S2	dept		
	S3	proj		
S2	S1	emp	$C = 1 + 4 + 10 = 15$ $S = 5 + 7 + 12 = 24$ $Sup = 4 + 7 + 11 = 22$	C
	S2	dept		
	S3	proj		
S3	S1	emp	$C = 4 + 7 + 10 = 21$ $S = 5 + 7 + 12 = 24$ $Sup = 3 + 6 + 11 = 20$	Sup
	S2	dept		
	S3	proj		

Figure 5. The percentage of local checking for update operations (a), (b), and (c) with various strategies

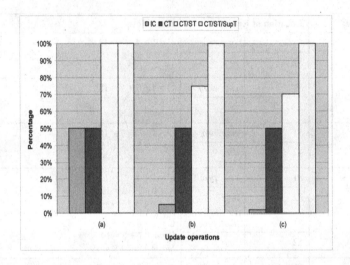

ing to reduce the cost and increase the efficiency of the constraint checking process.

We have conducted the following experiment. This experiment aims at analyzing the constraint checking process in terms of percentage of local processing for the update operation with/without ranking technique. This is shown through the type of integrity tests selected as follows.

- **IC:** the initial integrity constraints are checked to identify the existence of inconsistency in the distributed database. This method is termed the brute force checking.
- **CT:** only the complete tests are selected.

- ◦ Note that both *IC* and *CT* do not apply any ranking technique.
- **CT/ST:** either the complete test or the sufficient test is selected depending on which test will minimize the amount of data transferred as the main priority, for a given integrity constraint. This approach is as proposed by previous studies (Gupta, 1994; Ibrahim, Gray, & Fiddian, 2001; Ibrahim, 2006; Madiraju, Sanderraman, & Haibin, 2006; Mazumdar, 1993; Qian, 1989; Simon & Valduriez, 1989) which adopted the ranking technique proposed by Ibrahim, Gray, and Fiddian (2001).

- **CT/ST/SupT:** either the complete test, the sufficient test, or the support test is selected, as suggested by the ranking technique. This is what we proposed.

Figure 5 shows the constraint checking process with various strategies for the update operations (a), (b), and (c) discussed above. It is clear that the *CT/ST/SupT* strategy proposed by us has the highest percentage of local tests for both update operations (b) and (c) compared to the other three strategies due to all (for this example) the tests can be checked locally. However our strategy has the same ratio as *CT/ST* for update operation (a). This shows that the ranking technique proposed by us is able to select the most suitable test to be evaluated given a set of alternative tests.

Figure 6 illustrates the interface of the ranking and classifying integrity tests which is tailored for distributed database system that we have developed using the Visual Basic 6.0.

CONCLUSION

Integrity constraints are the most important and essential part in the modern database systems for enforcing consistency of data in database. Checking integrity constraints is a necessary process to check the update operation before performing it on the database. This process is the most critical task that influences the performance of database.

In this paper, we have proposed a model for ranking and selecting integrity tests. Selecting the suitable type of test which reduces the amount of data transferred across the network and minimizes the number of sites participated is important in distributed database. The novelty of this model is that local checking can be performed regardless the location of the submitted update operation. This is achieved by having several types of integrity tests and not focusing on certain type of integrity tests as suggested by previous researchers in this area. The cost which indirectly indicates the time taken

Figure 6. The interface of ranking and classifying integrity tests

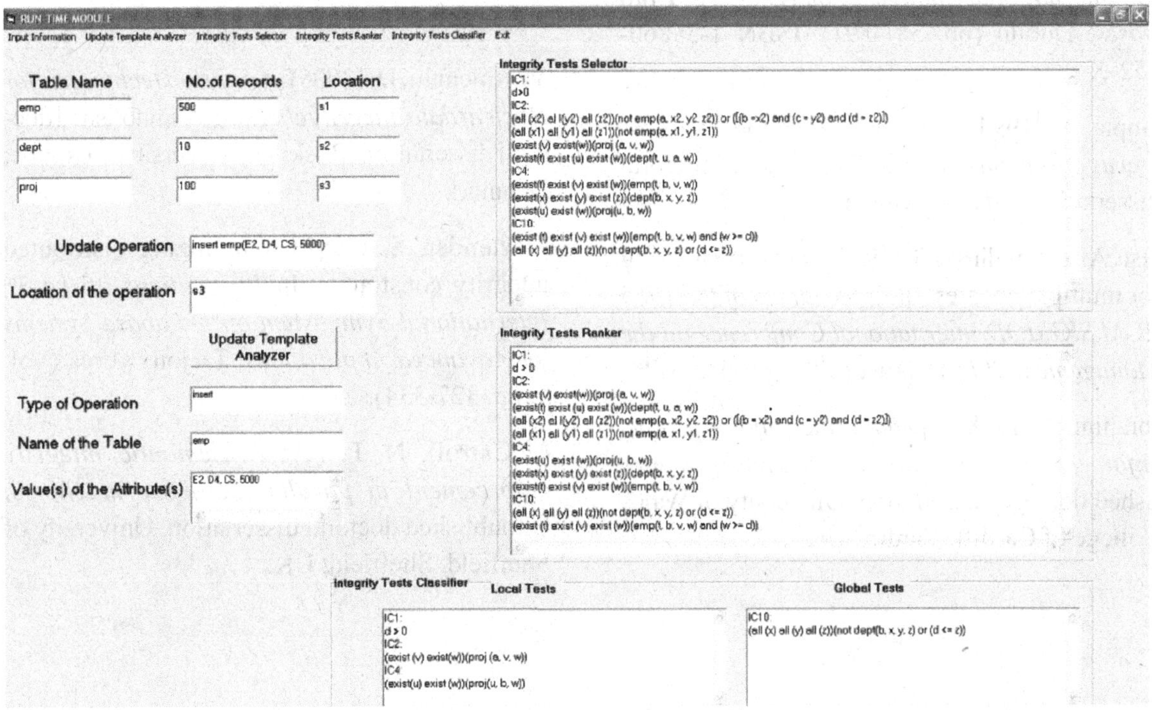

to check the consistency of distributed databases is reduced. Most importantly, we have proved that, in most cases, support tests can benefit the distributed database, where local constraint checking can be achieved. Thus, the efficiency of checking constraint process is increased.

REFERENCES

Ali, A. A., Hamidah, I., & Nur Izura, U. (2009). Improved integrity constraints checking in distributed database by exploiting local checking. *Journal of Computer Science and Technology, 24*(4), 665–674. doi:10.1007/s11390-009-9261-0

Feras, A. H. H. (2006). *Integrity constraints maintenance for parallel databases.* Unpublished doctoral dissertation, Universiti Putra Malaysia, Malaysia.

Grefen, P. W. P. J. (1993). Combining theory and practice in integrity control: A Declarative approach to the specification of a transaction modification subsystem. In *Proceedings of the 19th International Conference on Very Large Data Bases*, Dublin (pp. 581-591). ISBN 1-55860-152-X

Gupta, A. (1994). *Partial information based integrity constraint checking.* Unpublished doctoral dissertation, Stanford University, Stanford, CA.

Hsu, A., & Imielinski, T. (1985). Integrity checking for multiple updates. In *Proceedings of the 1985 ACM SIGMOD International Conference on the Management of Data*, Austin, TX (pp. 152-168).

Ibrahim, H. (1998). *Semantic integrity constraints enforcement for a distributed database.* Unpublished doctoral dissertation, University of Wales College of Cardiff, Cardiff, UK.

Ibrahim, H. (2002). A Strategy for semantic integrity checking in distributed databases. In *Proceedings of the Ninth International Conference on Parallel and Distributed Systems* (pp. 139-146). Washington, DC: IEEE.

Ibrahim, H. (2006). Checking integrity constraints – How it differs in centralized, distributed and parallel databases. In *Proceedings of the 17th International Conference on Database and Expert System Application - 2nd International Workshop on Logical Aspects and Applications of Integrity Constraints*, Krakow, Poland (pp. 563-568).

Ibrahim, H., Gray, W. A., & Fiddian, N. J. (2001). Optimizing fragment constraints – A Performance evaluation. *International Journal of Intelligent Systems – Verification and Validation Issues in Databases. Knowledge-Based Systems, and Ontologies, 16*(3), 285–306.

Madiraju, P., Sunderraman, R., & Haibin, W. (2006). A Framework for global constraint checking involving aggregates in multidatabases using granular computing. In *Proceedings of IEEE International Conference on Granular Computing*, Atlanta (pp. 506-509).

Martinenghi, D. (2005). *Advanced techniques for efficient data integrity checking.* Unpublished doctoral dissertation, Roskilde University, Roskilde, Denmark.

Mazumdar, S. (1993). Optimizing distributed integrity constraints. In *Proceedings of the 3rd International Symposium on Database Systems for Advanced Applications*, Taejon, Korea (Vol. 4, pp. 327-334).

McCarroll, N. F. (1995). *Semantic integrity enforcement in parallel database machines.* Unpublished doctoral dissertation, University of Sheffield, Sheffield, UK.

McCune, W. W., & Henschen, L. J. (1989). Maintaining state constraints in relational databases. A Proof theoretic basis. *Journal of the Association for Computing Machinery, 36*(1), 46–68.

Nam, Q. H. (1998). Maintaining global integrity constraints in distributed databases. *Constraints: An International Journal, 2*(3-4), 377–399.

Nicolas, J. M. (1982). Logic for improving integrity checking in relational data bases. *Acta Informatica, 18*(3), 227–253. doi:10.1007/BF00263192

Qian, X. (1989). Distribution design of integrity constraints. In *Proceedings of the 2ⁿᵈ International Conference on Expert Database Systems*, Vienna, VA (pp. 205-226).

Simon, E., & Valduriez, P. (1989). Integrity control in distributed database systems. In *Proceedings of the 19ᵗʰ Hawaii International Conference on System Sciences*, HI (pp. 622-632).

Soumya, B., Madiraju, P., & Ibrahim, H. (2008). Constraint optimization for a system of relation databases. In *Proceedings of the IEEE 8ᵗʰ International Conference on Computer and Information Technology*, Sydney, Australia (pp. 155-160).

This work was previously published in International Journal of Information Technology and Web Engineering, Volume 5, Issue 3, edited by Ghazi I. Alkhatib, pp. 65-84, copyright 2010 by IGI Publishing (an imprint of IGI Global).

Section 3
Web–Engineered Applications

Chapter 9

P2P-NetPay:
A Micro-Payment System for Peer-to-Peer Networks

Xiaoling Dai
Charles Sturt University, Australia

Kaylash Chaudhary
The University of the South Pacific – Laucala, Fiji

John Grundy
Swinburne University of Technology, Australia

ABSTRACT

Micro-payment systems are becoming an important part of peer-to-peer (P2P) networks. The main reason for this is to address the "free-rider" problem in most existing content sharing systems. The authors of this chapter have developed a new micro-payment system for content sharing in P2P networks called P2P-Netpay. This is an offline, debit based protocol that provides a secure, flexible, usable, and reliable credit service in peer-to-peer networks ensuring equitable participation by all parties. The authors have carried out an assessment of micro-payment against non-micro-payment credit systems for file sharing applications. The chapter reports on the design of our experiment and results of an end user evaluation. The chapter then discusses the performance of the credit model, comparing it to a non-micro-payment credit model. Through evaluation of the proposed system and comparison with other existing systems, the authors find that the new approach eliminates the "free-rider" problem. The chapter analyses a heuristic evaluation performed by a set of evaluators and presents directions for research aiming to improve the overall satisfaction and efficiency of this model for peers.

DOI: 10.4018/978-1-4666-0023-2.ch009

INTRODUCTION

A trend towards widespread use of peer-to-peer systems has become evident over the past few years for various content sharing, including files, music, videos, etc. However, these systems suffer from a common problem of individual rationality among peers (Shneidman & Parkes, 2003). This dilemma is also known as the "free-rider" problem. Peers become so self-absorbed that they participate in the peer-to-peer network only with the intent to download files rather than also making available some files for sharing with other peers. Thus these "free riders" exploit the peer-to-peer system for their own needs but do not contribute anything back for others.

Some content sharing systems provide a "credit" system encouraging or enforcing an equitable use of the system. This is usually some form of balancing of the amount of downloading allowed the degree of uploading or providing content to share. Such downloads and content sharing is very high volume, low cost transactions. Thus a micro-payment model may well be a suitable approach to implementing a credit-based scheme for peer-to-peer sharing networks. However, most current file sharing systems do not integrate a micro-payment approach. Most existing micro-payment systems implement a customer/vendor relationship which is suitable for client server and retail web applications but not for a P2P environment. These systems (Glassman et al., 1995; Hauser et al., 1996; Anderson et al., 1996; Pedersen, 1996; Liebau et al., 2006; Zghaibeh & Harmantzis, 2006; Vishnumurthy et al., 2003; Garcia & Hoepman, 2005) suffer from online processing and impose scalability and security issues on users.

We present our novel P2P-Netpay micro-payment model and its architecture. P2P-Netpay provides an offline micro-payment model that utilizes a light-weight hashing-based encryption approach. A "peer user" buys e-coins from a broker using macro-payment approach. These coins are stored in the peer's "e-wallet" stored on the peer's machine. The peer user pays for content as they download it by transparently passing e-coins to a "peer vendor". The peer vendor redeems their e-coins with the broker for coins of their own to do P2P downloads. E-coin information can be transparently exchanged between peer vendors when peer users download content from another peer vendor. Multiple sizes of coins and even multiple currencies are possible.

We give an overview of the current micro-payment schemes so far attempted for P2P networks. We describe our research methodology of assessing key requirements of content sharing application end users of micro-payment with non-micro-payment approaches. We describe the main aspects of the software architecture and design for P2P-Netpay. We describe three different evaluations we have performed on our P2P-Netpay prototype to compare micro-payment versus non-micro-payment usability, performance and conduct a heuristic assessment. We conclude with an outline of our further plans for research and development in this area.

MOTIVATION

It is all too easy for users of peer-to-peer content sharing networks to "free ride". This is a practice by which a user gains content but contributes little or nothing back. Contributions could be adding their own content to the network or allowing their machine to be a conduit for others to share content. To address this issue one common approach is a credit-based scheme. Users are given credit for contributions and this credit allows them to "pay" for content from others. Credit may be real money but more often is some form of virtual credit specific to the peer-to-peer network. Credit across peer-to-peer networks is currently very uncommon. Unfortunately, implementing such a credit-based scheme can severely impact the peer-to-peer network security, privacy, efficiency, scalability and robustness. One approach is to adopt micro-payment techniques developed for more traditional customer/vendor, client-server on-line applications. Micro-payment systems support very high-volume, low-cost transactions much more efficiently and effectively than macro-payment (e.g. credit card) or subscription services. Unlike ad-hoc credit-based systems they are designed with high security, integrity and performance in mind, being designed to facilitate real-money electronic commerce.

A number of P2P micro-payment systems have been developed for content sharing networks. These system includes XPAY (Chen et al., 2009), Huang's protocol (Huang & Zhao, 2009), Floodgate (Nair et al., 2008), WhoPay (Wei et al., 2006), and CPay (Zou et al., 2005). Unfortunately many of these suffer from problems with communication overheads, dependence on online brokers, lack of scalability and lack of coin transferability. The key requirements for P2P micro-payment system are generally agreed to be (Wei et al., 2006; Zou et al., 2005; Yang & Garcia, 2003):

- Ease of use for peers, ideally requiring nothing but point-and-click to purchase.
- Ease of addition to content sharing application software.
- Scalability: The load of either a peer or broker must not grow to an unmanageable size. This determines whether a system is an "online" or "offline" system.
- Security: The e-coins must be well-encrypted to prevent peers from double spending and fraud.
- Anonymity: peer user and peer vendor should not reveal their identities to each other or to any other third party.
- Transferability:
    1. E-coins must be spendable at any peers i.e. e-coins must not be peer specific; and

    2. the e-coin received by a peer can be spent at any other peer without contacting the issuer.

XPAY (Chen et al., 2009) is payword-based micropayment system. The peer vendor creates a connection to a group of peer customers. The peer vendor and the first peer customer run an "InitChain" operation to establish a micropayment chain for future use. All remaining peer customers in the group run an InitChain to establish other micropayment chains. The payword chain in XPAY is peer-specific and can only be spent with a specific peer vendor. The protocol features offline verification, aggregation, statistical overspending prevention and very low overheads. Throughputs of thousands of transactions per second are possible.

Huang's protocol (Huang & Zhao, 2009) is a scrip-based, secure and lightweight micropayment scheme. It requires a peer customer first purchase enough trivial scrip for service with little value and then pay a piece of scrip for each downloading. The peer vendor is responsible for the issue, signature, validation and double spending check of the scrips. There is a quorum of trusted peers to manage the peer's account information (increase and decease tokens) and several account holders to check the validity of the peer's account. The payment scheme is thus flexible for a peer's dynamic joining and departure of the network. The scrip is peer-specific and can only be used with a specific peer vendor. The centralized registry server (broker) does not participate in any transaction but trusted peers and several account holders must be online.

Floodgate is a token-based micropayment system (Nair et al., 2008) which provides monetary incentive to the clients to contribute their bandwidth to the system. Floodgate uses associated torrent file which includes the tracker information in a known public repository operated by a content provider. The tracker is a "middle-man" among the peers and the content provider manages micropayment tokens that can be used by peers to pay for content pieces. A provider peer needs to contact the content provider to verify a token received from requesting peer. The content provider and the trackers must be online.

A new micropayment protocol based on P2P networks, CPay (Zou et al., 2005), exploits the heterogeneity of the peers. CPay is a debit-based protocol. The broker is responsible for the distribution and redemption of the coins and the management of eligible peers called the Broker Assistant (BA). The broker does not participate in any transaction. Only payer, payee and the BA is involved. The BA is the eligible peer which the payer maps to and is responsible for checking the coin and authorization of the transaction. Every peer will have a BA to check its transaction. The performance will not be very high due to the involvement of the BA in every transaction.

WhoPay (Wei et al., 2006) inherits its basic architecture from PPay. Coins have the same life cycle as in PPay and are identified by public keys. A user purchases coins from the broker and spends them with other peers. The other peers may decide whether to spend the coin with another peer or redeem it with the broker. Coins must be renewed periodically to retain their value. Coins are renewed or transferred through their coin owners if they are online else through the broker. This system supports good anonymity, fairness, scalability and transferability. However, it is not very efficient because it uses heavyweight public key encryption operations on a per-purchase basis. In addition its downtime protocol is almost an online system.

PPay (Yang & Garcia, 2003) uses transferable coins and its main idea is to distribute broker workload onto peers. The concept of a floating and self-managed coin is introduced to reduce broker workload. Nevertheless, PPay also has its limitations. It does not take the heterogeneity of the peers into consideration. It also overlooks the simple fact that peers having low bandwidth, little on-line time or peers that are selfish and lazy are not appropriate to assume the role of "owner". The owner of the coin has too much authorization and can easily cheat other peers or collude with other peers. PPay has a downtime protocol which is almost an online micro-payment system. The use of layered coins introduces a delay in terms of fraud detection. The floating coins growing in size creates a scalability issue.

RESEARCH METHOD

We developed a new model for micro-payment in peer-to-peer networks and built a prototype of this system, P2P-Netpay. We then wanted to assess our new approach compared to non-micro-payment file sharing applications. We have also developed a file sharing application without a micro-payment credit system in order to compare these approaches.

We wanted to measure the characteristics of P2P-Netpay-based micro-payment systems from several different end user perspectives. We aimed to capture and understand customer views on P2P-Netpay, and advantages and disadvantages they saw with our system. Most currently used micro-payment credit systems in peer-to-peer environments are on-line systems (Dai et al., 2007), (Nielsen, 2005). These tend to suffer from dependence on online brokers and scalability and performance problems. We extended a micro-payment model that we had previously developed called NetPay. This uses light-weight, low cost e-coin encryption via hashing, offline micro-payment (i.e. the broker doesn't need to be involved in every transaction), protection from double spending and peer user and peer vendor forgery of coins or debits, and fully anonymous payment (Dai & Grundy, 2005). This is achieved by the use of a hashing mechanism embedded in the broker where peer users buy e-coins and peer vendors redeem spent e-coins.

To evaluate our P2P-NetPay prototype three types of evaluation were required. A usability evaluation was used to gain users' feedback on P2P-Netpay features. A heuristic evaluation was used to assess the overall user interface qualities. A performance evaluation was done to assess the scalability of the P2P-NetPay system. We approached assessing usability via a survey-based approach with representative target users of P2P-Netpay. A set of evaluators carried out a heuristic-based evaluation of P2P-Netpay using a set of well-adopted usability principles (Nielsen, 1994). A performance evaluation of our P2P-Netpay prototype was undertaken to determine the suitability of such a system in a large P2P network domain by assigning heavy loads to peers. We analysed the results from our three evaluations to assess whether (i) P2P-Netpay is usable in the opinion of our content sharing application target users; (ii) the performance of our P2P-Netpay prototype system would be acceptable in a complex P2P environment; and (iii) that our P2P-Netpay does meet the key requirements of a micro-payment system for content sharing in P2P networks. We describe each of these evaluations that we carried out, report on their key results, and discuss implications for micropayment usage in content sharing application domains.

OVERVIEW OF P2P-NETPAY

In this section we describe the main characteristics of our P2P-Netpay micro-payment protocol. We outline the key aspects of its architecture and discuss our prototype implementation.

NetPay Micropayment Protocol for use in Client-Server Networks

We developed a protocol called NetPay that provides a secure, cheap, widely available, and debit-based protocol for an off-line micro-payment system (Dai & Grundy, 2007). We developed

NetPay-based systems for client-server-based broker, vendor and customer networks. We have also designed three kinds of "e-wallets" to manage e-coins in our client-server-based NetPay micropayment systems. In the most common model the E-wallet is hosted by vendor servers. This e-wallet is passed from vendor to vendor as the customer moves from one site to another during e-commerce transactions. The second model we developed is a stand-alone client-side application resident on the client's PC. A third model we developed is a hybrid that caches E-coins in a web browser cookie for debiting as the customer spends at a site during e-commerce transactions.

The client-side e-wallet is a stand-alone application that runs on a client PC that holds e-coin information. Customers can buy article content using the client-side e-wallet at different sites without the need to log in after the e-wallet application is downloaded to their PC. Their e-coins are resident on their own PC and so access to them is never lost due to network outages to one vendor. The e-coin debiting time is slower for a client-side e-wallet than the server-side e-wallet due to the extra communication between vendor application server and customer PC's e-wallet application (Dai & Grundy, 2007). In a client-side e-wallet NetPay system, a Touchstone and an Index (T&I) of a customer's e-wallet are passed from the broker to each vendor. We designed that the broker application server communicates with vendor application servers to get the T&I to verify e-coins. The vendor application servers also communicate with another vendor application server to pass the T&I, without use of the broker. The main problem with this approach is that a vendor system cannot get the T&I if a previous vendor system goes down.

P2P-Netpay Micropayment Model

Using the client-side e-wallet NetPay protocol we adapted this for use as a P2P-NetPay protocol suitable for P2P-based network environments. P2P-

NetPay allows peer users to purchase information from peer-vendors on the web (Dai & Grundy, 2005). P2P-NetPay is a secure, cheap, widely available, and debit-based protocol. P2P-NetPay differs from previous protocols in the following aspects: P2P-NetPay uses touchstones signed by the broker and Index's signed by peer-vendors passed from peer-vendor to peer-vendor. The signed touchstone (T) is used for peer-vendor to verify the electronic currency. Paywords and a signed Index (I) is used to prevent double spending by peer-users and to resolve disputes between peer-vendors.

A P2P-Netpay micro-payment system comprises of peer-users, peer vendors and a broker. In our approach we make a fundamental assumption that the broker is honest and is trusted by both the peer users and peer vendors. Micro-payments only involve peer users and peer vendors. The broker is responsible for the registration of peers and for

crediting the peer vendors' account and debiting the peer users' account. Figure 1 outlines some of the key P2P-Netpay system interactions.

Initially a peer-user accesses the broker/Central Index Server (CIS) web site to register and buy a number of e-coins from the broker/CIS (1). The broker may provide credit as "virtual money" i.e. credit specific to this network only, or the P2P network may require peers to use real money to subscribe and/or to make use of the service. In this case, the broker uses a macro-payment e.g. credit card transaction with a conventional payment party to buy credit (2). The broker/CIS sends an "e-wallet" that includes the e-coin chain to the peer-user (3). When the peer-user selects content to download from peer-vendor1 site (4), the user's e-wallet sends e-coins to the peer-vendor1 (5). Then peer-vendor1 gets T & I from the broker and verifies the e-coins (6). The peer-user downloads content from the peer-vendor1 (7). The

Figure 1. P2P-Netpay component interaction

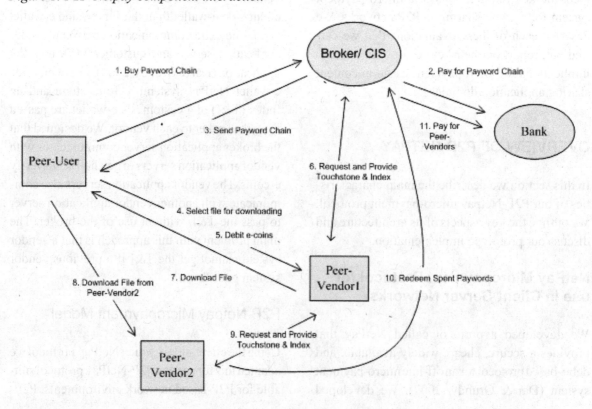

peer-user may download other content and their coins are debited. Different content may cost different amounts of e-coins and multiple denominations of e-coins are possible in a system. If coins run out the peer-user is directed to the broker/CIS's site to buy more. When the peer-user changes to a peer-vendor2 (8), peer-vendor2 contacts peer-vendor1 to get the T&I and then debits e-coins for further file downloading (9). At the end of each day, the peer-vendors send all the spent e-coins to the broker/CIS redeeming them (11) for their own credit to spend in the P2P network. In some P2P networks, peers may be able to cash in their credit for real money, again via a conventional macro-payment approach (12).

A peer user downloads a file from the peer-vendor1. The peer-vendor1 requests touchstone and index from broker/CIS and after verification, it allows the peer-user to download file. The peer-vendor1 sends the T&I to the broker/CIS. After browsing other peer-vendors, the peer-user requests for file download from the peer-vendor2 which contacts the peer-vendor1 for T&I. If the peer-vendor1 is offline then the peer-vendor2 requests the T&I from broker/CIS.

Software Architecture

We have developed a software architecture for implementing CORBA and socket-based P2P-Netpay micro-payment systems for content sharing in peer-to-peer networks. The interaction between the Broker/Central Index Server (CIS), Peer User and Peer Vendor is illustrated in Figure 2.

CORBA is a middleware platform standard developed by the Object Management Group (OMG) to support distributed object computing [13]. It uses an Object Request Broker (ORB) as an intermediary to handle distributed client and server object requests in a system. CORBA separates component interfaces from their implementation. It provides infrastructure for the programming of distributed systems using C, C++,

Smalltalk, Ada and Java. The ORB forwards operations on objects to the desired object and returns results to the client. We use such inter-ORB interactions authorize a peer user to communicate with peer vendor or peer user/vendor to communicate with Broker or vice versa.

The CORBA Interface Definition Language (IDL) describes the objects together with their methods and attributes. This is independent of the languages in which these interfaces will actually be implemented (e.g. C++, Java, etc). There are two IDL files used in P2P-Netpay, one for definition of Broker/CIS operations and other for peer operations. Peers use the Broker/CIS operations to buy e-coins, request Touchstones and redeem e-coins for real money. The peer operations IDL is used for downloading files from peers. The Broker/CIS operations IDL is compiled to generate interface code for both the peer and the Broker/CIS. The generated code for peers is in the form of object stubs. From the peer's perspective these stubs are function calls that go directly into the target object. In actuality the stubs forward the request to the remote object via the ORB. The compiler also generates skeleton code for the Broker/CIS (typically referred to as a servant in CORBA). This is fleshed-out with the implementation of the requested operations. The same happens for peer operations IDL file.

The peer user registers with the Broker/CIS website using Java Server Pages (JSP) in a web server layer. The JSP communicates with the application server layer using CORBA to register the user and to allow content downloads and uploads. When buying a file, e-coins are transferred using CORBA or a Socket from the peer user to Peer-Vendor1. This in turn requests a Touchstone from either Peer-Vendor2 or Broker/CIS depending on where the peer user has spent their last e-coin before this transaction.

The P2P-Netpay Broker/CIS system is built on top of a multi-tier web-based architecture as follows:

Figure 2. P2P-Netpay interactions

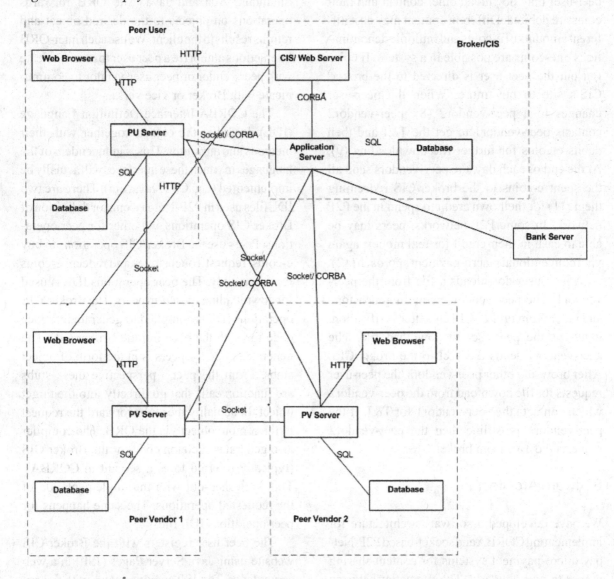

- **Client tier** (HTML Browser): The browser communicates with the web server which runs the JSPs to register peers.
- **Web tier** (Broker/CIS Web Server and JSPs): In the web tier, Java Server Pages (JSPs) and JavaBeans are used to service the web browser clients, process requests from the clients and to generate dynamic content for them. After receiving a client request, the JSPs request information via a JavaBean. The JavaBean can in turn

request information from an application server (CORBA). Once the JavaBean generates content the JSPs query and display the Bean's content. The Broker/CIS keeps track of online peers. It also stores the file names with the host and port of peers. The Broker/CIS is designed using CORBA in Java so that it can serve multiple clients at one time. The use of CORBA allows it to be language-independent. Thus non-Java peer to peer systems e.g. implemented in C# or C++ can be integrated using the same IDL-generated stubs as our Java implementation.

- **Application Server tier** (CORBA): In the P2P-Netpay Broker/CIS, CORBA is used as the middleware for the application server. In our prototype we have implemented the Broker/CIS functionality using the Java language using a CORBA IDL mapping.

- **Database Server tier:** On the back-end of the system our prototype uses a MS Access database to implement the Broker/CIS databases. The Access database is accessed using a Java Database Connectivity (JDBC) interface. JDBC, which is a multi-database application programming interface, provides Java applications with a way to connect to and use relational databases. When a Java application interacts with a database, JDBC can be used to open a connection to the database and SQL queries are sent to the database.

There are two types of the peer system that implement a three-tier architecture using CORBA or Sockets in the second tier:

- **Client tier** (Client Application): This client application communicates with the peer server to get requests from other connected peers or the Broker/CIS.

- **Peer Server tier** (Requesting peer server/ Supplying peer server): This is implemented in Java to handle the functionalities such as sharing files, communicating with server, checking balances, redeeming e-coins, searching the Broker/CIS and browsing peers.
  1. This server uses a CORBA IDL mapping for communication between peers.
  2. This server can also use sockets to listen on a port for requests from a peer or the CIS/Broker.

- **Database Server tier:** We use Ms Access to implement the database accessed via JDBC. Only the Java application can interact with the database. Peers cannot open the database manually and edit any data since this database is password protected.

CORBA-Based P2P-Netpay Implementation

The three main functionalities of P2P-Netpay are buying e-coins, downloading content and redeeming e-coins (Chaudhary & Dai, 2010).

Buy E-coins: Figure 3 shows how a peer buys e-coin in the P2P-Netpay system. The peer checks the amount left in the e-wallet and if wishing to buy e-coin, peer enters the amount and clicks buy button. The client application requests the e-coins through CORBA to Broker/CIS application server. The Broker/CIS application server debits from the peer's credit card, stores e-coins in the database and sends an e-wallet to java application (client application) in peer's computer.

Downloading a File: Figure 4 and 5 shows how a peer-user downloads a file using the P2P-Netpay micro-payment system. After browsing peers or searching the Broker/CIS, the peer-user clicks on a download popup menu on the title of the file name. The client application sends the request including file name and e-coins through the CORBA interface to the peer-vendor server (PVS) of PeerVendor1. If the touchstone and index (TandI) does not exist in its database, the server contacts Broker/CIS in order to obtain the required touchstone and index as shown in Figure 4. If the e-coins are valid, the PVS stores them in its redeem database and sends the file to the peer-user. The peer-user server (PUS) than debits the e-coin(s) spent to obtain this file. If the PUS wishes to buy file from peer-vendor2, the client application sends the request including file name and e-coins through the CORBA interface to the peer-vendor server of PeerVendor2. In order for peer-vendor2 to verify e-coin sent by PUS, peer-

Figure 3. Buy e-coin sequence diagram

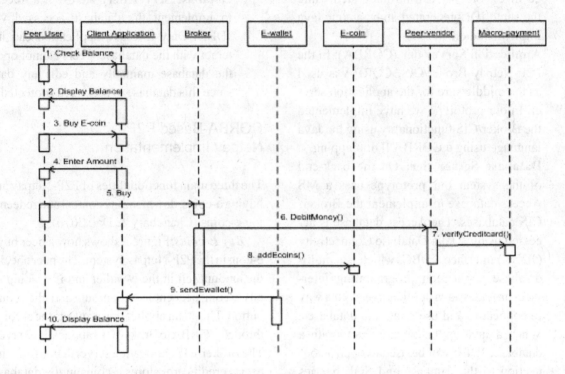

Figure 4. Download file sequence diagram (peer-vendor1)

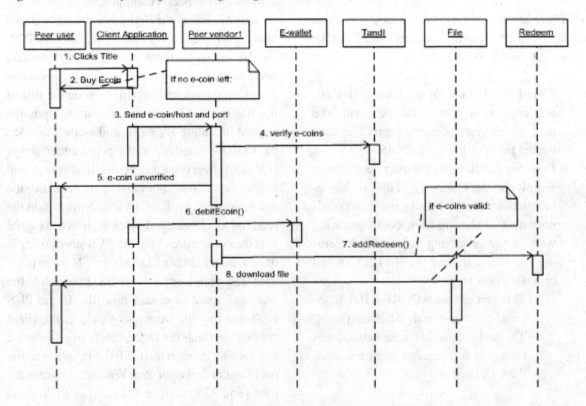

vendor2 needs to contact the peer-vendor1 for TandI. Once peer-vendor2 receives the TandI, e-coin is verified and if succeeded, file is sent to peer-user. This is illustrated in Figure 5.

Redeem Spending:Figure 6 illustrates how a peer-vendor redeems spent e-coins with the Broker/CIS. When redeeming spent e-coins, the peer-user clicks on the redeem button on the client application. The PVS aggregates all payments received so far and sends them to the Broker/CIS application server using the CORBA interface. If all redeem data are valid, the broker/CIS system generates a unique history_id and records the transaction history data. The Broker/CIS application server verifies the spent e-coins and sends the balance for this credit to the redeeming peer-user.

The basic user interfaces for these tasks are illustrated in Figure 7. A peer can be a user or a vendor. To use the services of P2P-Netpay, a peer must be registered with the broker and download the peer-to-peer content sharing application. A peer must remember the Peer ID generated by the system to login as described in Figure 7 (1). For the initial login the peer server connects to the Broker/CIS server to get the Peer ID and password using CORBA. After this it records these in its local application database. If the login was successful, the main interface of the application appears as shown in Figure 7 (3). The IP address and port is listed for the peers who are currently online. The IP and port is not that which a peer is listening to, as in socket communications, but it is the *RemotePeerManagerServer* which provides peers with a mechanism to invoke methods remotely. CORBA has its own Object Request Broker server(s) running to which peers can connect and use these services. There is a range of functionality provided such as checking balances, searching for files, uploading files and redeeming e-coins. A peer user can upload file for sharing as well (4). The file remains on the peer user computer, only the file name, file path and the cost is stored in the peer user database. The file

name and cost of the file is also supplied to CIS for storage.

For example, suppose a peer user would like to search for a file named "b". The user enters (part of) the file name and the search results are displayed (5). The results are obtained from the Broker/CIS application server. The application will only display results from the peers who are currently online and available for download. To download a file, the client application user right clicks on the desired file name and selects download from a popup menu (6). The application will connect to the peer hosting that particular file using CORBA and the download starts after e-coin verification.

EXPERIMENTAL DESIGN

We have carried out three evaluations of our P2P-Netpay micro-payment system to determine its suitability for providing credit support in peer-to-peer networks to discourage or prevent free riding. These included:

- A *usability evaluation* – this surveyed representative target end users of our prototype in order to assess their opinions about our approach when carrying out file downloading tasks using micro-payment, P2P-Netpay, and an alternative non-micro-payment file sharing application;

- A *heuristic evaluation* – this was used to gauge potential usability problems regarding the user interface design of our P2P-Netpay prototype using a range of common HCI design heuristics;

- A *performance evaluation* – this was used to gauge the likely performance of our P2P-Netpay prototype against a non-micro-payment file sharing application in regards to user response time to assess its potential scalability under heavy loading conditions.

Figure 5. Download file sequence diagram (peer-vendor2)

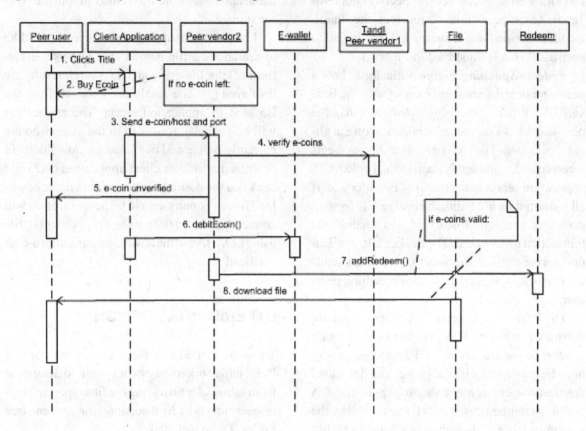

Figure 6. Redeem spending sequence diagram

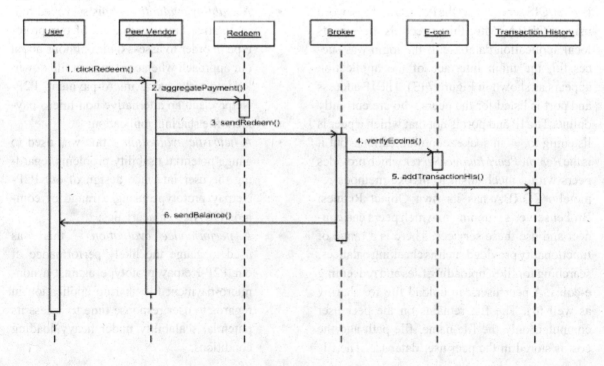

Figure 7. P2P-Netpay user interfaces

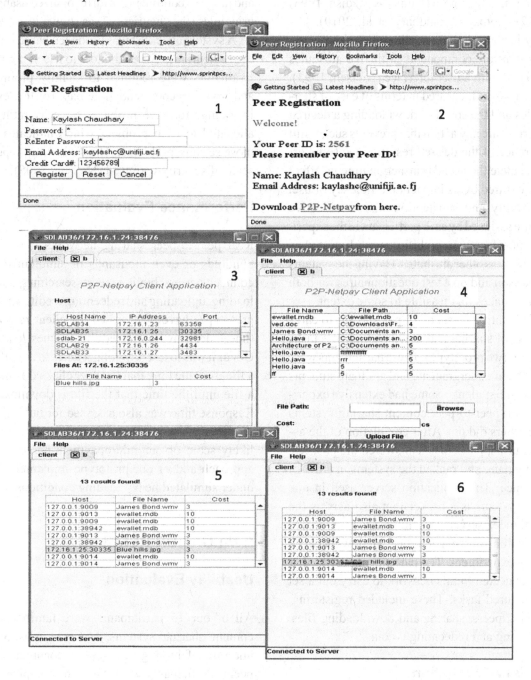

We outline the approach taken for each of these evaluations, report on the results of each experiment and draw some conclusions from these about the suitability of P2P-Netpay for credit management in a peer-to-peer environment.

Usability Evaluation

We evaluated participants' user satisfaction when sharing files, downloading files and general preferences about two systems – a non-micro-payment file sharing application (Dumas & Redish, 1993) and P2P-Netpay (Chaudhary et al, 2010). Usability is not a single, one dimensional property. It has multiple components with five attributes associated with a user interface (Nielsen, 1993). Efficiency was measured in terms of ease of sharing files and the speed of downloading a desired file. Errors are any action that prevents successful occurrence of the desired result. Since some errors escalate the users' transaction time its effect was measured by its impact on efficiency of use. Learnability and satisfaction was a subjective measure assigned by each participant in the experiment. Interface memorability is rarely tested as thoroughly as other attributes. Having the systems comparison and post test questionnaires for both systems makes this feasible to some extent.

We identified a set of 15 participants to carry out a set of sharing and downloading files tasks using our two prototypes. These participants were drawn from undergraduate and graduate students and non-IT students. Some had extensive experience with peer-to-peer content sharing systems while others did not. After completion of the assigned tasks, participants answered the post-test questionnaire and ranked the systems in order of preference. The application server used in one of the systems is a broker that also acts as the central index server. This application server was deployed on a host machine on a windows network for this experiment. The participants used other PCs connected to this network to carry out a set of structured tasks. These included registering, browsing peers, sharing and downloading files, and buying and redeeming e-coins.

Heuristic Evaluation

Heuristic evaluation is a usability engineering method for locating usability problems in a user interface design so that they can be attended to as part of an iterative design process (Nielsen, 1994). Based on Nielsen's heuristics (Nielsen, 1994), we chose 5 expert evaluators to examine the interface

and judge its compliance with recognized usability principles (the "heuristics") as shown in Table 1.

A system checklist was produced based on the above heuristics for evaluators to use as a guide. Evaluators were required to identify problems and provide recommendations based on the severity ratings for each problem. Severity rating is allocated to each problem which indicates the most serious problems. The following 5 point scale of severity was used (Nielsen, 2005):

Performance Evaluation

Real peer-to-peer sharing networks may have hundreds or even thousands of simultaneously connected users, buying credit, searching, downloading, uploading and redeeming credit. Our two prototypes have been tested for client response time under heavy simulated client request loading. Our aim was to test how long it takes to download a file from the time that the peer clicks the title of a file until the time that the file is downloaded. Response time was also assessed for buying and redeeming e-coins. This gives an indication of likely scale of our micropayment-based credit approach and of our prototype implementation under simulated heavy loading conditions.

EVALUATION RESULTS

Usability Evaluation

All of our 15 participants were familiar with content sharing systems, however most were unaware of buying and selling documents for peers. Participants were asked to complete the following tasks for the File Sharing Application without micro-payment:

- Download and install the File Sharing software
- Connect to the central index server
- Upload some files for sharing

Table 1. Usability principles (heuristics)

No.	Heuristics
1	Visibility of system status
2	Match between system and the real world
3	User control and freedom
4	Consistency and standards
5	Help users recognize, diagnose and recover from errors
6	Error prevention
7	Recognition rather than recall
8	Flexibility and minimalist design
9	Aesthetic and minimalist design
10	Help and documentation
11	Skills
12	Pleasurable and respectful interaction with the user
13	Privacy

Table 2. Severity rating

Scale	Description
0	I don't agree that this is a usability problem at all
1	Cosmetic problem only: need not be fixed unless extra time is available on project
2	Minor usability problem: fixing this should be given low priority
3	Major usability problem: important to fix, so should be given high priority
4	Usability catastrophe: imperative to fix this before product can be released

- Download two files from a peer who is online by browsing the files that particular peer has shared
- Download another file from a peer by searching the central index server

The following comparable tasks were performed by participants using our prototype P2P-Netpay system:

- Register and download P2P-Netpay software with the broker

- Install the software, login and connect to the central index server
- Buy e-coins from the broker
- Upload some files for sharing, gaining e-coin credits
- Download two files from a peer who is online by browsing the files that particular peer has shared
- Download another file from a peer by searching the central index server
- If e-coins run out, the user must either buy more e-coins from broker or upload more files for e-coin credits
- Redeem their e-coins with the broker

Our post-test questionnaire consisted of a 5-point rating scale to gauge each characteristic of both applications for their comparable features. The rating scale ranged from 1 to 5 where 1 is "Strongly Disagree" and 5 is "Strongly Agree". There were also open questions to gain further end user feedback. The bar chart shown in Figure 8 presents the average ratings for the tested features a, b and c. Figure 9 shows the number of participants out of 15 who preferred features d, e and f. The tested features were:

a. **Ease of use:** The applications are easy to use.
b. **Efficiency 1:** It is easy to share files using these systems.
c. **Efficiency 2:** The speed of downloading files is fast enough.
d. **Preference 1:** You preferred to use this system.
e. **Preference 2:** You preferred to upload files.
f. **Preference 3:** You preferred to download files.

Ease of use and Efficiency, which is mainly to do with the sharing of files, favoured the P2P-Netpay system. Participants mentioned that the speed of downloading files is better with the File Sharing Application without micro-payment

Figure 8. Usability test results on efficiency

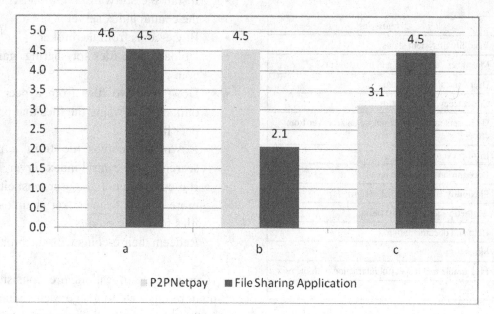

credits. This was essentially due to the way micro-payments in P2P-Netpay are actioned. Whenever a client requests that they download a file, the peer user sends the name of file, e-coins and port of host that has the index of the e-coins to the peer vendor. The host can be anyone, either the broker or another peer vendor. In both cases for the first download the peer vendor has to contact the host and request the index and touchstone of the e-coin to be spent, in order to validate it. Upon verification, the peer vendor than allows the peer user to download one or more files. Ease of use was almost the same but there was a vast difference in sharing files. In the feedback in the open questionaries, participants noted that it's better to share files in P2P-Netpay because it avoids free-riding and at the same time there is a gain in terms of credits to be spent elsewhere.

Participants preferred to use P2P-Netpay for sharing files and downloading files as shown in Figure 7. Though the speed of downloading a file is slower it encourages all peers to share files. If peers do not share files that others can download file then they themselves can not do downloads after their initial credit is used.

Heuristic Evaluation

Five evaluators evaluated the P2P-Netpay application with a set of heuristics. Three evaluators were IT specialists (one was software developer and others were network and server administrators) while two were graduate students majoring in other disciplines (accounting and mathematics student). The evaluators were asked to carry out this task independently so that a clear reflection of an evaluator can be captured. Different evaluators were selected because there will be different types of users with different area of expertise. Thus it was indeed necessary to select different evaluators. IT background evaluators can point out usability problem related to IT whereas non-IT background evaluator can give usability problems based on the usage of the software. Table 3 presents the results of these heuristic evaluations. It describes problems raised by the different evaluators with a summary of the heuristics violated and a severity rating out of 5. For example, three evaluators raised problem No.1 that violated heuristic No. 3 (User control and freedom) with severity of 2

Figure 9. Usability test results on preference

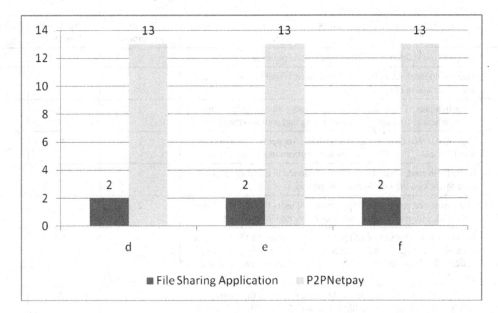

(Minor usability problem: fixing this should be given low priority).

There were four problems identified with a severity rating of three, which signifies that the problem is of high priority and it is important to fix. Other problems had a severity rating of one and two. These were minor problems identified. Table 4 discusses some recommendations to address the various problems with severity rating of three. All of the problems found by the evaluators were subsequently resolved in our prototype implementation.

Performance Evaluation

The performance results of downloading files with both systems, P2P-Netpay and the File Sharing Application without micro-payment, are shown in Table 5. The response delay time measures how long it takes for a file to be downloaded. The file was a picture and had a size of 27.8KB. All the ten tests download the same file. These tests were taken under a heavy concurrent load of forty other peers doing downloads at the same time. The client and server machines had WindowsXP and CPU speed was 3.2 GHz. The network speed was 100Mbs (LAN).

These results illustrate that when many simultaneous requests are made to peers or to the broker, it takes on average 2450.1ms to download a small file. The File Sharing Application without Micro-payment took 1994.7ms for the same file on average. There is a difference of 455.4ms on average. This is due to P2P-NetPay requesting an index/touchstone and e-coin verification. Several other timings were recorded and these are summarised in Table 6.

Browsing time was the same for both the systems. This is because browsing does not involve micro-payment. Redeeming e-coins takes on average 1867 ms and this is dependent on the amount and searching record in database in broker.

We have described above three kinds of experiments we have carried out on our P2P-Netpay prototype to assess usability, performance evaluation and carry out a heuristic evaluation of its interface. Usability and the performance evaluation were based on two prototypes which are P2P-Netpay and a content sharing application

Table 3. Results of heuristic evaluation

Problem No.	Issues	No. of Evaluators	Heuristics Violated	Severity
1	No keyboard shortcut key. What if there is no pointing device?	3	3	2
2.	There is no feedback to indicate that the users have run out of e-coins and files could not be downloaded.	2	1	2
3	No help topics. If users are confused or unsure about a menu/command, where to find information regarding that menu/command?	2	10	3
4	The login screen displays "Customer ID:". Are peers customers? It should display "Peer ID:" because this application is about Peers.	1	4	2
5	In the login screen, the cursor is not positioned in the id field; peers have to click on the field to enter the peer id.	3	8,11	2
6	In the main screen of P2P-Netpay, how will users know that doing a right click will give a popup menu for browsing/downloading?	2	1	2
7	In the file upload screen, when peers want to remove a file from sharing and if the system is not connected to server, there is no message to indicate that file could not be removed. Users are not aware what the problem is.	3	5	3
8	The upload command button should check whether a file exists or not. It should not upload the file unless and until the file path/ cost is correct.	2	5,6	3
9	The cost associated with file name doesn't specify the currency. E.g. distinguish between dollars and cents.	4	2	2
10	When peers redeem e-coins for real money, the window that shows the amount redeemed is not in center as other windows.	3	3	1
11	There is no clear command button in upload file screen to clear text in the text field.	2	3	3
12	There is no title (not window title) in the upload file screen to show that this screen is for uploading files.	2	3	2
13	No title in the main screen to indicate that this screen is for browsing peers and downloading file.	2	3	2
14	Menu item and command button have same name "Upload File" but there functionalities are different.	1	4	2

without a micro-payment system. The results mainly favor P2P-Netpay (usability) and show P2P-NetPay does not place undue scalability and performance problems on the peer-to-peer application (performance). Users were very satisfied with using of P2P-Netpay and they encouraged it for wide-spread use in file sharing applications to address the free rider problem. Through our user interface heuristic evaluation a number of problems were identified in our first prototype and these have been addressed.

DISCUSSION

We compare the features of our prototype P2P-Netpay protocol with other micro-payment protocols. Different architectures and associated middleware have been investigated for P2P-Netpay peers (Chaudhary & Dai, 2010). These include CORBA, sockets, Web Services and Java RMI. Each architecture and middleware have certain advantages and disadvantages. The releative advantages and disadvantages are in the context of P2P-Netpay and not in general. Currently, Web Services are another option for realising distrib-

Table 4. Recommendations to correct identified problems

Problem No.	Recommendation
3	Implement help topics as users may not be aware of the function of menu or command buttons.
7	Error messages should be improved to indicate that a file can not be removed with an appropriate reason.
8	Fields should be checked before passing information to the database. Apply error checking to fields and, if there is an error, display a pertinent explanation for the user.
11	The "Clear" command button should be implemented. This button should clear the text fields. Suppose the user has entered all the details in the text field and then the user decides not to share that file. Currently there are two options: either manually delete the information or click on "Upload" command button to clear the field. Neither of the two choices is relevant if the user has entered the information and then decides not to share.

uted computing infrastructure (Barai & Caselli, 2008). CORBA and web services are similar in that both support multiple programming languages and both are platform-indepdent middleware. CORBA interoperates with other middleware, is highly flexible and supports Remote Procedure Call and message-passing paradigms (Baker, 2010). Web services are a more recent development with advantages of generic XML payloads and greater dynamic discovery support. Web Services are designed for operation with heterogeneous connections at both ends. They also use XML schemas to define both interfaces and the format of payload messages (Baker, 2010). Table 7 summarizes our architecture comparison using the five criteria described below:

- *Easy to use*: less coding required in achieving our goals.
- *Programming language*: Some architectures support multiple programming languages while others do not. The benefit is that different systems implemented via different programming languages can communicate with each other. Why do we want implementation using different languages and not the same one? This is to suit other users expertise with coding using a particular language and legacy systems integration. It also allows extending and integrating the P2P system with other components e.g. social networking systems to rate and share content etc.

Table 5. Results of downloading file

Test	Response delay time with P2P-Netpay (ms)	Response delay time with File Sharing Application without Micro-payment(ms)
1	4018	3960
2	4002	1637
3	2281	2232
4	2437	2386
5	1753	1669
6	2007	1967
7	1950	1669
8	1867	1372
9	2094	1377
10	2092	1678
Average	**2450.1**	**1994.7**

Table 6. Results of browsing, buying and redeeming e-coins

	Average response delay time for P2P-Netpay (ms)	Average response delay time with File Sharing Application without Micro-payment (ms)
Browse peers	171	171
Buy E-coins	126	-
Redeem E-coins	1867	-

- *Platform*: Many architectures do not support multiple platforms and clients and servers need to be run on specific kinds of hosts. There may be a situation where a server may be running on certain types of machines and different clients on others. To enable these clients and servers to communicate, the chosen architecture should ideally be platform-independent.
- *File downloading*: large files may need to be broken into parts and then sent to peers for reassembling. File downloads may need to be paused and resumed or restarted. Some architectures enable this more easily than others.
- *Processing time*: the time taken to process remote requests. Benchmarked processing times are summarized in (Eggen & Eggen, 2001) and the results presented in the following table for time is based on (Eggen & Eggen, 2001).

The above comparison shows that the chosen CORBA architecture is recommended for P2P-Netpay. Sockets are simple to understand and program. Peer/server must listen to a port in order to communicate and have to cater for each message sent and received. Sometimes the connection times out which aborts downloading requiring establishing connection again. Web services, on the other hand, is a new technology which uses XML to define messages sent to peers by Broker/CIS. Only the Broker/CIS can be implemented as Web service which has a three-tier architecture when compared to multi-tier architecture for CORBA.

RMI is very similar to CORBA but it doesn't support multiple programming languages and platforms. The processing time for CORBA is higher when compared to RMI and sockets, as shown in (Eggen & Eggen, 2001). This is because the number of servers increase the processing time increases. In P2P-Netpay, only two servers will communicate so the processing time will be equivalent to socket and RMI.

P2P Micro-Payment Systems Comparison

We compare P2P-Netpay's characteristics to several well-known micro-payment systems. We also compare it to some more recent micro-payment systems in the peer-to-peer networks domain. The comparison criteria we use below are based on the set of key requirements outlined in Section 2: a need for an easy-to-use micro-payment system; a need for secure electronic coins and no double-spending; ensuring anonymity for customers; supporting transferable e-coins between vendors; and a robust, low performance impact, off-line micro-payment supported and scalable architecture for a very large number of peer end users. The comparison we use here is for the scenario of a peer customer (PC) downloading various content from peer vendors (PV), and micro-payment brokers (PB). Table 8 summarises this comparison of our P2P-NetPay protocol with these other systems.

XPAY provides offline verification with anonymity. PPay (Yang & Garcia, 2003) and Whopay (Wei et al., 2006) have a peer downtime protocol which is almost an online micro-payment system. In PPay, the use of layered coins introduces a

Table 7. P2P-Netpay architecture comparison

Property	Socket	Web Services	RMI	CORBA
Easy to use	**Low**, more code is required to handle a series of messages for downloading a file	**Medium-to-High**, messages are sent as SOAP; code is required to build and read SOAP messages and format payloads	**High**, most generated and less to code	**High**, functionality is achieved with less coding
Programming language	**No** direct support for multiple programming languages	**High**, supports multiple programming languages	**No** support for multiple programming languages	**High**, implementation for either Broker/CIS or peer may differ. Broker/CIS may be implemented in e.g. Java whereas client can be implemented as C++
Platform	**No** support for multiple platform – each implementation is platform-specific e.g. Windows vs LINUX	**Medium** – web services can run on different platforms but each implementation may be platform-specific. Needs web service-implementing server on every machine. This is challenging and opens up security challenges and problems.	**Medium** – clients and servers run on any Java-providing platform	**High**, currently P2P-Netpay has only windows based application; there is a possibility for different platform applications.
File downloading	**Low**, breaking down of file into parts and sent as series of messages. This may result in a download being aborted.Must code support in for restart and resume.	**Medium** - uses HTTP for communication – files may be split or sent as one large message.	**High**, a remote method is invoked – no breaking down of a file into parts.	**High**, a remote method is invoked – no breaking down of a file into parts.
Processing time	**Less** processing time	**Medium**	**Less** processing time	**Medium**, since it supports distributed system.

Table 8. Comparison of P2P micro-payment methods

System/ property	CPay	PPay	WhoPay	Floodgate	Huang	XPAY	P2P-NetPay
Security	High,	Medium,	High	Medium	High	Medium	Medium+
Anonymity	High	Low, Peers anonymity not supported	High	Low, Peers anonymity not supported	High	High	High
Transferability	**High**, The recipient of a coin can spend with other peers **through BAs**	**High**, The recipient of a coin can spend with other peers by using **layered coins**	**High**, The recipient of a coin can spend with other peers by using **public key operation**	**Medium**, the tokens can be spent to many peers with **the content provider and the trackers**	Low, scrip is peer-specific and can't be used with other peer	Low, scrip is peer-specific and can't be used with other peer	**Medium**, an e-coin chain of peer-user can be spent at many peer-vendors
Low-performance impact and robust	**Offline** for broker but BA peers are **almost Online**	**Online** down-time protocol causes **delay transactions**.	**Online** downtime protocol use of **public key operation** on every transaction.	The content provider and the trackers are **Online**.	The trusted peer & account holders are **Online**	**Offline** for bank, peers need to communicate with each other	**Offline** for broker, requesting peers only communicate with provider peers

delay in terms of fraud detection and the floating coins growing in size which creates a scalability issue. Whopay uses the expensive public key operation in every transaction in the downtime protocol. There are many BA peers must be online in every transaction in CPay. With P2P-Netpay downtime protocol, a PV contacts with PB in the first transaction with a PU to get T&I of a PU if a previous PV is not online. Floodgate and Huang's protocol are online systems with trackers or trusted peers.

Transferability is an important criterion which improves anonymity and performance of the P2P systems. The e-coin chain in our P2P-Netpay protocol is transferable between PVs to enable PUs to spend e-coins in the same coin chain to make numbers of small payments to multiple PVs. P2P-Netpay supports transferability between PUs without extra actions on the part of the PU and the PB. The e-coin chain in Xpay is not transferable. CPay, PPay, and WhoPay micro-payment protocols provide the transferability (2) that a peer's recipient coin can be spend to other peers similar with a real coin. However, they introduce scalability and performance problems in order to support the transferability (2). The e-coin chain in P2P-NetPay protocol is transferable between PVs to enable PUs to spend e-coins in the same coin chain to make numbers of small payments to multiple PVs. P2P-Netpay supports transferability (1) between PVs without extra actions on the part of the PU.

The aim of security in the payment protocols is to prevent any party from cheating the system. For peers, cheating security is specific to the payment scheme such as double spending coins and creating false coins i.e. forgery during payment. In CPay, double spending is detected timely while in PPay floating coins introduces delay in fraud detection. The security in Whopay, Huang's protocol is high. P2P-Netpay prevents double spending by using touchstones which are broker-generated and moderated.

CONCLUSION

Many file sharing systems suffer from a problem of many non-contributors. We have developed an approach to support efficient, secure and anonymous micro-payment for file sharing systems to encourage – or require – users to contribute more equitably. This incorporates a broker used to generate, verify and redeem e-coins, a peer e-wallet stored on peer machine and peer application server components. Our P2P-Netpay architecture provides for both secure and high transaction volume per item by using fast hashing functions to validate e-coin unspent indexes. P2P-Netpay is an offline protocol. The two evaluations (usability and performance) mainly favoured the P2P-Netpay-based system. Users were satisfied with their use of P2P-Netpay and they indicated they would adopt it for widespread content sharing use. Through our heuristic evaluation a set of problems was found with the current interface as it has been implemented.

We are investigating the use of XML-based interaction between peers and the broker using web services and ways to augment existing content sharing applications with P2P-Netpay support. This will allow us to conduct trials of the approach with much larger networks to gauge its wider impact on sharing behaviour. To date we have only prototyped the protocol and investigated its feasibilities for some cases. We need to find out the ways to overcome the need for downloadable clients that are installed on a peer's computer. We would like to investigate a web-based peer-to-peer file sharing service using our P2P-Netpay.

REFERENCES

Anderson, R., Manifavas, C., & Sutherland, C. (1996). Netcard - A practical electronic cash system. In M. Lomas (Ed.), *Proceedings of 1996 International Workshop on Security Protocols, Berlin, Germany, Lecture Notes in Computer Science, 1189,* (pp. 49-57). Berlin, Germany: Springer Verlag.

Baker, S. (2010). Web services and CORBA. *Lecture Notes in Computer Science, 2519,* 618–632. doi:10.1007/3-540-36124-3_42

Barai, M., & Caselli, V. (2008). *Service oriented architecture with Java.* Packt Publishing Ltd.

Chaudhary, K., & Dai, X. (2010). Architecture of CORBA based P2P-Netpay micro-payment system in peer-to-peer networks. *International Journal of Computer Science & Emerging Technologies, 1*(4), 208–217.

Chen, Y., Sion, R., & Carbunar, B. (2009). XPay: Practical anonymous payments for tor routing and other networked service. *WPES '09: Proceedings of the 8th ACM Workshop on Privacy in the Electronic Society,* (pp. 41-50). ACM Press.

Dai, X., Chaudhary, K., & Grundy, J. (2007). Comparing and contrasting micro-payment models for content sharing in P2P networks. *Third, International IEEE Conference on Signal-Image technologies and Internet-Based System (SITIS'07) China, 16 - 19 December 2007,* (pp. 347-354). IEEE Computer Society.

Dai, X., & Grundy, J. (2005). Off-line micro-payment system for content sharing in P2P networks. *2nd International Conference on Distributed Computing & Internet Technology (ICDCIT 2005) December 22-24, 2005, Lecture Notes in Computer Science, vol. 3816,* (pp. 297 –307).

Dai, X., & Grundy, J. (2007). NetPay: An off-line, decentralized micro-payment system for thin-client applications. *Electronic Commerce Research and Applications, 6*(1), 91–101. doi:10.1016/j.elerap.2005.10.009

Daras, P., Palaka, D., Giagourta, V., Bechtsis, D., Petridis, K., & Strintzis, G. M. (2003). *A novel peer-to-peer payment protocol.* IEEE EUROCON 2003, International Conference on Computer as a Tool (Invited paper), Ljubljana, Slovenia, September 2003.

Dumas, S. J., & Redish, J. C. (1993). *A practical guide to usability testing.* Norwood, MA: Ablex Publishing Corporation.

Eggen, R., & Eggen, M. (2001). *Efficiency of distributed parallel processing using Java RMI, Sockets, and CORBA.* Retrieved from www.imamu.edu.sa/dcontent/IT_Topics/java/paper3.pdf

Garcia, F. D., & Hoepman, J. H. (2005). *Off-line karma: A decentralized currency for static peer-to-peer and grid networks.* In 5th International Network Conference (INC2005), July 5-7 2005.

Glassman, S., Manasse, M., Abadi, M., Gauthier, P., & Sobalvarro, P. (1995). The millicent protocol for inexpensive electronic commerce. In *4th WWW Conference Proceedings,* (pp. 603-618). New York, NY: O'Reilly.

Hauser, R., Steiner, M., & Waidner, M. (1996). Micropayments based on IKP. *Proceedings of 14th Worldwide Congress on Computer and Communications Security Protection,* Paris-La Defense, France, 1996, C.N.I.T, (pp. 67- 82).

Huang, Q., & Zhao, Y. (2009). A secure and lightweight micro-payment scheme in P2P networks. *Proceedings of IEEE International Conference on Industrial and Information Systems (IIS '09),* (pp. 134 - 137).

Liebau, N., Heckmann, O., Kovacevic, A., Mauthe, A., & Steinmetz, R. (2006). Charging in peer-to-peer systems based on a token accounting system. *5th International Workshop on Internet Charging and QoS Technologies, LNCS 4033*, (pp. 49–60).

Nair, S. K., Zentveld, E., Crispo, B., & Tanenbaum, A. S. (2008), Floodgate: A micropayment incentivized P2P content delivery network. *Proceedings of 17th IEEE International Conference on Computer Communications and Networks, 2008,* (ICCCN '08), (pp. 1 – 7).

Nielsen, J. (1993). *Usability engineering.* Boston, MA: AP Professional.

Nielsen, J. (1994). Heuristic evaluation. In Nielsen, J., & Mack, R. L. (Eds.), *Usability inspection methods.* New York, NY: John Wiley & Sons.

Nielsen, J. (2005). Severity ratings for usability problems. Retrieved February 25, 2009, from http://www.useit.com/papers/heuristic/severity-rating.html

Pedersen, T. (1996). Electronic payments of small amounts. In M. Lomas (Ed.), *Proceedings of 1996 International Workshop on Security Protocols, Lecture Notes in Computer Science, vol. 1189,* (pp. 59 - 68). Berlin, Germany: Springer Verlag.

Shneidman, J., & Parkes, D. (2003). Rationality and self-interest in peer-to-peer networks. In *Proceedings of 2nd International Workshop on Peer-to-Peer Systems (IPTPS '03),* Berkeley, CA, USA, February 2003.

Vishnumurthy, V., Chandrakumar, S., & Sirer, E. G. (2003). KARMA: A secure economic framework for P2P resource sharing. *Proceedings of the First Workshop on Economics of Peer-to-Peer Systems (P2PEcon '03).*

Wei, K., Smith, A. J., Chen, Y. R., & Vo, B. (2006). WhoPay: A scalable and anonymous payment system for peer-to-peer environments. In *Proceedings of 26th IEEE Intl. Conf. on Distributed Computing Systems,* (p. 13). Los Alamitos, CA: IEEE Computer Society Press.

Yang, B., & Garcia-Molina, H. (2003). PPay: Micropayments for peer-to-peer systems. In *Proceedings of the 10th ACM Conference on Computer and Communication Security,* (pp. 300-310). ACM Press.

Zghaibeh, M., & Harmantzis, F. C. (2006). Lottery-based pricing scheme for peer-to-peer networks. *ICC APOS; 06. IEEE International Conference on Communications, 2006,* (vol. 2, pp. 903–908).

Zou, E. J., Si, T., Huang, L., & Dai, Y. (2005). A new micro-payment protocol based on P2P networks. *Proceedings of the 2005 IEEE International Conference on e-Business Engineering (ICEBE'05),* (pp. 449 – 455). IEEE Computer Society Press.

Chapter 10
Virtual Telemedicine and Virtual Telehealth:
A Natural Language Based Implementation to Address Time Constraint Problem

Shazia Kareem
The Islamia University of Bahawalpur, Pakistan

Imran Sarwar Bajwa
University of Birmingham, UK

ABSTRACT

Telemedicine is modern technology that is employed to provide low cost, high standard medical facilities to the people of remote areas. Store-and-Forward method of telemedicine suits more to the progressive countries like Pakistan as not only is it easy to set up but it also has a very cheap operating cost. However, the high response time taken by store & forward telemedicine becomes a critical factor in emergency cases, where each minute has a price. The response time factor can be overcome by using virtual telemedicine approach. In virtual telemedicine, a Clinical Decision Support System (CDSS) is deployed at rural station. The CDSS is intelligent enough to diagnose a patient's disease and prescribe proper medication. In case the CDSS cannot answer a query, the CDSS immediately sends an e-mail to a medical expert (doctor), and when the response is received, the CDSS knowledge-base is updated for future queries. In this chapter, the authors not only report a NL-based CDSS that can answer NL queries, but also present a complete architecture of a virtual telemedicine setup.

INTRODUCTION

In last few decades, the concept of e-Health has made easy to provide health facilities such as curative care and preventive care to the patients of remote areas such as rural areas (Kensaku, 2005). Two most common types of e-Health are telemedicine and telehealth. The telemedicine (Khalid, et al., 2008) has been emerged into one of the most pertinent applications of telecommunication and information technologies (Ackerman & Carft, 2002). Telemedicine has been proved to

DOI: 10.4018/978-1-4666-0023-2.ch010

be an effective and efficient solution to provide curative care to patients in the remote areas, where medical experts are not available in physical. Similarly, telehealth has been come into sight as an expansion of telemedicine as telehealth also provides the preventive care besides the curative care (Berner, 2007). In telehealth, patient's data is sent to the physicians or medical experts for diagnosis and medical prescription. Following are the details of both the technologies.

Telemedicine

In current times, one of the most developing applications in the field of clinical medicine is telemedicine (Puskin, et al., 2006). In the telemedicine, the modern technologies such as telecommunication & information technology are employed (Puskin, 1995) to deliver clinical care at remote areas where advanced health facilities like hospitals are not available (Perednia, 1995). With the help of telemedicine, the improved medical facilities are made available to the people of remote areas at low cost (Kensaku, 2005). The key attribute of the telemedicine based healthcare is easier and cost-effective solution for medical consultation as patient's medical information is transmitted through telephone, internet or satellite and a medical specialist can examine the patient's report, diagnose the diseases and prescribe medication (Ackerman & Carft, 2002).

Telehealth

In contrast with telemedicine, telehealth provides both curative and preventive care facilities. The real concept involved in telehealth is the providence of health care facilities to the people of remote areas (Puskin, et al., 2006). The successful application of telehealth in developed countries such as USA, UK, China, etc the telehealth technology has been proved to be an easier and cost-

effective solution for medical consultation. Using telehealth technology, a physician communicates with the patient using modern technologies such as e-mail and voice-chat, and video conferencing (Berner, 2007).

As telehealth has been used successfully in developed countries, telehealth can be a possible solution in Pakistan to provide better medical facilities in remote areas. Real-time telehealth is not feasible in Pakistan due to its high cost. However, the store and forward method based telehealth can be a cost-effect solution. In (Ackerman & Carft, 2002) a major problem with store-and-forward method is highlighted that time-factor can be very high in certain cases. To address this issue a concept of virtual telemedicine is proposed in (Bajwa, 2010) and an improved framework is presented in (Kensaku, 2005). Typically, telemedicine and telehealth basically work in two different ways:

Store-and-Forward Method

This is simple and cheapest way of using telehealth technology (Khalid, et al., 2008). In store-and-forward method of telehealth, a physician communicates with a patient using e-mail or fax facilities (Houston, 1999). A patient's medical report containing various bio-signals such as blood pressure, body temperature, weight, pulse rate, blood sugar level, etc are sent to the physician for diagnosis (Rashid, 2003). The physician may also ask for additional reports such as blood test, urine test, X-Rays, etc. All the patient reports are sent to the physician in the form of medical images (see Figure 1).

However, a problem with store-and-forward clinical method is the time factor involved that is typically from 1 minute to 48 hours (Perednia, 1995), depending upon the availability of the physician. The situation may become more verse if patient is in serious condition and requires immediate medication.

Figure 1. A typical telehealth care system providing primary health care, expert opinion and e-learning facilities

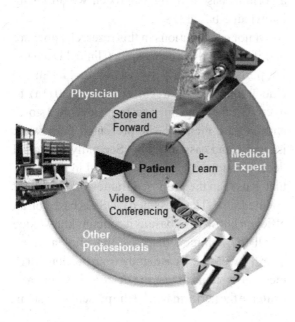

Real Time Method

In this method of telemedicine and telehealth, doctors monitor patient remotely using various modern technologies such as video conferencing (Ackerman & Carft, 2002). In the real time method based telemedicine is quite expensive for progressive countries like Pakistan due to high cost of hardware and communication channels for audio and video streaming involved in video conferencing (Perednia, 1995). There is online communication between doctor and patient such as video conferencing which is costly for the countries like Pakistan. High bandwidth is required for data transmission.

Telemedicine in Pakistan

Pakistan is a developing country with high population and low resources. Moreover, 60% population of Pakistan lives in rural and distant areas (W.H.O., 2010). However, this is not possible to provide high quality medical facilities in distant areas due to shortage of medical experts and financial resources. Health care report of 2005-2006 shows that there are 113,937 doctors (H.M., 2006) in Pakistan and doctor-people ratio is one doctor for 1400 people. Similarly, there are 968 hospitals and 4813 dispensaries (H.M., 2006) and one hospital/dispensary has to support 6000 people. In addition to scarcity of resources, the qualified doctors do not prefer to work in remote areas and such attitude of medical experts make this scenario more complicated.

Figure 2. A typical telehealth care network communicating in a patient/local health provide and a medical expert

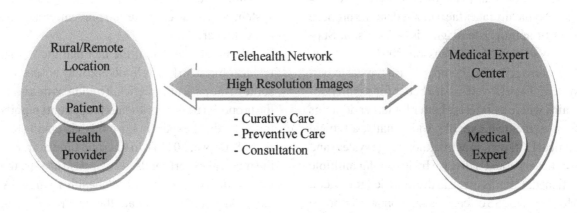

To provide better medical facilities in remote areas of Pakistan, telemedicine can be a solution. Due to shortage of financial resources, store and forward method based telemedicine can be a cost-effect solution. As the response time of store-and-forward method is very high, solution like virtual telemedicine (Bajwa, 2010) can be used. Virtual telemedicine proposes the use of a virtual physician at the remote medical center to provide immediate response to the patients. An expert system based telemedicine solution presented in (Bajwa, 2010) that provides a base for improved and efficient virtual telemedicine systems. However, the presented system demands following improvements:

- Accuracy (Bajwa, 2010) of the diagnostic algorithms is needed to be improved to improve usability.
- A better and improved analysis of the patients' medical data.
- A better and enhanced medical explanation (Bajwa, 2010) is required to make medical system more useful.
- Expert Systems do not provide automated support for updating medical expert's opinion

To address these issues and provide a better and improved framework for virtual telemedicine we propose the use of a Clinical/Diagnosis Decision Support System (CDSS/DDSS) (Tatnall & Burgess, 2007) in place of an expert system. A CDSS typically takes the patients data and propose a set of appropriate diagnosis. A Decision Support System (DSS) (Johansson, 2001) is better in many respects when compared to a typical expert system. Expert systems are better for smaller domains where as DSS are better for larger domains such as medical experts. Additionally, a typical clinical DSS uses an active knowledge system and diagnosis patient's disease by involving multiple patient data items given in the knowledge system. Moreover, a DSS offer decision support to improve

patient outcomes and reduce errors (Klonoff, et al., 2005). A DSS can be knowledge-based or algorithm based. In our approach we are using knowledge-based DSS.

Major contributions in this research paper are manifold. Firstly, we present a Clinical Decision Support System (CDSS) based virtual telemedicine framework (Kareem & Bajwa, 2011a) to overcome the long time constraint in typical store-and-forward method. The presented framework is simple and easy to set up. Secondly, a clinical decision support system is reported in this paper that is used in the proposed framework of virtual telemedicine. The presented CDSS is based on an efficient knowledge base as it contains knowledge of all the common diseases and implemented. Thirdly, we solve a case study from the domain of medical systems to test the designed system and evaluate the performance of the presented system.

VIRTUAL TELEMEDICINE & TELEHEALTH CARE SYSTEM

In store-and-forward telemedicine and telehealth, we are facing the same problem of time factor. To address this problem, we propose a concept of virtual telemedicine (Kareem & Bajwa, 2011a) and telehealth framework (Kareem & Bajwa, 2011b). In virtual telehealth, a virtual physician is used at the rural/remote medical center so that the patients can get benefit of telehealth facility on immediate bases. The proposed health care system covers the curative and preventive aspects of health care.

We are also aware of the fact that the virtual telehealth system has its limits and may not answer some of the patient's queries. To address this issue the proposed framework sends the patients reports to the medical expert in hospitals via email (Kareem & Bajwa, 2011a). In a certain time of frame, the medical expert sends his response against the patient query. The virtual telehealth framework has robust ability to update the new response in the existing system so that similar patient queries in future may be responded locally. The presented virtual telehealth system uses a Clinical/Diagnosis Decision Support System (CDSS/DDSS) to answer the curative care or preventive care queries of the users. In (Tatnall & Burgess, 2007) it is explained that a Decision Support System (DSS) for answering health care queries is efficient and effective. In the implementation of our virtual telehealth system, the knowledge-based DSS has been used.

The presented research contributes to the knowledge in multiple aspects. Firstly, we address a challenge of long time factor in store-and-forward telehealth system and propose a solution of virtual telehealth to get immediate response. Secondly, an enhanced and improved version of CDSS presented in (Klonoff, et al., 2005) is used to complement the needs of a virtual physician in a virtual telehealth system. The presented CDSS is based on an efficient knowledge base as it contains knowledge of all the common diseases. Thirdly, a small case study is also solved to evaluate the performance of the presented telehealth care system.

Rest of the chapter is structured as next section presents the related work. Then the architecture of the designed medical decision support system is presented following with the implementation details. A case study is solved with the discussion on results is also presented. Chapter is concluded with the future work.

RELATED WORK

This section presents the related work to presented research project. The major areas covered are such as Telemedicine, virtual telemedicine and Decision Support System (DSS).

Field of telemedicine is being proved the technology of the electronic age. In 1959 telemedicine was first introduced by Douglas (1959) was the first researcher, who worked to make the efficient use of available resources by using store & for-

ward telemedicine method. Dena (1995) used the telemedicine for the rural areas of America and gave its benefits. For the help and care of diabetic patients DIAB Tel (Gomez, et al., 1996) named telemedicine service is working efficiently.

Telemedicine unit for patient and doctor, which is low in cost was given by Khalid (2008) and covers the main issues of high speed network forms. In 2007 remote display protocol (RDP) was used in telemedicine on two different platforms WAN and over wi-fi (Albert, 2007). Home telemedicine architecture was given. UCD Health system (Rashid, 2003) for pharmacy using video conferencing was used by UC Davis. The paper highlights the role and benefits of advance technology in psychiatric and medical education. Initial results after using telepsychiatry are valid and reliable. Usage of telepsychiatry is increased by secure email and telephone. Technical designs and recommendations are given to develop telemedicine devices with help of advance telecommunication technology (Ullah, 2009).

Another example of a telemedicine project to provide a low cost solution with facilities like antiretroviral therapy (ART) and to test Acquired Immune Deficiency Syndrome (AIDS). Telemedicine Network model is presented for the developing countries specially Pakistan, which would improve the quality of medical facilities in the rural areas of Pakistan. Paper briefly discusses the present situation of the health sector of Pakistan and the improvements which could be seen after the use of Telemedicine Network Model (Klonoff, 2009).

Proposed solution to over come the time constraint of telemedicine is virtual telemedicine. Virtual physician is developed which is basically an artificial intelligent based expert system. Heart of expert system is knowledge base medical expert; its task is to diagnose the correct disease from symptoms given as input, after that give proper treatment. Expert medical system gives proper treatment immediately and efficiently. Experiments show that 90% cases were treated locally and accuracy of results was 85.5% (Bajwa, 2010).

The concept of Decision Support Systems (DSS) was first introduced in late 1960s and with the passage of time DSS gained importance in almost all field of science. In 2007, Tatnall (2007) build and used simple Decision Support Systems (DSS) for two postgraduate subjects at Victoria University in Melbourne. It enables students to look inside how system works with a feature to customize it. Klonoff (2009) declared decision support software as a necessary component which could accelerate the advancement and acceptance of telemedicine. To demonstrate innovative practices, it was suggested that decision support software should be used in U.S. MHS. New generation of physicians should know how to work with decision support software for diabetes and other diseases (Johansson, 2001).

USED METHODOLOGY

A CDSS based virtual telemedicine framework has been discussed in this section. Typically, a CDSS is a DSS that is primarily designed to achieve clinical advices for patients by incorporating knowledge management (historical patient data). The CDSS is an important component of the presented framework of virtual telemedicine. The proposed framework of virtual telemedicine works as a dispenser from remote location gathers patient data and processes data using a clinical decision support system. The CDSS processes the patient's report and diagnosis the diseases and prescribe the medication. If the CDSS can not diagnose the disease (in a case it is new disease, not updated in medical knowledge base), an email is sent to a medical expert (doctor). When the medical expert gives his response, the response is updated in CDSS medical knowledge base. The working of overall system is shown in Figure 3. There are multiple issues involved in

the development of a conventional telemedicine (Lai, 2007) system as following:

- First of all there is need of infrastructure that is based on hardware, software and connectivity mechanism of multiple nodes (patient and doctor).
- A dispenser who is enough expert to record patient's medical data such as X-rays, bio signals (such as blood pressure, body temperature, weight, pulse rate, blood sugar level, etc), blood test, etc and prepare a patient's symptoms report.
- Typical medical equipment is required at the patient end where a dispenser (literate person) can transmit patient's information to the medical expert (if required).
- A natural language processing system that can understand and analyze input patient report and extract symptoms and signs.
- A clinical decision support system (CDSS) is required that can diagnose a disease by reading symptoms (and signs) of disease and prescribe appropriate medication.
- There are important issues like accurate information exchange, security, transmission bandwidth, protocols, data sets etc.

A CDSS has been embedded in typical telemedicine system due to the high time constraint of conventional telemedicine system. A CDSS based virtual telemedicine is presented to cover up this time constraint and make telemedicine more effective and efficient. Virtual telemedicine is the extension of conventional telemedicine. A new component 'clinical decision support system' is the major contribution. The CDSS is able to understand and process natural language patient reports. Second major contribution is that Stanford parts of speech (POS) Tagger to improve accuracy in diagnosis. A patient's symptoms report is processed in multiple phases and all these phases are explained in detail in following.

Figure 3. A working scenario of the virtual telehealth care system used for providing health care services to a patient in a rural or remote area

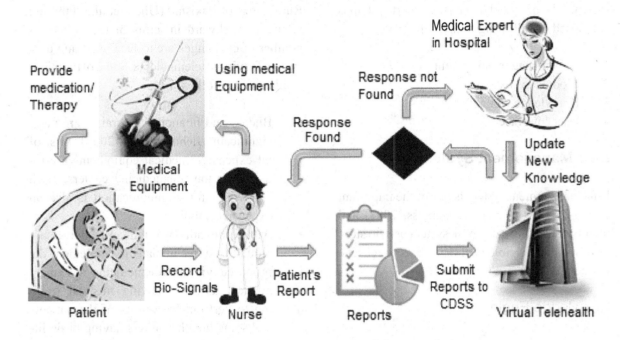

Process Patient's Symptom Natural Language Report

Natural language processing is performed to read the given text and extract the related information. Used NLP steps to analyze text are as follows.

Preprocessing NL Text Report

First of all the sentences are split by reading the input text document and separately stored in an array-list. Each input sentence is handed over to Stanford POS tagger (Toutanova & Manning 2000) to tokenize the input English text and identify the particular parts of speech of each token. The POS tagged text is further processed to identify and analyze morphemes. For example a verb "feels" is stored as "feel+s" and a noun "vegetables" is stored as "vegetable+s".

Syntacticaly Analysing NL Report

The pre-processed text is further processed to identify their particular role in the sentence and grammatical rules also assist this type of analysis such as subject, verb, object, etc.

Semantically Analysing NL Report

The input text is further semantically analyzed by doing roll labeling to extract symptoms and signs those are further used to identify patient's disease.

Used Clinical Decision Support System

This virtual medical expert based system is shown in Figure 3. This system has robust ability of reading the patient's symptoms and immediately diagnosing the disease and also prescribing the appropriate medication for the patient. A natural language processing (NLP) based medical expert system is the base of the proposed health care system. The designed rule based expert system is composed of following four components:

1. Data management system
2. Module system
3. Knowledge engine
4. User Interface

Data Management System

Data management system is one of the important parts of CDSS (Tatnall & Burgess, 2007) and mainly consists of a database system used to store the patient's physiological data so as to retrieve it when required.

Knowledge Engine

Facts, rules, structures and procedures based on expert knowledge are included in Knowledge Engine (Klonoff, et al., 2005). It contains decision making rules, which are based on the previous experiences and the expertise of the doctors. Accuracy of the diagnosis depends on the information contained in Knowledge Engine.

Module System

The module system is the key unit that is actually responsible for processing the input queries and answer. To answer a patient's query, the module system evolves a health model by analyzing the patient's data in the knowledge base system with the used model for formulating an advice. Module system and knowledge engine works as a unit to diagnose the patient and to take decisions.

User Interface

User interacts with user interface which allows verifying the correctness of the diagnosis and decisions of the CDSS.

IMPLEMENTATION DETAILS

Rural areas of Pakistan (Ullah, et al., 2009) are relatively backward in terms of technology. A number of challenges are to face in setting up a system for virtual telemedicine. Some of the major challenges are following:

- Budget and financial constraints are more significant (Jiehui & Jing, 2007). First of all expensive medical equipments are required at the telemedicine centers. High bandwidth for communication is also an expensive solution.
- At the site, adequate human resources are required i.e. technicians to implement the proposed virtual Telemedicine system, a medical assistant having medical training to perform basic tests of the patients and some health workers having basic literacy of computer and capable of using computers.

A complete infrastructure is required to actually set up the proposed virtual telemedicine framework. A satellite based wireless internet work or WiFi system supporting speed of 1.0 Gbps or above is required. 3G cellular technology is also getting very popular these days in the field of telehealth. This technology can help out in fast video sharing, video male and video conferencing. On the other side, a telemedicine center at the remote area needs basic eequipments i.e.

- Virtual telemedicine software
- Camera (s), lights, projector
- Digital X-Ray System
- UPS system
- Computer hardware, system and application software and accessories

RESULTS AND DISCUSSION

To evaluate the performance of the designed CDSS, a sample case study is presented. The case study was consisted of a NL patient's report containing nursing data that was originally generated by a dispenser. Following is the problem statement of the solved case study.

The patient's age is 32 years. The patient's weight is 61 Kg. The patient's body temperature is 103.5C°. The blood pressure of patient is 71 and 124. Patient is feeling sudden weakness/ numbness in the face, arm or leg on one side of the body. There is also an abrupt loss of vision, strength, coordination, sensation, speech. There is a sudden dimness of vision, especially in one eye. There is a sudden loss of balance, possibly accompanied by vomiting, nausea, fever, hiccups or trouble with swallowing. There is a sudden and severe headache with no other cause followed rapidly by loss of consciousness an indication of a stroke due to bleeding.

In the solved case study, there were 10 sentences. Smallest sentence was composed of 4 and the largest sentence was composed of 12 words. Average length of all sentences was 9. Similar to this case study four other case studies were solved with the designed health care system. A medical assistant was involved to use the system. A multiple step procedure is involved to use the designed medical health care system. The steps are following:

1. Registration of a patient
2. Recording patient's data
3. Processing patient's report
4. Diagnosis and Medication

Brief description of all these phases with the help of a case study has been provided in the later part of the section.

Registration Patient

A patient is needed to register with his personal details i.e. name, age, sex, address, family history, previous cases, etc for using the proposed virtual telehealth system. Figure 4 shows the form that is used to register the patient first.

Recording Patients' Data

After registration, the medical expert performs basic tests of a patient to get the reading of temperature, blood pressure, blood group, sugar level and ESG (if required). Then he records the common symptoms of the patient in the system. Besides these tests, the data i.e. color of tongue, color of eyes, heartbeat, face color; etc is also captured and is updated in the system. The output of this step was a patients report in English.

Processing Patient's Report

Medical assistant can also use digital stethoscope and electrocardiograph file with ECG recorder or images with the examination camera (Gomez, 1996). A text file containing the patient's case details is prepared. A prescription will be generated after the patient's data is submitted. If the knowledge base cannot reply then the patient's data will be emailed to expert. The correctness of the decision made by the software and the medical expert is based on the accuracy of the data captured by the medical assistant. The quality and accurateness of the images and video of the patient is also quite important.

Diagnosis and Medication

Finally, CDSS diagnosis the patient's disease and prescribes medication (including dose details and side effects, if possible) with diet details and exercise details. At this phase, facility of explaining and reasoning of the system to the user is

Figure 4. A CDSS based framework for virtual telemedicine and virtual telehealth

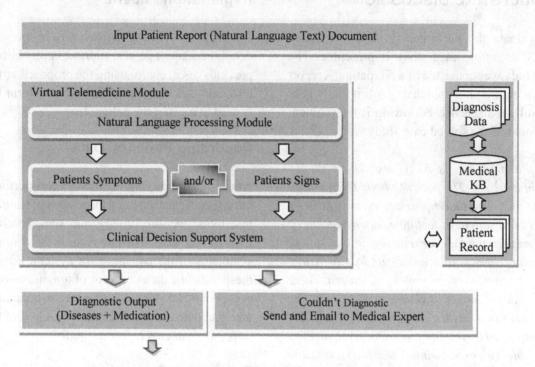

also provided. User can make different queries regarding the system domain and system.

To validate the precision and affectivity of the designed system symptom reports of three groups of ten patients were defined. For each group three reports i.e. easy, average and difficult were generated for each group. The symptom reports were carefully prepared and processed for each patient using the designed health care system. For correct and wrong diagnosis of a symptom report various points were given. Table 1 shows the details of the results and graphical representation of the results is shown in Figure 5.

Following are some benefits over using the proposed framework virtual telehealth.

- Improved and immediate access the specialty care
- Upgraded emergency medical services
- Reduction in un-necessary duplication of services
- Less dependency on the medical expert

- Easier diagnostic consultation
- Expanded disease cure education
- More patient health queries
- Remote medical consultation
- Reduction in health care cost
- Automated patient record keeping

CONCLUSION AND FUTURE WORK

Virtual Telemedicine is the new concept which actually works faster than that of the traditional telemedicine systems. An expert system has been deployed in place of a medical expert that has ability to immediate respond. This immediate response can help to treat patients in time and more effectively. 95% queries can be entertained locally. The accuracy achieved with the designed system is 91.00%. The Virtual expert system becomes more robust and intelligent with the passage of time as the knowledge-base grows and the level accuracy will also improve. For the

Table 1. Virtual telehealth system representing the accuracy of used CDSS in the proposed virtual telehealth care system

#	Symptoms	Diagnosis	Average Accuracy
1	92.00%	89.00%	90.50%
2	89.00%	85.00%	86.50%
3	84.00%	81.00%	82.50%
4	93.00%	89.00%	91.00%
5	89.00%	87.00%	88.00%
6	87.00%	82.00%	84.50%

Figure 5. Architecture of used clinical DSS

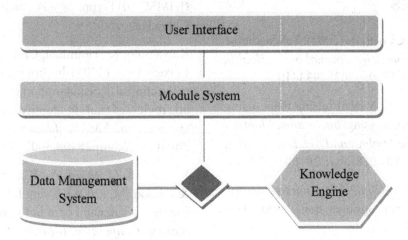

Figure 6. Results of the solved case study with the presented CDSS based virtual telehealth care system

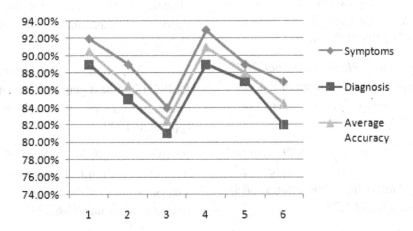

under developed and developing countries like Pakistan, Bangladesh, Sri Lanka etc the usability of the Virtual telemedicine will be more useful and beneficial. The experiments were performed on a simulator and it is acceptable that these results may vary when the system will be run real time. In future enhancements the algorithms is needed to be improved to increase the accuracy level of the system. Medical explanation module is also needed to enhance its usability.

REFERENCES

Ackerman, M., & Carft, R. (2002). Telemedicine technology. *Telemedicine Journal and e-Health*, *8*(1).doi:10.1089/15305620252933419

Bajwa, I. S. (2010). Virtual telemedicine using natural language processing. *International Journal of Information Technology and Web Engineering*, *5*(1), 43–55. doi:10.4018/jitwe.2010010103

Berner, E. S. (2007). *Clinical decision support systems*. New York, NY: Springer. doi:10.1007/978-0-387-38319-4

Gomez, E. J., Pozo, F. D., & Hernando, M. (1996). Telemedicine for diabetes care: The DIABTel approach towards diabetes telecare. *Informatics for Health & Social Care*, *21*(4), 283–295. doi:10.3109/14639239608999290

Health Ministry. (2006). *Health year book 2005-2006*. Health Ministry, Government of Pakistan, 2007. Retrieved January 28, 2011, from http://www.health.gov.pk/

Houston, M. S., & Myers, J. D. (1999). Clinical consultations using store-and-forward telemedicine technology. *Mayo Clinic Proceedings*, *74*(8), 764–769. doi:10.4065/74.8.764

Jiehui, J., & Jing, Z. (2007). Remote patient monitoring system for China. *IEEE Potentials*, *26*(3), 26–29. doi:10.1109/MP.2007.361641

Johansson, B. G. (2001). *Design and implementation of a clinical decision support system based on open standards*.

Kareem, S., & Bajwa, I. S. (2011a). *A virtual telehealth framework: Applications and technical considerations*. In IEEE International Conference on Emerging Technologies 2011 (ICET 2011)

Kareem, S., & Bajwa, I. S. (2011b). Clinical decision support system based virtual telemedicine. *3rd International Conference on Intelligent Human-Machine Systems and Cybernetics* (IHMSC 2011) (pp. 78-83). Zhejiang University, China Hangzhou.

Kawamoto, K., Houlihan, C. A., Balas, E. A., & Lobach, D. H. (2005). Improving clinical practice using clinical decision support systems: A systematic review of trials to identify features critical to success. *British Medical Journal*, *330*(7494), 765. doi:10.1136/bmj.38398.500764.8F

Khalid, M. Z., Akbar, A., Kumar, A., Tariq, A., & Farooq, M. (2008). Using telemedicine as an enabler or antenatal care in Pakistan. *2nd International Conference: E-Medical System*, Tunisia, (pp. 1-8).

Klonoff, D. C., & True, M. W. (2009). The missing element of telemedicine for diabetes: Decision support software. *Journal of Diabetes Science and Technology*, *3*.

Lai, A. M., Nieh, J., & Starren, J. B. (2007). REPETE2: A next generation home telemedicine architecture. *AMIA 2007 Symposium Proceedings*, (pp. 1020-1022).

Perednia, D. A., & Allen, A. (1995). Telemedicine technology and clinical applications. *Journal of the American Medical Association*, *273*(6), 483–488. doi:10.1001/jama.273.6.483

Puskin, D. S. (1995). Opportunities and challenges to telemedicine in rural America. *Journal of Medical Systems, 19*(1), 59–67. doi:10.1007/BF02257191

Puskin, D. S. (2006). *Telemedicine, telehealth, and health Information Technology.* An ATA Issue Paper, The American Telemedicine Association, May 2006.

Rashid, E., Ishtiaq, O., & Gilani, S. (2003). Comparison of store and forward method of teledermatology with face-to-face consultation. *Journal of Ayub Medical College, Abbottabad, 15*(2), 34–36.

Tatnall, A., & Burgess, S. (2007). Experiences in building and using decision- support systems in postgraduate university courses. *Interdisciplinary Journal of Information, Knowledge, and Management, 2*.

Thomas, S. N. (2007). Meeting the health care needs of California's children: The role of telemedicine. *Digital Opportunity for Youth Issue Brief, Number 3,* September 2007.

Toutanova, K., & Manning, C. D. (2000) Enriching the knowledge sources used in a maximum entropy part-of-speech tagger. In the *Joint SIGDAT Conference on Empirical Methods in Natural Language Processing and Very Large Corpora* (pp. 63-70).

Ullah, N., Khan, P., Sultana, N., & Kwak, K. S. (2009). A telemedicine network model for health applications in Pakistan: Current status and future prospects. *International Journal of Digital Content Technology and its Applications, 3*(3).

WHO. (2010). *World health statistics*. World Health Organization (WHO). Retrieved December 27, 2010, from http://www.who.int/whosis/whostat/2007/en/

Chapter 11
Predicting Temporal Exceptions in Concurrent Workflows

Iok-Fai Leong
University of Macau, Macau

Yain-Whar Si
University of Macau, Macau

Robert P. Biuk-Aghai
University of Macau, Macau

ABSTRACT

Current Workflow Management Systems (WfMS) are capable of managing simultaneous workflows designed to support different business processes of an organization. These departmental workflows are considered to be interrelated since they are often executed concurrently and are required to share a limited number of resources. However, unexpected events from the business environment and lack of proper resources can cause delays in activities. Deadline violations caused by such delays are called temporal exceptions. Predicting temporal exceptions in concurrent workflows is a complex problem since any delay in a task can cause a ripple effect on the remaining tasks from the parent workflow as well as from the other interrelated workflows. In addition, different types of loops are often embedded in the workflows for representing iterative activities, and presence of such control flow patterns in workflows can further increase the difficulty in estimation of task completion time. In this chapter, the authors describe a critical path based approach for predicting temporal exceptions in concurrent workflows that are required to share limited resources. This approach allows predicting temporal exceptions in multiple attempts while workflows are being executed. The accuracy of the proposed prediction algorithm is analyzed based on a number of simulation scenarios. The result shows that the proposed algorithm is effective in predicting exceptions for instances where long duration tasks are scheduled (or executed) at the early phase of the workflow.

DOI: 10.4018/978-1-4666-0023-2.ch011

INTRODUCTION

In a highly dynamic business environment, simultaneously executing departmental workflows form digital business ecosystems (Boley & Chang, 2007) which are designed to share limited resources. In this chapter we address the temporal exception prediction problems of a digital business ecosystem involving concurrent workflows.

A workflow specification defines how a business process functions within an organization (Son, Kim, & Kim, 2005). Based on these specifications, Workflow Management Systems (WfMS) allocate and dispatch work to users (Li & Yang, 2005). A workflow instance is an execution of a workflow specification. During the execution of a workflow instance, some events that are not defined in the workflow specification may occur. These events are typically considered as exceptions in workflow management systems.

In the area of programming languages, exceptions may interrupt or abort the execution of a program. Exceptions in WfMS are also similar to those of programming languages. Two types of exceptions (Casati, 1998) can be defined in WfMS; expected exceptions which are the results of predictable deviations from the normal behavior of a process and unexpected exceptions which are the outcomes of inconsistencies between the business process in the real world and its corresponding workflow description.

A workflow specification may consist of several workflows that execute concurrently and share the same pool of resources. In large organizations, a number of different workflows could be executed simultaneously. These concurrent workflows are often interdependent since they are required to share limited resources. For example, two concurrent workflows designed, respectively, for an inventory management process and a delivery process may share the same human resources (e.g. a group of technicians) who are responsible for managing a storage facility and a

fleet of vehicles. Delays occur when only a limited number of resources are available during a given interval. We denote deadline violations caused by such delays as temporal exceptions.

In workflow management systems, control-flow patterns are used to describe the order of tasks that make up a process and the relationship between them (Van der Aalst, Ter Hofstede, Kiepuszewski, & Barros, 2003). In this chapter, we focus on predicting temporal exceptions for workflows which include iteration patterns (loops). Iteration pattern refers to a repeated execution of one or more tasks within a workflow. Although iteration patterns are extensively used in workflow specifications, less attention has been devoted to understanding their implications for temporal exceptions. Specifically, rapid changes in the number of iterations can cause deadline violations as well as conflicts in resource usage.

Our approach can be divided into two phases: *design time* phase and *run time* phase. During the design time, temporal and resource constraints are calculated for each task within the workflow specifications. During the run time, an algorithm is used to predict potential deadline violations of workflow instances by taking into account constraints calculated at design time.

The remainder of this chapter is organized as follows. In the following section we give a brief introduction to resource and temporal constraints in a workflow. Then we describe the critical path based method for exception prediction. Next we illustrate an example of temporal exception prediction, and evaluate the proposed method based on a simulation experiment. We then review recent related work and finally summarize our ideas.

OVERVIEW

A workflow management system can be used to host several simultaneously executing workflows. Conflicts usually occur when tasks from instances of different workflows compete for limited resources (e.g. human resources, financial resources, etc).

Figure 1 (a) shows an example of conflict involving tasks from two concurrent workflows when only four units of resources are available for sharing. In this example, resource conflicts occur at tasks T_{22}, T_{23}, T_{24} and T_{25}. Furthermore, these resource conflicts are linked to the underlying temporal constraints of a task within the workflow.

During the design time, each task T within a workflow can be specified with a value $D(T)$ (see Figure 1 (b)) which is the maximum allowable execution time of task T. Based on $D(T)$, we can derive two temporal constraints: start time $ST(T)$ and end time $ET(T)$.

These constraints are crucial in determining the deadline of a workflow. Basically, temporal constraints of a task are related to its predecessor and successor tasks. For instance, if a task is scheduled to execute immediately after another task, its start time will be the end time of its predecessor task. However, depending on the control flows (Van der Aalst et al., 2003) within the workflow, derivation of these constraints can be different.

A workflow may also have loops. A loop is a repeated execution of one or more tasks within a workflow. Such repetition has a significant impact on temporal constraints of a task which may belong to one or more loops. Since tasks within a loop can be repeated a number of times, calculating temporal constraints of a task also needs to take into account possible repetition of its ancestors and descendants as well. Such calculation can be even more problematic when a task belongs to one or more nested loops. Based on the definition from (Van der Aalst et al., 2003), loops can be classified as structured loops and arbitrary loops. In this chapter, we further divide arbitrary loops into nested loops and crossing loops. In a nested loop, all tasks within a loop appear in another loop. In a crossing loop, only a certain number of tasks within a loop appear in another loop.

Let I_N be the maximum number of iterations that a loop N within a workflow wf can be executed, and let S_N be the set of tasks in loop N in wf. We assume that $I_N > 0$. We also assume that each task within a workflow is atomic, i.e., a task must be executed entirely until it is completed and cannot be interleaved with other tasks.

Structured Loop. The simplest form of a loop is the Structured Loop (Van der Aalst et al., 2003). An example of a structured loop is shown in Figure 2.

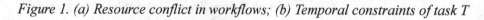

Figure 1. (a) Resource conflict in workflows; (b) Temporal constraints of task T

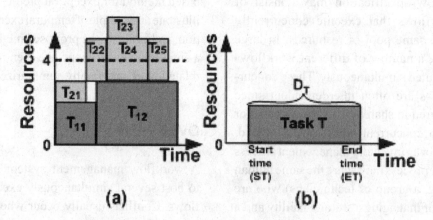

Figure 2. Structured loop

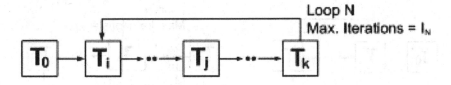

Figure 3. Nested loop

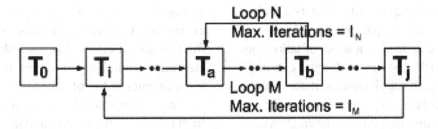

Definition 1: Structured Loop. A loop N in a workflow *wf* is a Structured Loop if for all tasks T in S_N, T does not exist in any other loop in *wf*.

Nested Loop. In a nested loop, all tasks within a loop are also members of another loop. Such loops are also classified as Arbitrary Loops (Van der Aalst et al., 2003). For instance, in Figure 3, a loop N is nested within another loop M.

Definition 2: Nested loop: A loop M in a workflow *wf* is a nested loop if there exists another loop N such that all the tasks in N are also members of M and both M and N do not share the same initial task. If M and N share the same initial task, we consider M as a crossing loop.

Crossing loop. Crossing loops are also classified as Arbitrary Loops (Van der Aalst et al., 2003). In Figure 4, an example of a crossing loop is depicted with two loops N and M crossing each other at T_a and T_k.

Definition 3: Crossing loop. Loop N and M in a workflow *wf* are crossing loops if (1) both N and M share at least one task and (2) in each of N and M there exists at least one task of *wf* which is a member of only one but not both of the loops N and M. Both loops may begin at the same task, i.e. both flows from T_k and T_b may point back to either T_i or T_a in Figure 4. In this case, the loop whose last task ends earlier is considered as loop N.

In this chapter we define workflow specifications in XML Schema based on YAWL (Van der Aalst & Ter Hofstede, 2005). A sample XML schema is shown in Figure 5. The XML code defines the duration of each of the workflow specifications and the resource usage for each of them. It also defines the loops within the workflow specifications.

CRITICAL PATHS IN WORKFLOWS

In Figure 6, we describe the calculation of the execution time for the control flow patterns defined in Li and Yang (2005). The resulting execution

Figure 4. Crossing loop

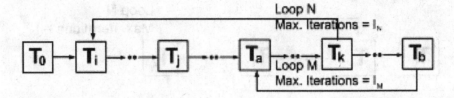

time is then used for calculating the critical path of the workflow. A critical path in a workflow specification can be defined as a series of tasks from the first task to the last with the longest execution time. Critical paths can be used to determine the hard deadline of the workflow after resolving any conflict in resource usage. In this research, we apply a breadth-first-search algorithm to identify all the possible paths of a workflow specification. We assume that the workflow is well-formed and free from structural errors. The breadth-first-search algorithm recursively traverses the workflow and generates a list of all possible paths. The total execution time of a path can be calculated from the sum of the execution times of each task within the path. In case there are loops defined in

Figure 5. Workflow schema defined using XML schema

```xml
<?xml version="1.0" encoding="UTF-8"?>
<workflowSchema>
  <resource size="4"/>
  <workflow name="wf1">
    <task name="T11" duration="5" resource="2">
      <flowsInto task="T12"/>
      <join type="or"/>
      <split type="and"/>
    </task>
    ...
    <lastTask name="T16"/>
    <firstTask name="T11"/>
    <loop name="wf1_n" maxExecution="3">
      <member name="T11"/>
      <member name="T12"/>
      <member name="T13" loopTo="T11"/>
    </loop>
  </workflow>
  <workflow name="wf2">
    ...
  </workflow>
</workflowSchema>
```

Figure 6. Execution time for different control patterns (Adapted from Li and Yang (2005))

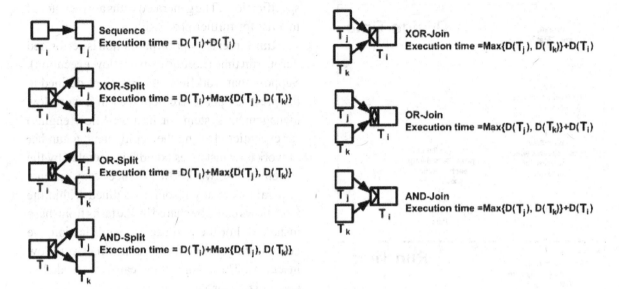

the workflow, the algorithm traverses the loops in all possible numbers of iterations and treats each of them as a different path. The algorithm for calculating all possible paths within a workflow is described in Algorithm 1 below.

The path(s) with the longest execution time are then selected as the critical path(s) of the workflow. In addition, the shortest path(s) in the workflow are also recorded. Later in this chapter, we will make use of the longest and shortest paths for calculating the Hard Deadline (*HD*) and Earliest Completion Time (*ECT*) of the workflow.

PREDICTING TEMPORAL EXCEPTIONS

The algorithm for predicting temporal exceptions is divided into two phases (see Figure 7). The first phase can be performed during the design time and second phase is performed during run time. These phases are shown in the upper and lower areas of Figure 7, respectively.

Design Time: During the design time, each workflow specification is defined with corresponding temporal and resource allocation pa-

Algorithm 1. Generation of all possible paths within a workflow

Input: Workflow specification *wf*   **Output:** List of all possible paths for *wf*
1. Initialize *pathlist* as an empty list of task lists   2. Initialize *path* as an empty list of tasks   3. Start with the first task *T* of *wf*   4. Call Algorithm 1-(a) with *T, path, pathlist* as parameters
Algorithm 1-(a) – Recursive procedure that adds a task to a path
Input: Task *T,* List of tasks *path,* list of paths *pathlist*   **Output:** Updated list of paths *pathlist*

1. Add *T* to *path*
 2. If *T* has an outgoing flow to another task (i.e., *T* is not the end task of the workflow)
     a) For each next task T' from T, call Algorithm 1-(a) with *T', path, pathlist* as parameters
     Else if *path* is not in *pathlist*, add *path* to *pathlist*

Figure 7. Steps for predicting temporal exceptions

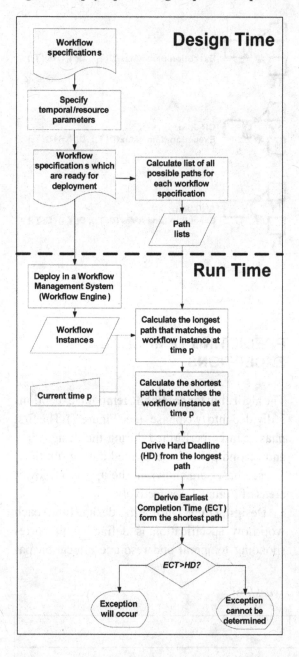

derive all the possible paths from these workflow specifications. The generated paths are then stored in a list for further processing.

Run Time: Exception prediction is performed during run time (i.e. during workflow execution). Suppose that workflow specifications defined in the previous phase are deployed in a workflow management system (or in a workflow engine) for execution. During the deployment, a number of workflow instances could be generated by the system. These workflow instances may execute in parallel and any resources defined within the workflows could be shared by the tasks from these instances. For the purpose of simplification, we define our algorithms for two concurrent workflow instances. These algorithms can be extended or generalized for cases including more than two workflow instances.

Let p be the current time point during the execution. To predict exception at time p, all paths generated from the previous phase and the detailed information of the ongoing workflow instances at time p are input into four algorithms (algorithms II, III, IV, and V).

First, Algorithm 3 is used to estimate the end times of two ongoing concurrent workflow instances for worst case scenarios. Such estimation can be done by calculating/projecting the longest possible execution path of each workflow instance beyond current time point p. First, we identify a longest path from the path lists which contain the identical sequence of tasks from the paths of each ongoing workflow instance. Note that the path lists are generated during design time based on Algorithm 1. The identified two longest paths can be considered as the sequence of tasks which can be executed during the worst case scenario.

However, when the two ongoing workflow instances progress (according to the identified longest paths in the worst case scenarios), these instances may need to share limited resources and therefore any conflicts resulting from the tasks within these workflow instances need

rameters. Specifically, every task in the workflows is configured with parameters such as maximum allowable execution time and the number of required resources. These parameters should be defined to reflect the actual constraints of the business process. Next, Algorithm 1 is used to

Algorithm 2. Generation of Execution Trace from two workflow instances

Input: Workflow instances $wi1$, $wi2$ **Output:** Conflict free Execution Trace
1. For each occurrence of task $t_{(i)}$ in the workflow instance, where i represents the i-th occurrence of task t in the workflow instance, check if any other task t' has overlapping executing time and resource conflict. If it does, put t' into a list of tasks called $dep(t_{(i)})$. For each task $t_{(i)}$, each task in $dep(t_{(i)})$ is said to have a "resource conflict" with $t_{(i)}$. 2. For each task t' in $dep(t_{(i)})$, modify $[ST(t'), ET(t')]$ as $[\mathbf{MAX\{ST(t'), ST(t_{(i)})+D(t_{(i)})\}}, \mathbf{MAX\{ET(t'), ET(t_{(i)})+D(t')\}}]$. Modify the ST/ET values of tasks t'' where all tasks t'' are reachable by t' accordingly.

to be resolved. In Algorithm 2, we resolve the resource conflicts between any two workflow instances by interleaving and ordering of tasks from each instance. The result of Algorithm 2 is called an *Execution Trace* which represents the conflict-free ordering of tasks from two workflow instances. By using Algorithm 2, we can resolve any conflicts among two longest paths identified from the previous step. The end time of the resulting execution trace is called the *Hard Deadline* (HD- shown in algorithm 3) which is the latest time by which both ongoing workflow instances must complete their execution without causing an exception. HD can be also considered as the outermost boundary of the acceptable time limit for both workflow instances.

Next, Algorithm 4 is used to estimate the end times of two ongoing concurrent workflow instances for best case scenarios. In contrast to Algorithm 3, such estimation can be done by calculating/projecting the shortest possible execution path of each workflow instance beyond current time point p. First, we identify a shortest path from the path lists which contain the identical sequence of tasks from the paths of each ongoing workflow instance. The path lists are generated during design time based on Algorithm 1. The identified two shortest paths represent the sequence of tasks which can be executed during the best case scenario.

However, when the two ongoing workflow instances progress according to the identified shortest paths in the best case scenario, these instances may need to share limited resources and therefore any conflicts resulting from the tasks within these workflow instances need to be resolved. By using Algorithm 2, we can resolve any conflicts among two shortest paths identified from the previous step. The end time of the resulting execution trace is called the Earliest Completion Time (*ECT*) which is the earliest time by which both ongoing workflow instances can complete their execution.

Algorithm 3. Calculation of hard deadline (HD)

Input: Workflow specifications $wf1$, $wf2$; Workflow instances $wi1$, $wi2$; time point p **Output:** Hard Deadline (*HD*) for the workflow instances
1. For workflow specification $wf1$, call Algorithm 1 to generate the list of all possible paths for $wf1$. From the list, identify the longest path $wi1'$ that contains the same sequence of tasks from $wi1$. 2. Set maximum allowable duration for each task in $wi1'$ based on workflow specification $wf1$. 3. For workflow specification $wf2$, call Algorithm 1 to generate the list of all possible paths for $wf2$. From the list, identify the longest path $wi2'$ that contains the same sequence of tasks from $wi2$. 4. Set maximum allowable duration for each task in $wi2'$ based on workflow specification $wf2$. 5. Call Algorithm 2 with $wi1'$, $wi2'$ and time point p as input to generate the execution trace. 6. The end time of the execution trace is the Hard Deadline (*HD*) of the workflow. 7. If there is more than one longest path (with the same path cost) from either $wf1$ or $wf2$, repeat steps 1 to 6 for each longest path to calculate *HD* and choose the largest one as the final result.

Algorithm 4. Calculation of earliest completion time (ECT)

Input: Workflow specifications $wf1$, $wf2$; Workflow instances $wi1$, $wi2$; time point p
Output: Earliest Completion Time (ECT) of the workflow instances

1. For workflow specification $wf1$, call Algorithm 1 to generate the list of all possible paths for $wf1$. From the list, identify the shortest path $wi1'$ that contains the same sequence of tasks from $wi1$.
2. Update the duration of tasks from $wi1'$ with the actual execution time (actual duration is used to reflect the real execution progress of the workflow instance) recorded in $wi1$. In this step, only tasks before and up to time point p will be updated.
3. Set maximum allowable duration for all remaining unexecuted tasks in $wi1'$ based on workflow specification $wf1$.
4. For workflow specification $wf2$, call Algorithm 1 to generate the list of all possible paths for $wf1$. From the list, identify the shortest path $wi2'$ that contains the same sequence of tasks from $wi2$.
5. Update the duration of tasks from $wi2'$ with the actual execution time recorded in $wi2$. In this step, only tasks before and up to time point p will be updated.
6. Set maximum allowable duration for all remaining unexecuted tasks in $wi2'$ based on workflow specification $wf2$.
7. Call Algorithm 2 with $wi1'$, $wi2'$ and time point p as input to generate the execution trace for the two workflow instances.
8. The end time of the resulting execution trace is the Earliest Completion Time (ECT) of the workflow.
9. If there is more than one shortest path (with the same path cost) from either $wf1$ or $wf2$, repeat steps 1 to 8 for each shortest path to calculate ECT and choose the smallest one as the final result.

Algorithm 5. Predicting exception

Input: Workflow specifications $wf1$, $wf2$; workflow instances $wi1$, $wi2$; time point p
Output: Either exception is predicted or cannot be determined

1. Call Algorithm 3 with $wf1$, $wf2$, $wi1$, $wi2$, and p as input to calculate HD.
2. Call Algorithm 4 with $wf1$, $wf2$, $wi1$, $wi2$, and p as input to calculate ECT.
3. If $ECT > HD$, then ongoing workflow instances are predicted to have an exception. Otherwise an exception cannot be determined.

Based on Algorithms 3 and 4, we can derive the Earliest Completion Time (ECT) from the shortest paths and Hard Deadline (HD) from the longest paths. HD is the latest time by which both workflow instances must complete their execution while ECT is the earliest possible time for completion. By comparing ECT and HD we can predict whether there will be an exception or not in the future. Algorithm 5 shows the steps for exception prediction.

EXAMPLE

In this section we outline a sample exception prediction for two workflow instances from specifications $wf1$ and $wf2$ (see Figure 8 and Table 1, respectively).

In $wf1$, loops m and n form a crossing loop. In $wf2$ loops m and n form a nested loop.

For the purpose of illustration, in this example, we assume that each loop can iterate for at most one time. Suppose that $wf2$ starts later than $wf1$ and both workflows are required to share 4 units of identical resources. The durations and resource usages for each of the tasks in both workflows are specified during the design time and are shown in Table 1. $D(T)$ represents the maximum allowable execution time and $R(T)$ represents the required number of resources for a task T.

First, we apply Algorithm 1 to find all the possible paths for the two workflow specifications (see Figure 9). Figures 10 (a) and (b) show the list of all possible paths generated for $wf1$ and $wf2$, respectively.

Suppose two workflow instances $wi1$ and $wi2$ are executing at time point p (see Figure 11) based on workflow specifications $wf1$ and $wf2$. Our aim is to perform Algorithm 5 at different time points to predict potential exceptions for both workflow instances (Figures 11 (a) and (b)). In this example, two prediction attempts are made at $p = 20$ and $p = 40$.

At time point = 20, $wi1$ has completed $T11 \rightarrow T12 \rightarrow T14 \rightarrow T15 \rightarrow T16$ (the last one was partially executed), and $wi2$ has executed $T21 \rightarrow T22 \rightarrow T25 \rightarrow T26 \rightarrow T27$. First we apply Algorithm 3 to find their longest paths. The longest path for $wf1$ that starts with $T11 \rightarrow T12 \rightarrow T14 \rightarrow T15 \rightarrow T16$ is $T11 \rightarrow T12 \rightarrow T14 \rightarrow T15 \rightarrow T16 \rightarrow T12 \rightarrow T13 \rightarrow T11 \rightarrow T12 \rightarrow T13 \rightarrow T15 \rightarrow T16$; whereas the longest path of $wf2$ that starts with $T21 \rightarrow T22 \rightarrow T25 \rightarrow T26 \rightarrow T27$ is $T21 \rightarrow T22 \rightarrow T25 \rightarrow T26 \rightarrow T27 \rightarrow T28 \rightarrow T21 \rightarrow T22 \rightarrow T23 \rightarrow T24 \rightarrow T23 \rightarrow T24 \rightarrow T27 \rightarrow T28$ (see Figure 12 (a)). These two longest paths are fed into Algorithm 3 and a Hard Deadline HD (65) is calculated from the resulting conflict-free execution trace.

Next we apply Algorithm 4 to find their shortest paths. For $wf1$, the shortest path that starts with $T11 \rightarrow T12 \rightarrow T14 \rightarrow T15 \rightarrow T16$ is $T11 \rightarrow T12 \rightarrow T14 \rightarrow T15 \rightarrow T16$. For $wf2$, the shortest path that starts with $T21 \rightarrow T22 \rightarrow T25 \rightarrow T26 \rightarrow T27$ is $T21 \rightarrow T22 \rightarrow T25 \rightarrow T26 \rightarrow T27 \rightarrow T28$ (see Figure 12 (b)). Note that $T16$ has not finished execution at $p = 20$. These two shortest paths are fed into Algorithm 4 and the Earliest Completion Time ECT (27) is calculated from the resulting conflict-free execution trace. Since $ECT < HD$, at time point $p = 20$, we cannot determine whether an exception will occur in the future.

Figure 8. Workflow specifications wf1 and wf2

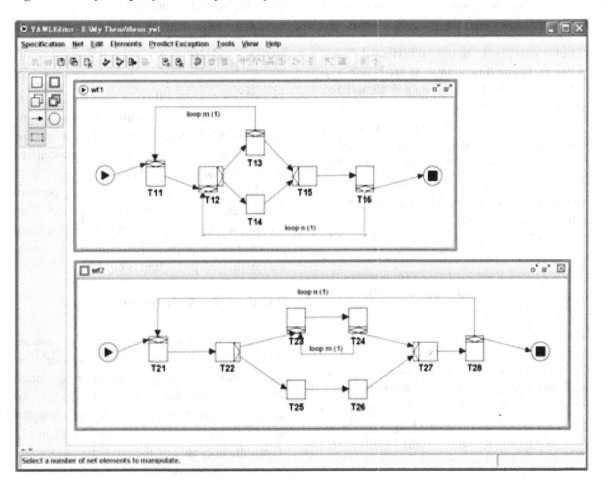

Table 1. Durations and resource usage for the tasks in both workflows

Task T	Duration D(T)	Resource Usage R(T)
T11	5	2
T12	2	1
T13	5	4
T14	3	2
T15	1	2
T16	5	1
T21	2	1
T22	3	2
T23	5	3
T24	4	3
T25	2	1
T26	2	1
T27	1	1
T28	6	4

Since exception can not be predicted at $p = 20$, we make another attempt at prediction at $p = 40$ (see Figure 11 (b)). When $p = 40$, $wi1$ has executed $T11 \rightarrow T12 \rightarrow T14 \rightarrow T15 \rightarrow T16 \rightarrow T12 \rightarrow T14 \rightarrow T15 \rightarrow T16$ (the last one was partially executed), and $wi2$ has executed $T21 \rightarrow T22 \rightarrow T25 \rightarrow T26 \rightarrow T27 \rightarrow T28 \rightarrow T21 \rightarrow T22 \rightarrow T23$ (the last one was also partially executed). First we apply Algorithm 3 to find their longest paths. The longest path for $wf1$ that starts with $T11 \rightarrow T12 \rightarrow T14 \rightarrow T15 \rightarrow T16 \rightarrow T12 \rightarrow T14 \rightarrow T15 \rightarrow T16$ is $T11 \rightarrow T12 \rightarrow T14 \rightarrow T15 \rightarrow T16 \rightarrow T12 \rightarrow T14 \rightarrow T15 \rightarrow T16$; whereas the longest path of $wf2$ that starts with $T21 \rightarrow T22 \rightarrow T25 \rightarrow T26 \rightarrow T27 \rightarrow T28 \rightarrow T21 \rightarrow T22 \rightarrow T23$ is $T21 \rightarrow T22 \rightarrow T25 \rightarrow T26 \rightarrow T27 \rightarrow T28 \rightarrow T21 \rightarrow T22 \rightarrow T23 \rightarrow T24 \rightarrow T23 \rightarrow T24 \rightarrow T27 \rightarrow T28$ (see Figure 12 (c)). Notice that the two longest paths are different from those at $p = 20$ since both workflow instances have progressed up to time point 40. These two longest paths are fed into Algorithm 3 and a Hard Deadline *HD*

(53) is calculated from the resulting conflict-free execution trace.

Next we apply Algorithm 4 to find their shortest paths. For $wf1$, the shortest path that starts with $T11 \rightarrow T12 \rightarrow T14 \rightarrow T15 \rightarrow T16 \rightarrow T12 \rightarrow T14 \rightarrow T15 \rightarrow T16$ is $T11 \rightarrow T12 \rightarrow T14 \rightarrow T15 \rightarrow T16 \rightarrow T12 \rightarrow T14 \rightarrow T15 \rightarrow T16$. For $wf2$, the shortest path that starts with $T21 \rightarrow T22 \rightarrow T25 \rightarrow T26 \rightarrow T27 \rightarrow T28 \rightarrow T21 \rightarrow T22 \rightarrow T23$ is $T21 \rightarrow T22 \rightarrow T25 \rightarrow T26 \rightarrow T27 \rightarrow T28 \rightarrow T21 \rightarrow T22 \rightarrow T23 \rightarrow T24 \rightarrow T27 \rightarrow T28$ (see Figure 12 (d)). These two shortest paths are fed into Algorithm 4 and the Earliest Completion Time *ECT* (55) is calculated from the resulting conflict-free execution trace. Since $ECT > HD$ we can predict that an exception may occur in the future. In the next section we will perform an experiment to show how different prediction time points can affect the prediction accuracy.

EXPERIMENT

To validate our algorithm we perform a simulation experiment by generating a set of workflow instances using Algorithm 6. The workflow instances are generated based on randomized execution times for each task.

Algorithm 7 shows how we validate our prediction result. In this algorithm, we check if an instance will have an actual exception or not, and compare it with the predicted result. If they match, we say that the prediction is correct.

Random Task Duration

In our experiment, we use Algorithm 6 to generate 6000 workflow instances (3000 for each workflow specification $wf1$ and $wf2$). Exception prediction is then performed for each pair of workflow instances at every 20 time units (except when $p=0$ and $p >=$ end time of these instances). Figure 13 shows the accuracy of the prediction algorithm after running Algorithm 7. The x-axis shows dif-

Figure 9. Applying algorithm 1 to find all possible paths

ferent time points (*p*) that we have chosen, and the y-axis indicates the accuracy of the prediction result. Since the latest end time of the generated instances is 138, we perform prediction at $p = 20$, 40, 60, 80, 100 and 120 respectively. The result shows that the accuracy of the proposed method reaches approximately 60% when $p = 60$. We can observe that the prediction accuracy increases with the time point. This makes sense since the longer the workflow instance executes, the more information we can get from the instance for predicting exceptions.

In addition, we further conduct six simulation experiments to analyze the influence of execution time of tasks on the prediction accuracy. In the following six scenarios we use various randomization procedures to generate the execution time of the tasks.

1. Early "long-task" scenario
2. Late "long-task" scenario
3. High resource usage "long-task" scenario
4. Low resource usage "long-task" scenario
5. Long duration "long-task" scenario
6. Short duration "long-task" scenario

For each scenario, we generate 6000 instances (3000 instances for each workflow specification *wf*1 and *wf*2) and perform exception prediction at various time points for each pair of workflow

Figure 10. (a) List of all possible paths for wf1; (b) List of all possible paths for wf2

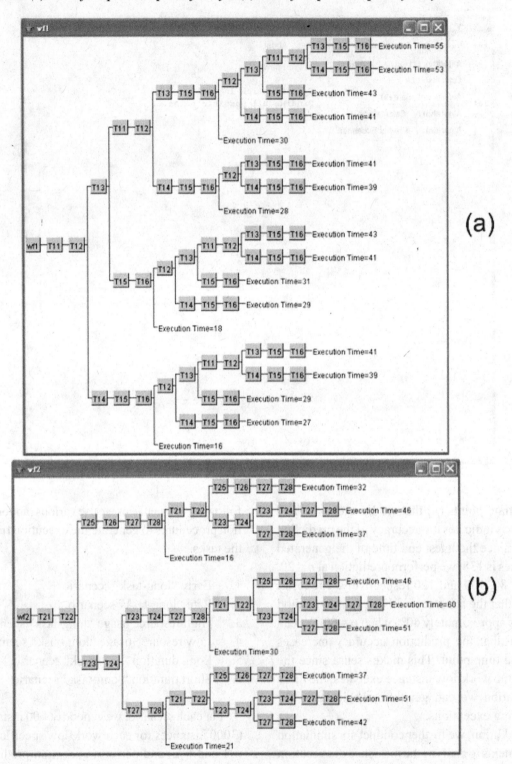

Figure 11. (a) Workflow instance at time point = 20; (b) Workflow instance at time point = 40

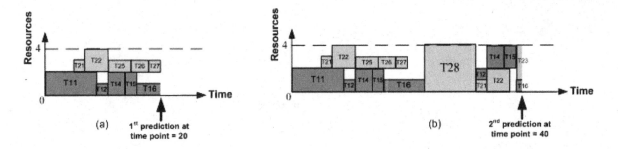

Algorithm 6. Generating workflow instances for simulation

Input: Workflow specification $wf1$, $wf2$ **Output:** Workflow instances $wi1$ and $wi2$ with random durations for tasks
1. Extract temporal and resource parameters from workflow specifications $wf1$ and $wf2$.   2. Create workflow instances $wi1$ and $wi2$ corresponding to $wf1$ and $wf2$. Set starting time of $wi1$ to 0, and randomize the starting time of $wi2$.   3. Continue the execution of $wi1$ and $wi2$ according to the control flows defined in the respective workflow specifications. During the execution, interleave and resolve any conflicts for tasks in $wi1$ and $wi2$ whenever necessary. Randomize the execution time of each task in $wi1$ and $wi2$.   4. For any loop within the workflow, a random number (which is less than the maximum number of allowable iterations) is chosen as the number of iterations.

Algorithm 7. Validating the accuracy of prediction

Input: Workflow specification $wf1$, $wf2$; Workflow instances $wi1$, $wi2$; time point p **Output:** Validity of the predicted result compared with the actual result
1. For Workflow instances $wi1$, $wi2$, pick a time point p and calculate the Hard Deadline (HD) using Algorithm 3.   2. Retrieve the actual end time $aet\text{-}wi1$ and $aet\text{-}wi2$ of $wi1$, $wi2$ from the audit trail.   3. Set Actual End Time (AET) = max($aet\text{-}wi1$, $aet\text{-}wi2$).   4. If $AET > HD$, then this workflow instance has an exception.   5. Perform the exception prediction using Algorithm 5, and record whether this workflow instance is predicted to have an exception or not.   6. Compare results from steps (4) and (5). If the actual result is the same as the predicted result, then we say the prediction is correct.

instances. Table 2 shows the statistics about these scenarios. The total number of prediction points (attempts) varies from one scenario to another since some simulated workflow instances have shorter end times.

In the following sections, we summarize the experiment results from these scenarios.

Early "Long-Task" Scenario

In this scenario, certain tasks are randomly selected to have longer execution times (compared to the predefined maximum allowable duration). Selection is done for those tasks which are executed before $p=30$. Figure 14 shows the accuracy of the proposed prediction algorithm. The result shows that nearly 50% accuracy is achieved at an early stage ($p=40$) compared to the result from Figure 13. It also shows that the performance of the proposed algorithm increases when tasks exceed their maximum allowable duration in the early phase of the workflow execution.

Figure 12. (a) Longest paths when p = 20, (b) Shortest paths when p = 20; (c) Longest paths when p = 40, (d) Shortest paths when p = 40

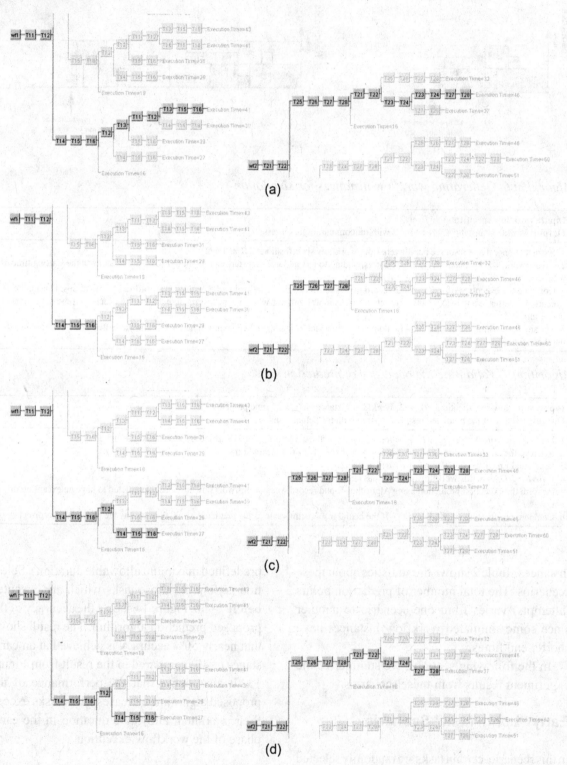

(a)

(b)

(c)

(d)

Figure 13. Prediction accuracy for workflow instances with randomized task duration

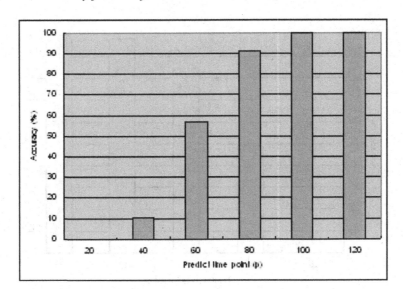

Table 2. Statistics on different scenarios

Scenario	No. of instances	No. of prediction points (attempts)	No. of actual exceptions	Max. end time of workflow instances
1	6000	5914	640	102
2	6000	5492	552	100
3	6000	7233	2542	137
4	6000	6815	1314	121
5	6000	7892	3138	148
6	6000	5977	801	97

Late "Long-Task" Scenario

In this scenario, certain tasks are randomly selected to have longer execution times (compared to the predefined maximum allowable duration) at the later phase of the workflow. Selection is done for those tasks which are executed after $p=50$. Figure 15 shows the accuracy of the prediction algorithm.

The result reveals the lower accuracy in prediction compared to the previous scenario. The maximum accuracy only reaches 60% when $p=80$. The low accuracy result is caused by the late scheduling of tasks which have longer execution time in the simulated workflow instances. Due to the delayed execution of "long-task" at later

stages of the workflow, the prediction algorithm has less information about the future execution situation. As a result, the accuracy of prediction is affected. It is also interesting to see that 20% accuracy in prediction is achieved at $p=40$ even though "long-tasks" are introduced at a later stage of the workflow (when $p>50$).

High Resource Usage "Long-Task" Scenario

In this scenario, certain tasks with resource usage of more than 2 units are randomly selected and assigned with longer execution times (compared to the predefined maximum allowable duration).

Figure 14. Prediction accuracy on early "long-task" scenario

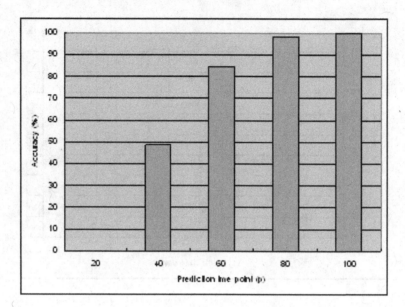

Figure 15. Prediction accuracy on late "long-task" scenario

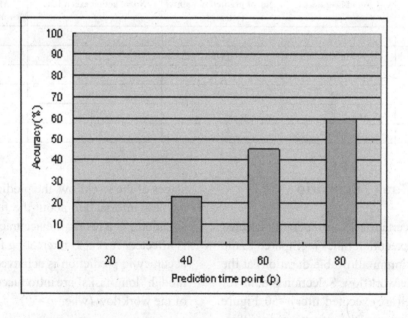

Figure 16 shows the accuracy of the prediction algorithm. In concurrent workflows, tasks with a high resource demand are more likely to affect other competing tasks since any delay in these tasks can severely reduce the level of available shared resources in the workflow for an extended

period. Such a situation can significantly increase the chance of resulting in an exception.

Our simulation result also confirms that a high number of actual exceptions is generated in this scenario (see Table 2). However, the prediction accuracy is similar to the experiment result from

Figure 16. Prediction accuracy on high resource usage "long-task" scenario

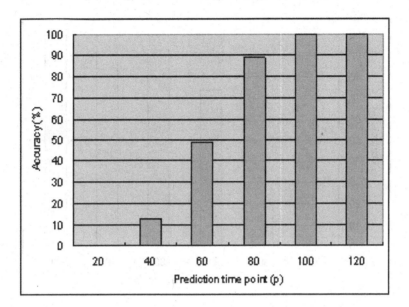

Figure 13. It shows that an increase in the number of actual exceptions does not affect the accuracy of the prediction algorithm.

Low Resource Usage "Long-Task" Scenario

In this scenario, certain tasks with resource usage less than 3 units are randomly selected and assigned with longer execution times (compared to the predefined maximum allowable duration). In contrast to the previous scenario, Figure 17 shows the increase in the accuracy of the prediction algorithm. Our simulation result also shows that a low number of actual exceptions is generated in this scenario (see Table 2). This is in line with our expectation since any delays in tasks with low resource usage have a lower chance of generating an exception.

Long Duration "Long-Task" Scenario

In this scenario, tasks with predefined execution time greater than 3 units are randomly selected and assigned with longer execution times (compared

to the predefined maximum allowable duration). Figure 18 shows the accuracy of the prediction algorithm. The result is similar to that of the random task duration experiment (see Figure 13).

Short Duration "Long-Task" Scenario

In this scenario, during the simulation, tasks with execution time less than 4 units are randomly selected and assigned with longer execution times (compared to the predefined maximum allowable duration). Figure 19 shows the accuracy of the prediction algorithm. In this scenario, short tasks are favored for creating a delay and as a result, a significantly lower number of actual exceptions is generated (see Table 2). The maximum end time of the workflow instances is also shorter compared to other scenarios. The result also shows that the prediction accuracy is similar to that of the low resource usage "long-task" scenario.

In summary, our experiment results show that the proposed algorithm is able to predict more accurately at $p=80$ (at slightly over half-way to completion). The experiment result also reveals

Figure 17. Prediction accuracy on low resource usage "long-task" scenario

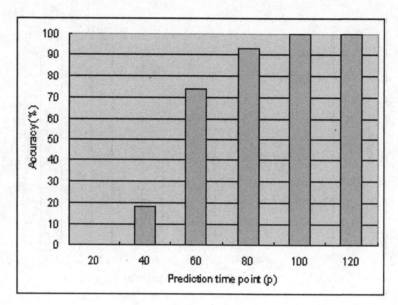

Figure 18. Prediction accuracy on long duration "long-task" scenario

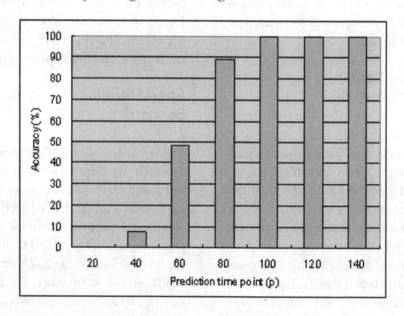

that the execution of long duration tasks and their positions within the workflow instances have a significant impact on the accuracy of the prediction algorithm. The result shows that the proposed algorithm is more effective in predicting excep-

tions for instances where long duration tasks are scheduled (or executed) at the early phase of the workflow.

Figure 19. Prediction accuracy on short duration "long-task" scenario

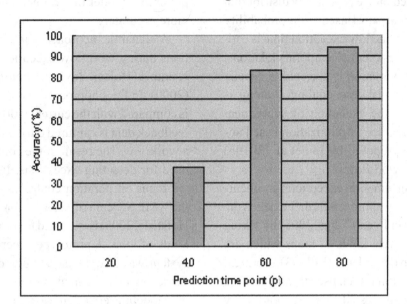

RELATED WORK

Critical paths are widely used in predicting as well as in handling exceptions. For detecting artifact anomalies in temporal structured workflows, Hsu and Wang (2011) unfold loops into a decision structure with three branches. The first branch represents the case when there is no iteration. The second branch represents the case when there is a single iteration. The last branch represents the case for maximal iterations. The resulting decision structure is then used for the analysis of structure and temporal relationships between artifact operations to reveal any anomalies buried in the workflow. In the algorithm for handling temporal exception (Xie, Yu, & Kuang, 2009), the longest and shortest time to complete is used to detect potential exceptions at the process instantiation or during run time. When a potential exception is detected, the slack time of remaining activities is reduced so that the process can complete before the overall deadline. Son and Kim (2001) also propose a method for determining the minimum number of servers for the activities within the critical path of a given workflow specification.

Their approach maximizes the number of workflow instances that satisfy the deadline and hence improve the performance of time-constrained workflow processing.

The M/M/1 queuing network based approach for identifying the critical path of a workflow model is detailed by Son et al. (2005). In their approach, loops are transformed into sequence control constructs for finding the longest execution paths. The main difference between their approach and ours is that in their approach, each activity is considered as an independent M/M/1 queuing system and the average execution time of a workflow instance in an activity is derived from the sum of the average servicing time and the average waiting time of the activity, whereas in our approach, an exhaustive search (breadth-first search) is used to identify the critical path based on the maximum allowable execution time of activities. In addition, the waiting time at the queue of the activity is implicitly taken into account when the conflict-free execution trace is derived from the resulting critical paths.

For preventing temporal violations in scientific workflows, Liu et al. (2010a) proposed an estimation method based on the probability distribution model of the critical path. Based on the probability distribution, the workflow execution time is estimated for detecting potential temporal violations. Once a potential violation is detected, a genetic algorithm based local rescheduling strategy is used to speed up the execution of subsequent activities. An ant colony optimization based approach was also proposed by Liu et al. (2010b) for rescheduling workflows.

For predicting temporal exceptions in concurrent workflows, our approach further extends the algorithm given in Li and Yang (2005) by taking into account different kinds of loops which are identified by Van der Aalst et al. (2003). A critical path based algorithm is then used to calculate the duration of tasks in nested loops with arbitrary depth. In resource sharing, our approach allows sharing of identical resources whereas the approach proposed in Li and Yang (2005) only considers sharing of single unique resources by concurrent workflows.

Besides temporal exceptions, algorithms for predicting other types of exceptions are also proposed in a number of research reports. For instance, Yuan, Ding, and Sun (2008) propose a Support Vector Machine (SVM) to forecast potential exceptions in a workflow. In their approach, the state and operation instances of the workflow captured in a Label Petri Net are used for training the SVM.

A knowledge-based approach is also proposed by Klein and Dellarocas (2000) to determine when a workflow is going to fail. In their approach, rules are created for detecting exceptions during the workflow definition stage. The approach proposed mainly focuses on prediction of exceptions at the design time of the workflow. However, in our approach, audit trail information of the workflow instance is used to dynamically generate the execution trace which is then used for exception prediction. Therefore our approach produces a prediction model that reflects the real situation more accurately.

A dynamic approach for detecting exceptions during workflow execution is proposed in Kammer, Bolcer, Taylor, Hitomi, and Bergman (2000). In their approach, historical information is compared with the current workflow status and feedback data is captured during the execution of a workflow. The result of the comparison is then used for detecting exceptions. In our approach, we focus on iteration control flow patterns and the XML Schema of YAWL (Van der Aalst & Ter Hofstede, 2005) is extended for modeling different kinds of loop. A prototype exception prediction system was also developed based on the YAWL workflow management system.

Modeling errors in workflow specifications can often lead to exceptions. An algorithm for validating a workflow model represented in the form of a graph is proposed in Lu, Bernstein, and Lewis (2006). The algorithm identifies the pre-conditions and post-conditions of a workflow and checks if the input and output edges of the workflow follow a valid pattern or not. In contrast, our approach does not validate the correctness of the workflow specification during design time. Instead, our algorithm utilizes predefined temporal constraints and run time information for predicting exceptions.

Exception prediction based on data warehousing and mining techniques was proposed in Grigori, Casati, Dayal, and Shan (2001). In their approach, audit trail data from a workflow management system is extracted and stored in a relational database. The audit trail data is then used in generating rules and decision trees for exception prediction in future workflow instances. The main difference between our approach and the approach proposed by Grigori et al. (2001) is that our approach allows prediction of exceptions in concurrent workflows which are required to share identical resources.

CONCLUSION

In this chapter, we describe a critical path based approach for predicting temporal exceptions in resource constrained concurrent workflows. These workflows form a highly dynamic element of a larger digital business ecosystem. During design time, we use a brute-force approach to find the paths with maximum and minimum path length. During run time, we create conflict-free execution traces for concurrent workflow instances to resolve both time and resource constraints. The conflict-free execution traces are calculated based on the critical paths as well as the shortest paths of the workflow. These execution traces are then used for predicting exceptions at selected time points. The accuracy of the proposed prediction algorithm is analyzed based on a number of simulation scenarios.

The proposed exception prediction algorithm was developed based on YAWL (Van der Aalst & Ter Hofstede, 2005) Editor 1.4 and Java 5.0. YAWL Editor is an open source tool written in Java and allows modeling of workflow specifications with different workflow patterns (Van der Aalst et al., 2003). It also supports the export of workflow specifications to XML formats.

The main contributions of our research work are two-fold; from the theoretical standpoint, we contribute to the calculation of critical paths for concurrent workflows which contain iterative control-flow patterns. The proposed critical path based approach can be effectively used to calculate the hard deadlines of the workflows. In addition, the proposed approach takes into account identical resources and real time audit trail data for predicting temporal exceptions in multiple stages. From the practical standpoint, our research opens the door to the further development of mechanisms for deriving temporal constraints and predicting exceptions for workflows which are designed with complex control-flow patterns (Van der Aalst et al., 2003).

ACKNOWLEDGMENT

This research was funded by the University of Macau under grant RG074/09-10S/SYW/FST "Resource Assignment in Business Process Simulation, Performance Monitoring During Process Enactment, and Change Management for Process Improvement."

REFERENCES

Boley, H., & Chang, E. (2007). Digital ecosystems: Principles and semantics. In *Proceedings of the Inaugural IEEE International Conference on Digital Ecosystems and Technologies* (pp. 398-403). Cairns, Australia.

Casati, F. (1998). *Models, semantics, and formal methods for the design of workflows and their exceptions.* Unpublished doctoral dissertation, Politecnico di Milano.

Grigori, D., Casati, F., Dayal, U., & Shan, M. C. (2001). Improving business process quality through exception understanding, prediction, and prevention. In *Proceedings of the 27th International Conference on Very Large Data Bases* (pp. 159-168). San Francisco, CA, USA.

Hsu, H. J., & Wang, F. J. (2011). Detecting artifact anomalies in temporal structured workflow as reusable assets. In *Proceedings of the 35th IEEE Annual Computer Software and Applications Conference Workshops* (pp. 362-367). Munich, Germany.

Kammer, P. J., Bolcer, G. A., Taylor, R. N., Hitomi, A. S., & Bergman, M. (2000). Techniques for supporting dynamic and adaptive workflow. *Computer Supported Cooperative Work, 9*(3-4), 269–292. doi:10.1023/A:1008747109146

Klein, M., & Dellarocas, C. (2000). A knowledge-based approach to handling exceptions in workflow systems. *Journal of Computer Supported Collaborative Work, 9*(3-4), 399–412. doi:10.1023/A:1008759413689

Li, H., & Yang, Y. (2005). Dynamic checking of temporal constraints for concurrent workflows. *Electronic Commerce Research and Applications, 4*(2), 124–142. doi:10.1016/j.elerap.2004.09.003

Liu, X., Chen, J., Wu, Z., Ni, Z., Yuan, D., & Yang, Y. (2010b). Handling recoverable temporal violations in scientific workflow systems: A workflow rescheduling based strategy. In *Proceedings of the 10th IEEE/ACM International Conference on Cluster, Cloud, and Grid Computing* (pp. 534-537). Melbourne, Australia.

Liu, X., Yang, Y., Jiang, Y., & Chen, J. (2010a). (Accepted for publication). Preventing temporal violations in scientific workflows: Where and how. *IEEE Transactions on Software Engineering.*

Lu, S., Bernstein, A., & Lewis, P. (2006). Automatic workflow verification and generation. *Theoretical Computer Science, 353*(1-3), 71–92. doi:10.1016/j.tcs.2005.10.035

Son, J. H., Kim, J. S., & Kim, M. H. (2005). Extracting the workflow critical path from the extended well-formed workflow schema. *Journal of Computer and System Sciences, 70*(1), 86–106. doi:10.1016/j.jcss.2004.07.001

Son, J. H., & Kim, M. H. (2001). Improving the performance of time-constrained workflow processing. *Journal of Systems and Software, 58*(3), 211–219. doi:10.1016/S0164-1212(01)00039-5

Van der Aalst, W. M. P., & Ter Hofstede, A. H. M. (2005). YAWL: Yet another workflow language. *Information Systems, 30*(4), 245–275. doi:10.1016/j.is.2004.02.002

Van der Aalst, W. M. P., Ter Hofstede, A. H. M., Kiepuszewski, B., & Barros, A. P. (2003). Workflow patterns. *Distributed and Parallel Databases, 14*(1), 5–51. doi:10.1023/A:1022883727209

Xie, T., Yu, Y., & Kuang, G. (2009). A time exception handling algorithm of temporal workflow. In *Proceedings of the 2009 IEEE International Symposium on Parallel and Distributed Processing with Applications* (pp. 641-646). Chengdu, China.

Yuan, H. T., Ding, B., & Sun, Z. X. (2008). Workflow exception forecasting method based on SVM theory. In *Proceedings of the 2008 International Symposium on computational Intelligence and Design* (pp. 81-86). Hefei, China.

Chapter 12
Geographic Information Retrieval and Text Mining on Chinese Tourism Web Pages

Ming-Cheng Tsou
National Kaohsiung Marine University, Taiwan

ABSTRACT

The World Wide Web (WWW) offers an enormous wealth of information and data, and assembles a tremendous amount of knowledge. Much of this knowledge, however, comprises either non-structured data or semi-structured data. To make use of these unexploited or underexploited resources more efficiently, the management of information and data gathering has become an essential task for research and development. In this paper, the author examines the task of researching a hostel or homestay using the Google search web service as a base search engine. From the search results, mining, retrieving and sorting out location and semantic data were carried out by combining the Chinese Word Segmentation System with text mining technology to find geographic information gleaned from web pages. The results obtained from this particular searching method allowed users to get closer to the answers they sought and achieve greater accuracy, as the results included graphics and textual geographic information. In the future, this method may be suitable for and applicable to various types of queries, analyses, geographic data collection, and in managing spatial knowledge related to different keywords within a document.

1. INTRODUCTION

The Internet has an abundance of information and has become one of the most significant resources in our daily lives. These resources contain an enormous number of elements, the retrieval of

which depends mostly on webpage search engines. However, most of this knowledge comprises either non-structured or semi-structured data (Mitra et al., 2003), and at present the ability of regular search engines is limited to the retrieval of basic keywords, rather than analysis of the subject matter and content of the webpage itself; these

DOI: 10.4018/978-1-4666-0023-2.ch012

applications then, are still far from perfection. For the reasons mentioned above, much research on efficient message and data extraction has been focused on the effective management of data.

This research has resulted in many developments in information retrieval and data mining strategies. However, these strategies are mostly aimed at semantic data only. Based on a recent estimation, about twenty percent of web-users enquire for spatial context, such as searching for restaurants, theatres or academic institutions; in addition, eighty percent of those web-users type their queries in services with a location orientation (Kornai & Sundheim, 2003; Souza et al., 2005), such as searching for New York delicacies or hostels in Portland. However, they are hindered not only by the current state of development of search engines, but also by the limitations imposed by differences between written English and Chinese. Because it is not possible to put a space between written Chinese characters, this has caused perplexities in relation to word segmentation. For regular search engines therefore, this is quite a barrier to effectively indexing web-content written in Chinese. For example, if the user is browsing for "Portland 民宿" [read as 'min su'], this means hostels in Portland; a regular search engine might only perform some search on either Portland or on 民宿 separately from the database. This inquiry therefore lacks thematic and spatial context, and this creates a big gap between the demand and supply of the query (Buyukkokten et al., 1999).

Although the Geographic Information System (GIS) has the capability to handle geographic data, access to spatial data is mostly limited to coordinates created by geometric space expressions. Even though Web GIS is presently available, it is quite difficult to combine GIS analysis and text analysis, since the former's usability does not surpass traditional GIS. Generally, people express their knowledge of geographic locations by using spatial content such as place names, labels, addresses or even telephone numbers, instead of using geometric coordinates (Jones et al., 2001).

The content of a webpage is a concrete example of this phenomenon, where people express location data in a spatial context. However, by using this implicit spatial data, the connection between web pages and any particular location can be established.

In regard to the above-mentioned requirement, neither text mining nor GIS are sufficient in themselves; instead, a new search engine needs to be built with the capability to manage thematic and spatial contexts, not separately but simultaneously. Geographical Information Retrieval (GIR) has become one means of satisfying such queries, and it is starting to receive some attention from academic and commercial communities (Byrd & Ravin, 1999; Jones et al., 2002). GIR combines text mining, information retrieval (IR) and geographic metadata with spatial cognition related to the research area. The main purpose of this is to retrieve spatial-related information from text documents more accurately.

Roughly seventy-five to eighty percent of human activity, especially traveling, is related to geographic location (Lee et al., 2007). Following the increasing trend to holiday outings in Taiwan, there has been a noticeable expansion in agricultural leisure farms and rapid growth in the number of hostels and homestay facilities. At a conservative estimate there are more than ten thousand homestays in Taiwan, both listed and unlisted and about two hundred new hostels are being opened every year. Consequently, the number of related web pages is also increasing tremendously. Nevertheless, some of these web pages are not mainly related to this subject, or they offer only text descriptions, rather than concrete spatial information display. Such information may confuse users during their searches, and this is where GIR can provide searches with better functionality.

This paper focuses on the topic of hostel or homestay searches within Chinese-based webpages, using the Google Search Web Service combined with the Chinese Word Segmentation

System developed by Academia Sinica, and Text Mining technology. It examines the results of web searches, mining, retrieving and ranking of spatial contexts and semantic data in relation to the purpose of finding the webpage that is most closely related in context and geography to the topic 'hostel or homestay'. Furthermore, by using geographic information retrieval (GIR) and *regular expression*, these results were filtered to obtain the most usable results in terms of geographic information. By means of geocoding within the Google Map API (Application Programming Interface), geographic information and text data could be featured and mapped directly into Google Map for the end user. This method was used to enhance the otherwise inadequate function of regular search engines by obtaining results that include graphic and text data tailored to satisfy the user's demand. Additionally, in the future, this kind of method could be used for data mining and spatial knowledge management.

2. RELATED RESEARCH

In the past, research on Geographical Information Retrieval was based on two categories: context entity-based and content-based (Amitay et al., 2004; Martins & Silva, 2005). Both forms use a webpage's actual location as the main pedestal, based on the webpage's IP address and domain name service in order to find the computer server's engine and the possible location of the original webpage. The Gtrace tool (Periakaruppan & Nemeth, 1999) is one research tool available, and Buyukkokten et al. (1999) indeed found that based on accessing a domain name server database, information on the domain manager could be found. By using the manager's telephone area code or postal code number, we could determine a possible location for the webpage and thereby determine spatial relationships between webpages. This particular method could only identify a possible location, not spatial contexts within the

webpage. Additionally, as mobile telecommunication starts to become available, webpage locations can function as a reference, but without actually reflecting spatial contexts (McCurley, 2001; Vogel et al., 2005). The original location of a web page therefore, can only provide very limited spatial information, whereas our research emphasizes the retrieval of spatial information within the content of a website.

2.1 Research Planning

GIPSY (Georeferenced Information Procession System) is considered to be the first research planning tool that automatically indexes documents based on place names and phrases (Vestavik, 2008). GIPSY incorporates a gazetteer that includes place names and the names of other significant landmarks (Woodruff & Plaunt, 1994). SPIRIT (Spatially-aware Information Retrieval on the Internet) is a scheme that involves geographical thesauri and spatial hierarchies for use in geographic information retrieval. This method is similar to the concept developed by Larson (1995), improving search methods within the WWW (Purves et al., 2002). GeoXwalk is a project aimed at providing a British and Irish gazetteer service. This project uses a middleware component comprising APIs supporting open protocols to issue spatial and/or aspatial search queries. Within a query, a semi-automatic document scanner can parse around non-geographically indexed documents for place names, relate them to the gazetteer and return appropriate coordinates for confirmed matches (Reid, 2003). GeoVSM (Geographic Vector Space Model) combines coordinate-based geographic indexing with the key-word based vector space model in representing an information space. Relevancy measures are based on matching the similarity of spatial context and thematic measures, which can be combined into one single measuring system to fit the space and location results (Cai, 2002).

2.2 Current Search Engines Provided with GIR

At the present moment, there are many search engines, such as Google, Yahoo!, Ask Jeeves, and MSN, which provide a Local Web Search service using or displaying mapping ability to show more accurate search results. For example, Google Local (http://local.google.com), uses the Google Map function to show a map that allows the user to search for hotels, stores, restaurants and so on in the requested area, in a much simpler way. This function is different from the original Google search engine in that it depends on the website providing a city information database, for example CitySearch (http://www.citysearch.com) or WCities (http://www.wcities.com). It can be concluded that Google Local is more likely a webpage that can provide the end user with city guide information, but it is not a GIR website. MSN City Guides (http://local.msn.com) and Ask Jeeves Local also provide the same services as mentioned above. Another, similar service provider to Google Map is Yahoo! Local Maps (http://local.msn.com). AOL Local (http://local-search.aol.com), on the other hand, incorporates all of its abundant information such as local news and events, movie information and so on, within specific regions.

2.3 Related Applications

Some Taiwanese scholars are using Chinese Word Segmentation and WEB-GIS to gather information from semi-structured Chinese tourism data and then transferring it into longitude and latitude formats. Then, based on traveling schedules, a tourism map is produced (Lee et al., 2007). However, this process is not based on thematic analysis. Tezuka et al. (2006) notes that from the content of traveling blogs, spatial information such as location, timing, action and activity content can be collected. By using this information or data, a traveling experience map can be built,

and this particular system could be used to gather the places or locations into a webpage using the parsing of the gazetteer and ontology. McCurley (2001) used the content of geospatial information within a webpage, including telephone numbers, addresses, place names, locations of links between web addresses and contents of webpages, in order to find out the way and results of corresponding geospatial mapping. Tezuka and Tanaka (2005) use Cognitive Geography as their frame of reference, where landmarks are cognitively significant geographic objects. Landmarks, significant geographic objects have always enjoyed outstanding space recognition by human beings. Although the World Wide Web (WWW) provides a rich source of region-related information, this is also becoming a drawback for human societies. It can also represent the understanding of a general, non-professional public on various matters. However, by querying combinations of spatial relationships and by data mining, landmark information within a webpage can be found. Furthermore, identifying the important knowledge on each landmark enables replacement of the previous questionnaire method with cognitive geography, thereby obtaining a better result.

3. RESEARCH METHOD

The data this research needs to manage includes two major groups: semantic and spatial data. How to derive semantic information, spatial data, spatial information, and the integration between semantic and spatial information from the original text data, is our focus. In contrast to the past, the obtaining of data, managing and analysis of data, and finally the resulting demonstration of data, are all completed through the Internet and Web 2.0. In regard to text semantics, not only traditional Chinese text mining techniques, such as word segment processing in the Chinese Word Segmentation System, the vector space model representing documents, and utilizing the BM25 function to calculate document similarities, were utilized, but a subject keyword database was also established for reference. In regard to spatial data management, most past research has focused on the analysis and retrieval of spatial semantics; without including the Geographic Information System's management of spatial data. This has limited the integration of semantic and spatial information and actual spatial location. Firstly, a database for matching place names is established in this research. The content includes not only place names and assigned numbers, but also something different: that is, spatial objects' minimum bounding rectangle (MBR) coordinate data, which is often used during spatial inquiries in the Geographic Information System. This is not just a single point coordinate. It also establishes the hierarchical relationship between place names and spatial objects, to retrieve address data through regular expression. It utilizes Regional Co-occurrence Summation (RS), to verify the relationship between semantics and space. Finally, though geocoding and presentation of maps, semantic information and spatial information are visually integrated for demonstration. We believe through this research, the effect of data mining of text data, visual data, and spatial data together can be achieved. In the following sections, the methodologies of text data and space data management, and their roles, relationship, and integration in this research are explained. The research design structure is shown in Figure 1.

3.1 Text Data Management

The text data retrieval is based on Chinese semantics, and through data mining, similarities between the retrieved webpage content and the searched subject are calculated. Webpages that rank higher in similarity with subject semantics are identified and then, within the content of the webpages, spatial information is retrieved and filtered. The research methodology includes not only relevant text mining theories and methods

such as vector space model and BM25 document similarity calculation function, but also Chinese Word Segmenting processing.

3.1.1 Thematic Keyword Database

The first action is to plan different kinds of subjects, such as hostel or homestays, restaurants, hotels and so on. Through interviewing experts and the collection of correlated literature, we can reorganize and generalize the groups of keywords that best represent and typify each subject, and then enter the established keywords one by one into the database. For example, for hostel or homestay, these might include hostel, transportation, accommodation, traveling, scenic area, board and lodging, reservation, and so on. When the user selects a subject to query, it is also this group of keywords to which this subject corresponds, that provides document and query results.

3.1.2 Vector Space Model – VSM

Generally, encoding work is done to each article before text mining is begun. The term-weighting approach is applied in order to find important terms within a document. In addition, term extraction and natural language handling are used (Boguraev & Neff, 2000; Gey et al., 2005). Using VSM to represent one group of articles or webpages means that each article or webpage uses one group of weighted keyword vectors as representative (Salton & Buckley, 1988). The documents would be represented in the form:

$$D = (t_i, t_j, ..., t_p)$$

where D identifies the document being addressed and each t_k identifies a content term assigned to the sample document D. Similarly, a typical query might be formulated as follows:

$$Q = (q_a, q_b, ..., q_r)$$

Figure 1. Framework of geographic information system

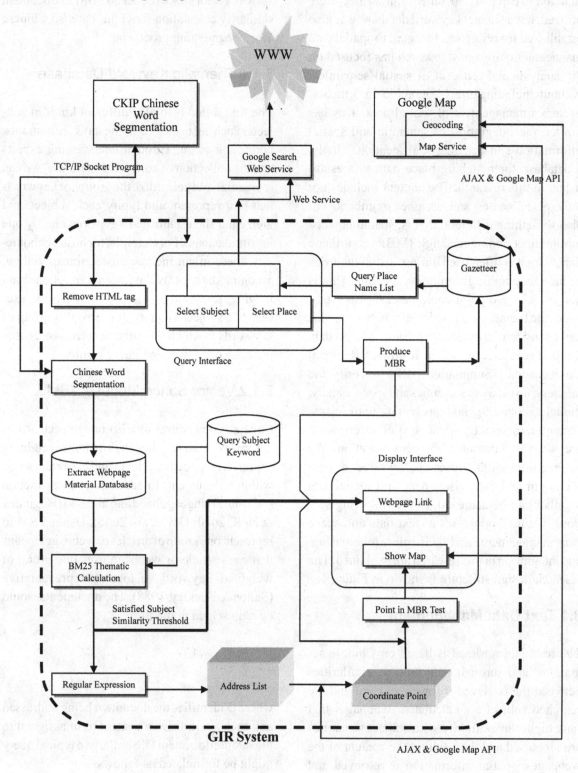

where Q identifies the query and each q_k represents a term assigned to query Q.

Calculations on Term Frequency (TF), Document Frequency (DF) and Inverse Document Frequency (IDF) are then carried out. TF identifies the frequency with which t_k appears within document D. The larger the TF, the more important the word is to the document being searched. Additionally, if we are using frequency of words in order to determine the keywords, we might find some words that are commonly used but not significant, such as "the", "a", "it". These words are not usually important words within the document; therefore we need the IDF to make a revision. IDF is calculated by the following formula:

$$IDF(q_i) = \log \frac{N - n(q_i) + 0.5}{n(q_i) + 0.5} \quad (1)$$

Where N represents the total number of documents in the collection, $n(q_i)$ represents the number of documents containing term q_i. When the IDF is large, this means that the term emerges less often within the documents. After calculating the number mentioned above, similarity level calculations can be done between document queries and criteria queries. Based on the similarity and its ranking function, the semantic similarity document is chosen and treated as the search result. The most common corpus weighting is TF × IDF as the cosine of VSM and the BM25 weighting scheme, in order to calculate schematic similarity.

3.1.3 BM25 Similarity Calculation Function

BM25 was introduced by Robertson (1994). The difference between this formula and others with a cosine function is that it is based on the probabilistic retrieval framework. Using the retrieval function, it ranks a set of documents based on the query terms appearing in each document. BM25 and its newer variants have represented and con-

tinue to represent state of the art retrieval functions used in document retrieval. Since this research is based on search queries and keywords, similarity calculations are applied to the retrieval document. Since the condition of the query is comparatively shorter, it is more complicated if we firstly need to compare the relationship between one complete document and another and then calculate the term weightings. By using BM25, we can avoid this particular difficulty and its associated ineffectiveness. Therefore, within this particular research, in regard to similarities between thematic and query criteria, BM25 can be used for this calculation, as described below:

Given a query Q, containing keywords $q_1,...,$ qn, the BM25 score of a document D is:

$$\text{score}(D,Q) = \sum_{i=1}^{n} IDF(q_i) \cdot \frac{f(q_i, D) \cdot (k_1 + 1)}{f(q_i, D) + k_1 \cdot (1 - b + b \cdot \frac{|D|}{avgdl})} \quad (2)$$

Where $f(q_i, D)$ is q_i's term frequency in the document D, $|D|$ is the length of the document D (number of words), and *avgdl* is the average document length in the text collection from which documents are drawn. k_1 and b are free parameters. $IDF(q_i)$ is the IDF (inverse document frequency) weight of the query term q_i. By using a BM25 calculation, every document can be allocated a BM25 score; ranking and filtering to find the matching documents according to their relevance to the search query are the next and final step.

3.1.4 CKIP Chinese Word Segmentation System

In European languages there is a clear gap or segmentation between each word or phrase, which simplifies data collection. This is unlike Asian languages such as Chinese or Japanese, where there is no gap or segmentation between words. In these languages, identification of words is more challenging. In this particular case, the information retrieval task becomes very difficult as preliminary processes such as VSM, TF, DF, IDF and BM25 need to find the words or phrases within the document before any other process. In order to increase accuracy in data extracting, Academia Sinica researched and developed a tool called the Chinese Word Segmentation System. The strong point of this particular system is combining heuristic and statistical rules, as in mixing the segmentation rules. According to experimental results, this system includes approximately 100,000 words within its database and the segmentation accuracy reaches and exceeds ninety percent (May & Chang, 2003). CKIP not only provides an online service for its users, but also opens up API access for calling up the service program, and XML access for data sharing applications. This particular research chose a TCP/IP Socket connection as intermediary to deliver the information packet and text to the server database.

3.2 Spatial Data Management

Geographic names and concepts that are stored in documents but have not been processed through spatial information are merely general vocabulary, and cannot be connected with space concepts. Therefore, spatial information is hidden within the documents. In order to highlight spatial meaning within documents, we perform the following spatial data processing for documents to satisfy semantic similarity requirements.

3.2.1 Gazetteer

This research relates to a pre-established geographic information bank (geographic database). This information database's column contains the place's serial number, place name, the top left-hand corner X coordinate of the place's Minimum Bounding Rectangle - MBR, the top left-hand corner Y coordinate of MBR, the right bottom X coordinate of MBR, the right bottom Y coordinate of MBR, the place's position respective to the spatial structure hierarchy (e.g., county city is 1, villages and town areas are 2 and so on). It also contains the serial number of spatial structure addresses located on the level above, in that particular structure hierarchy. After having this information database, the geographic name may be used to look up its MBR, or perhaps using the present inquiry range of MBR to find out what is included within the area of search and to use the WWW to look for what is included within the related area from the webpage information. At present, this research is only aimed at the county and city levels; however, in the future, more complete information databases may be established.

3.2.2 Regional Co-Occurrence Summation (RS) and Spatial Sentence

Tezuka et al. (2006), during the analysis of place names using DF or TF found that they might not be able to differentiate whether or not these place names are applicable within the spatial context. Consequently, it is often the case that company subdivisions, schools or chain stores receive a high rank in DF but actually do not have any significance in relation to the query. In order to avoid such inaccurate results, it is necessary to measure the frequency with which a place name is actually used in a spatial context. In calculating a regional co-occurrence summation (RS), we assume that when two neighboring place names appear in the same document, it is likely that these two place names were both used in a spatial context. In terms of text mining, we consider a co-occurrence of two neighboring place names as an indicator of spatial context. Co-occurrence is a commonly used measure for term relationships in text mining (Salton, 1968; Rijsbergen, 1979). According to this measurement, not only can we grasp spatial context, but we can also analyze the following ambiguities in place names. If the document can find neighboring place names or geographical relationships, therefore spatial contexts can be differentiated. Additionally, if a geographical term is combined with a spatial phrase trigger, spatial significance and the accuracy of search results can become more certain.

3.2.3 The Retrieval of Spatial Location Information

Word processing is always one of the most important matters in information retrieval research because the content of a document and the information being inputted by the user are not merely based on text or characters. The style of text materials such as XML, CSV, and others, can use simple functions and parsing. However, researchers still do not want to face free style situations. Therefore, when they are unable to input the search, the format needs to be changed. In order to face a loose string, we need more complicated algorithms to parse the content of the loose string and thereby support elements of fuzzy handling. The algorithm called the Regular Expression method was used in the 1950s by a neurophysiologist. Afterwards, a mathematician used the same method, in the parsing of text data. It supports a specific character to limit certain ideas from the subject group. In addition, after an extension, master combined filtering machines will be perfected in order to match the needs of the string (Jeffrey, 2006). This algorithm has become one of the standard features in a wide range of popular software development tools.

Using the number measured by the RS and the spatial sentence, the Chinese address material in the document may be discovered to contain the target characteristics. Within addresses, a geographic name usually follows a closely related place according to the respective geographical hierarchy. For example, Portland follows Oregon; this kind of hierarchy may be a very accurate spatial sentence. Furthermore, we may consider the road, lane, alley and house numbers within an address, as spatial

triggering words. The appearance of such a glossary can also deepen our grasp of some geographic information. Therefore, we may say that address material within the document is treated as one of the best data sources for recognizing geographic information such as a hostel or homestay facility. A hostel webpage inevitably contains a hostel's address so that by parsing the address material, the document's most direct geographic information can be grasped. The only issue is that the address material usually belongs to one kind of free format, but it may also conform to fuzzy rule logic. For example, in Taiwan, where one standard format is unable to be found, by penetrating the regular expression, address material can be provided to carry on fuzzy parsing for evidence of implication in geographic material.

This research was based on the Microsoft.Net Framework, utilizing Regular Expression as part of the tools, and was aimed at Chinese addresses by using the following fuzzy search pattern:

$$\text{"}\backslash w\{2\}[縣|市](\backslash d\backslash d\backslash d)?\backslash w\{2,3\}[市|鎮|鄉|區]\backslash w\{1,20\}\backslash d*\text{-}?\backslash d+號\text{"}$$

Where 縣 [read as 'xian'] – means county, 市 [read as 'shi'] – means city, 鎮 [read as 'tzhen'] – means town, 鄉 [read as 'xiang'] – means village, 區 [read as 'qu'] – means area, and 號 [read as 'hao'] – means number.

3.2.4 The Matching of Relationships between Documents and Space

After spatial information has been obtained from documents, merely representing this information through texts is not enough. Most past research was restricted to just this. Since it is spatial information, mapping is certainly the best platform. If it can be integrated with a Geographic Information System, the first benefit is that documents and spatial location can be visually matched; secondly, through showing a map, spatial distribution of the documents can also be seen. A proper visual presentation is more valuable than many words of description, for retrieving meaningful information from images is strength of human perception. If a Geographic Information System can also be utilized to conduct spatial analysis, space patterns of this information may be further explored. This has great potential for future application in similar business promotions, commercial zone analysis, and location-based service (LBS) provision.

In this research, Google Map fills a role similar to a Geographic Information System. Its geographic information functions are carried out on the Internet through Google Map and its map service application programming interface (Google Map API) (Davis, 2006). In this research, it provides the following functions:

1. **Development Platform:** As one of the Web 2.0 platforms, it should integrate with the system's AJAX program rapidly. It should also provide a comparatively traditional Internet geography information system and thus be more convenient and effective in terms of operational experience. The geography information function should be binding through the Internet. It recognizes an application program interface (Google Map API) developed from the map service station, Google Map. It may use general webpage composition technology development different to that utilized in the experience of Web GIS's in the past. It does not require the user to install a map browsing program in the first browsing occurrence, in order to simplify installation and operation.

2. **Operation and Demonstration Platform:** Within the map service, it should contain the newest settings of national base map information and fine resolution satellite pictures. It should not need to purchase the map separately—moreover, this service should continuously update revision map material.

3. **Geocoding:** It provides the function of geocoding where the user only needs to enter the address in text. This service can act according to the street address to carry on spatial interpolation and to calculate this address's possible latitude and longitude.

3.3 Data Gathering

3.3.1 Thematic and Spatial Query

So-called focused searches will search for the subject limited within a specific spatial scope and subject range, instead of in a comprehensive universal search. This system aims to promote search efficiency and integrate spatial and subject semantic queries. Based on the subject semantic query, the user chooses the subject to match the place as a condition, and then the Internet gathers the information. The search format is "place + subject keywords", for example, "Portland home-stay". From the subject keywords database, the group of related keywords is found. This keyword group is matched with the collection of documents found on the Internet in order to proceed with text mining on similar measures, while screening for the closest thematic document.

Speaking of spatial inquiry, the user may choose the following three input modes:

1. **Place text:** The user chooses the county or city of inquiry directly under the menu choice. The system will discover the particular MBR of the queried county or city, and treat this county or city name as another kind of keyword. The geographic name and subject keywords are grouped together within the keyword search group.

2. **Current map's extent (De Floriani et al., 1996; Laurini & Thompson, 1992; Larson, 1995):** sometimes the place that the user wants to inquire about is near several counties or cities, in which case a single

pull-down menu is quite insufficient. For example, the user possibly has an interest in Yushan national park (located in Taiwan's middle national park), but Yushan national park actually stretches across counties like Nantou, Hualien, Chiayi and Kaoshiung, and so it is not suitable to inquire by writing this input. As mentioned before, the user may use the current map range and adjust it to the Yushan national park scope, then using the map scope (Extent) query for the region. This also means that it is equal to the current display area of the MBR. In this way rectangular coordinates can be obtained. Furthermore, we can match rectangular coordinates and MBR within the geographic name database, and if both MBRs overlap, this query contains the county or city being looked for. However, if it overlaps with many counties or cities, this query will contain many county or city geographic names that become spatial keywords in the query. Additionally, the combination of the related keywords being searched might divide into query conditions, for example:

(subject X, county or city name 1) OR (subject X, county or city name 2) OR…

3. **Query on designated range (figure seven):** Basically this query method is similar to the second method, providing the function of range query but still using a query format. The operator may use the present map demonstration scope by dragging the query rectangle on the screen to establish the MBR query scope.

3.3.2 Hyperlink Traversal

There are many web pages that might satisfy the search conditions of the subject, but which do not offer spatial information. Sometimes, in order to obtain more explicit spatial information, we can use hyperlinks to link to the related information, or an existing webpage that clearly contains spatial information (McCurley, 2001). Based on such an idea, this system enables the user the possibility of continuing to visit (traverse) the hyperlink layer and to discover the spatial information.

3.4 Execution Procedure

Figure 2 explains the pseudo code of the system framework. First, based on the user's query, when the choice is a range query, it penetrates Google Map by using the current map's query range (extent) to gain the MBR. Then, comparison between the MBR and various geographic names in the thesaurus within the MBR is carried out; when both MBRs overlap, allocation of the geographic name into the geographic query name list will be done (Figure 2 lines 02~04). If a selection is made from a place name in a pull-down menu, then it will directly allocate this geographic name into the query geographic name list. The geographic name will be retrieved from within the gazetteer and based on this geographic name, a query on the corresponding MBR can also be done (Figure 2 lines 06~07).

Subsequently, based on a subject chosen by the user as a query criterion—for example, hostel or homestay, the query keywords list finds the related keyword group which represents the hostel or homestay and establishes a set of query subject keywords (Figure 2 line 09). At the same time, making a query on a collection of place names and on subject keywords as query conditions, automatic search results can be obtained by using a program to call up the Web Service format and the Google Search Web Service (Mueller, 2004). Furthermore, the above-mentioned web pages can be tallied and built into an HMTL document list. Such effects are actually equal to manually inputting the search keywords into the Google search webpage (Figure 2 line 10).

Afterwards, we can use Microsoft HTML Object Library to remove the HTML tags from

Figure 2. Geographic information system pseudo code

```
01  //Geographic Range Query Selection
02  if Query By Map Extent then
03    QueryMBR = Current Map Extent ;  //MBR: Minimum Bounding Rectangle
04    QueryPlaceNameList = LookupGazetteerForPlaceNameList(QueryMBR);
05  else  //Query By Place Name
06    QueryPlaceNameList .Add(UserInputQueryPlaceName);
07    QueryMBR = LookupGazetteerForMBR(UserInputQueryPlaceName);
08  end if
09  QueryKeywordList = LookupThematicKeywordDB(UserInputQueryTheme);
10  HTMLDocumentList = CallGoogleWebSearch (QueryTheme, QueryPlaceNameList);
11  PureTextList = RemoveHtmlTag(HtmlDocumentList);
12  DocumentList = CallCKIPWordSegmentationAndBuildVSM(PureTextList);
13  for each Document in DocumentList do
14    AddressStringList = empty;
15    Score = BM25ThematicCalculation(DocumentList, Document, QueryKeywordList);
16    if Score > User SetupThreshold then
17    DocumentAddressList = RegularExpression(PureText, AddressPattern);
18    if DocumentAddressList is empty then
19      recursively trace the hyperlink in HtmlDocument;
20    end if
21    for each AddressString in DocumentAddressList do
22      if AddressString not exist in AddressStringList then
23        AddressStringList.Add(AddressString);
24        CoordinatePoint = CallGoogleMapAPIGeocoding(AddressString);
25        if CoordinatePoint in QueryMBR then
26          Get Google Map and Draw CoordinatePoint;
27          Associate CoordinatePoint Icon with HtmlDocument;
28        end if
29      end if
30    end for
31    end if
32  end for
```

within the HTML document in order to obtain the pure text within the document list (Figure 2 line 11). Then, the TCP/IP Socket connection method is used to reach Academia Sinica's CKIP Chinese Word Segmentation, in order to segment each pure text document. This way, character word selection within the document can be obtained and this can present a character phrase set and vector space model within the document (Figure 2 line 12).

BM25 subject similarity computations are then carried out on the character word compilation and the query subject keywords are set for each pure text document. This is done in order to obtain a score for each document, matching scores on the query subject (Figure 2 line 15). If the similarity score surpasses the User Setup Threshold value, then the word string on the address extraction pattern will become the basis of a fuzzy comparison through Regular Expression. Address information can then be extracted from the document in order to create a document address list (Figure 2 line 17). If it is not possible to find any address list within the document, hyperlink tags within the HTML document can be surveyed by using the recursion method to continue the searches on the linked webpage to determine whether or not it contains the tallying material (Figure 2 line 18~20).

The next step is to judge whether or not each address in the address string already existed and had been indicated in the set Address String List (Figure 2 line 22~23). If not, it is necessary to transform the address string into the address's latitude and longitude coordinates by using the AJAX call method. The AJAX call method uses a geocoding function provided by Google Map API (Figure 2 line 24). This particular system, other than comparing similarities within the thematic aspect, also needs to continue screening for proximity. The result of this is that the position queried will be placed within the query region. Therefore, the "point in a polygon" test will be carried out in regard to the address' latitude, longitude and query scope (MBR) to judge whether

the coordinate points are situated within the query scope (Figure 2 line 25). If the address' latitude and longitude are situated within the query scope, this address location will be drawn on the Google Map and indicated by an icon. Furthermore, it will establish hyperlink relations connecting the map to the hyperlinks in the HTML document, thus facilitating the inspection of search results (Figure 2 line 26~27).

4. EXPERIMENT DESIGN AND RESULT EVALUATION

ASP.Net is the building block of this experiment. Relevant major modules include a semantic retrieval module, a geographic information retrieval module, a TCP/IP Socket connection module in a CKIP Word Segmentation System, the Google Web Service, and Google Map API. Map is the center of operation and inquiry interface in the whole system. Document content and geographic information on spatial position are connected to each other through hyperlinks on the map, which is also the concept of Geospatial Web promoted by Scharl (2007, 2008), where all information and operations are connected with map (figure six). The design of the experiment uses Google search engine and Google Local Search as platforms for semantics and space testing and comparison, setting multiple search keyword strings and having Google Search and Local Search conduct searches and experiments as well as result comparisons.

4.1 The Setting of BM25 Parameters

In regard to the setting of k_1 and b in formula 2, because the content in webpage documents is relatively less, the most searched Term Frequency is less than 7. Through a set number of articles (1000 items) and a number of articles which contain individual search keywords (20 items), assuming tested articles are of average length to conduct

Figure 3. BM25 score of searched keywords under different k_1 values and different TF values

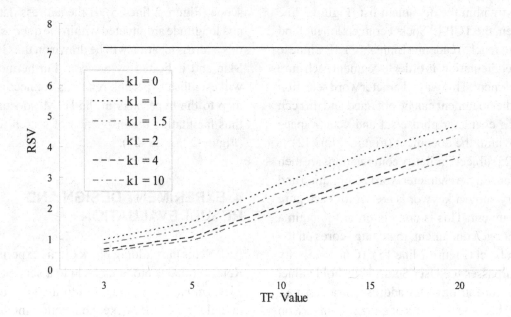

the experiment (in f 3), and with reference to the relevant implementation (Hawking et al., 2004; Lin, 2005; Andrade & Silva, 2006; Pérez-Iglesias, 2008), it is found that in relation to individual keyword BM25 score value, the settings of k_1 and b will reach near convergence (when $k_1 = 0$), and there is not much difference in BM25 scores when Term Frequency is low. Therefore k_1 is set at 2.0, and b is set at 0.75. In the future, further studies are needed with documents of higher Term Frequency and of different kinds or subjects, such as non-webpage documents or reports.

4.2 Semantic Retrieval and Evaluation

In regard to the evaluation of semantic retrieval, in this research keyword strings of "County/City, Subject" of different districts and subjects were combined, to compare with Google search. The searched webpages were also saved, to facilitate experimentation evaluation. Since there are many combinations, and the focus of this research is on hostels/homestays in southern Taiwan, especially

Kaohsiung county, Kaohsiung county was selected as an example for the purpose of demonstration and explanation. We utilized Rijsbergen's (1979) evaluation indicators Precision rate (P), Recall rate (R), and evaluation methods. After changes in Precision and Recall rates are sorted with webpages' BM25 similarity score, the retrieved front k value for judgment purpose is relevant. In other words, the lower the k value, the higher the Precision rate and the lower the Recall rate; the higher the k value, the lower the Precision rate and the higher the Recall rate. The k value can be set by users. Given the limited labor resources and time, we manually judged semantic consistency. 500 webpages searched through Google Search were put into testing first. Thus the k value was set between 1 and 500; experiment results are each shown using P-R curve lines in Figures 4 and 5. The evaluation indicator calculation methods were as follows:

$$\text{Precision} = \frac{N_P \cap N_S}{N_S} \qquad (3)$$

Figure 4. Experiment results with "Kaohsiung County Homestays" as the searched subject, shown through the P-R curve line

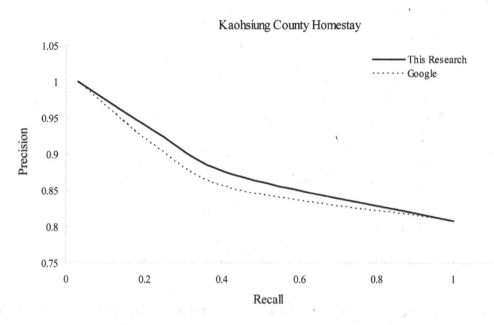

$$Recall = \frac{N_P \cap N_S}{N_P} \qquad (4)$$

N_S: Number of k articles in the front after the system's sorting, k value is between 1 and 500.

N_P: 500 articles manually judged as being consistent with subject semantics.

$N_P \cap N_S$: Number of articles judged both by the system and human as being consistent with subject semantics.

Since travel is a popular trend nowadays, most search results from this research or Google are highly related to the subject matter. Further analysis and comparison of the searched results generate the following understanding (See Figures 6 and 7):

1. The results obtained from Google search were merely hyperlinks on the content of the document. They did not include any processing or demonstration of geographic information. Our system not only provides document hyperlinks, but also has the map demonstration and stores spatial information contained in the webpage. By such a method we can visualize spatial distribution as the reference point of knowledge discovery.

2. Some results in the Google search contain very high-ranked webpages, which might not appear in our system's search results because our particular system unified appraisals of semantic similarity. Therefore, some tabulations that show only homestays or only webpages providing a query interface, were automatically discarded by our system.

3. Webpage that contains only Flash animations or pictures were also rejected by this particular system. There are now many web pages with address information embedded in pictures instead of pure text and this will affect the search results.

4. The results of searches on this system have a more detailed text explanation and explicit spatial information. From the user's point of

Figure 5. Experiment results with "Kaohsiung County Hotels" as the searched subject, shown through the P-R curve line

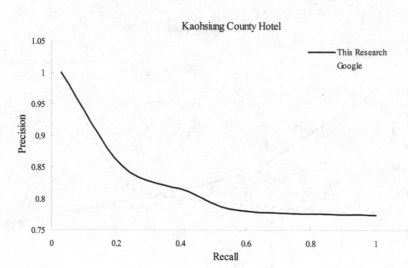

view, they may attain additional, precious information. Since the results include a spatial para-position based on information gleaned from the address, there may be some errors. This is probably unavoidable, but content comparison in the semantic area will reduce the number of errors.

5. Pages searched by Google often contain excess information on un-related subjects and irrelevant geographic information. This does not favor the user's ability to absorb information, whereas our system's search results are drawn closer to the user's needs.

6. Since Google does not process spatial information, it is unable to query the range between cross region and concept region. Our system may achieve such an effect with the aid of a gazetteer and the MBR query area.

7. Webpages found within Google include many blog sites. This kind of article is generally based on the travelers' own experience and feelings; the content may be crucial and have little commercial advertising value. It may be more useful if each article could be matched to a particular time indication,

especially in regard to seasonally influenced destination. However, the majority of these articles contain implicit rather than explicit spatial information and will be omitted from our search. All things considered, having omitted such articles is not seen as being problematic.

4.3 Spatial Data Retrieval Evaluation

Google Local Search has Chinese-related material so we can compare Local Search against it. The query condition is, using the keywords "homestay or hostel" roughly around Kaohsiung County's map range. Just as the second article states, it relies on the webpage providing a city information database or a yellow pages blend. In the Kaohsiung County's rough map scope, Google search shows 18 homestays, which is less than were found by our search. According to material within the webpage of the Taiwanese Ministry of Communications' Sightseeing Bureau, on the hotel industry and homestay management (http://hscc.tbroc.gov.tw), there was a total of forty four homestay registrations within the Kaohsiung County area alone. The effectiveness of Local

Figure 6. Experimentation testing platform

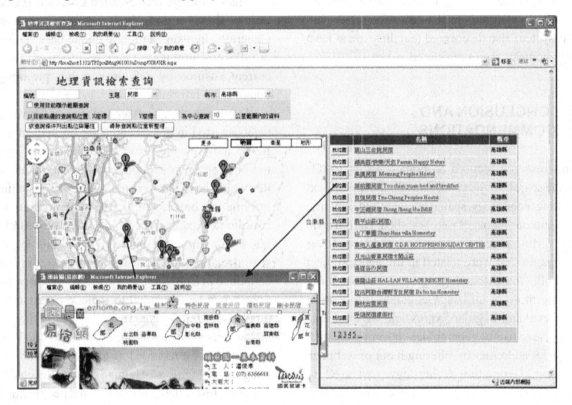

Figure 7. Query on designated range or present map range

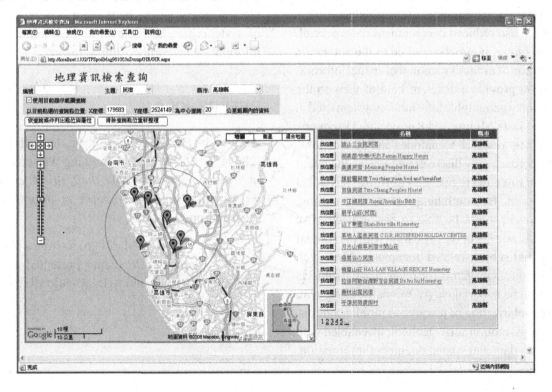

Search is therefore quite limited, since it is mainly focusing on commercial unification. However, this system also discovered that there are at least fifty homestays not registered as yet.

5. CONCLUSION AND RECOMMENDATIONS

This research focused on homestays or hostels as the search subject and, through the use of Text Mining technology, spatial contexts and semantic contents on webpages were searched. The retrieval and sorting processes were designed to discover webpages that were related to the query subject in both content and geographic information. Furthermore, by utilizing Geography Information Retrieval and Regular Expression, we were able to retrieve useful geographical information such as street addresses by filtering it out of webpage content. Geocoding technology was used to retrieve geographic data and text contents that could then be integrated and displayed on a map.

This method of using the GIR improved search results, especially in spatial-related subjects. This method also reduced development costs overall. In the future, it may be applied to the automatic collection of a massive amount of spatial information and provide a different kind of view to the traditional geographic information system used in Spatial Data Mining and Knowledge Discovery. As to new research techniques and applications, this approach may discover patterns in document lists, or rules and regularities as well as statistical tendencies. By combining it with data mining, this system might be able to perform retrieval, analysis, sorting or classification with the content of spatial subject-related documents. Moreover, knowledge management on related spatial subjects can also be undertaken. Additionally, along with the popularization of ubiquitous mobile communications, motion surfer technology enabled with location data and spatial document information,

will also stimulate the so-called Location-Based Service (LBS) or drive the Local Search service. Generally, by using a mobile phone base station or GPS positioning, the user will transmit their current position by default, and thus promptly obtain local information and service.

At present, the gazetteer and subject keyword database used in this research are limited. In the future, a more widespread and standardized spatial ontology may be constructed. The result of this would be to discover not only explicit spatial information, but also implicit spatial information, in order to expand and promote effective searching. Furthermore, in regard to the discussion of spatial relations, this research mainly tested points contained in a polygon. In the future, spatial relations could be expanded to include orientation, topology, proximity, and so on. Because of the day by day accumulation of online material, the Web has come to be regarded as a kind of miniature image of the opinions of human society. It includes many cognitive descriptions of what people think in spatial terms as well as other aspects of a subject. Therefore, from a Cognitive Geography point of view, to carry on the betterment and extension of systems for the retrieval and exploration of information is worthy of further study.

REFERENCES

Amitay, E. Har'E, N., Sivan, R., & Soffer, A. (2004). Web-a-Where: Geotagging Web content. In *Proceedings of SIGIR-04, the 27th Conference on Research and development in information retrieval* (pp. 273-280).

Andrade, L., & Silva, M. (2006) *Relvance Ranking for Geographic IR*. Paper presented at the Workshop on Geographical Information Retrieval (SIGIR'06).

Boguraev, B., & Neff, M. S. (2000). Discourse segmentation in aid of document summarization. In *Proceedings of the Hawaii International Conference on System Sciences (HICSS)* (pp. 10-17).

Buyukkokten, O., Cho, J., Garcia-molina, H., Gravano, L., & Shivakumar, N. (1999). Exploiting geographical location information of web pages. In *Proceedings of the ACM SIGMOD Workshop on the Web and Databases (WebDB'99)* (pp. 91-96).

Byrd, R., & Ravin, Y. (1999). Identifying and extracting relations in text. In *Proceedings of the Applications of Natural Language to Information Systems (NLDB)* (pp. 149-154).

Cai, G. (2002). GeoVSM: An integrated Retrieval model for geographic information. In M. J. Egenhofer & D. M. Marks (Eds.), *GIScience 2002* (LNCS 2478, pp. 65-79).

Davis, S. (2006). *Google Map API, V2: Adding Where to You Applications*. The Pragmatic Programmers LLC.

De Floriani, L., Marzano, P., & Puppo, E. (1996). Spatial Queries and Data Models. In A. Frank & I. Campari (Eds.), *Spatial Information Theory: A Theoretical Basis for GIS* (LNCS 716, pp. 113-138).

Gey, F., Larson, R., Sanderso, M., Joho, H., & Clough, P. (2005). GeoCLEF: the CLEF 2005 cross-language geographic information retrieval track. In *Proceedings of CLEF 2005: Working Notes for the CLEF 2005 Workshop* (pp. 908-919).

Hawking, D., Upstill, T., & Craswell, N. (2004). Toward Better Weighting of Anchors. In *Proceedings of the 27th Annual International ACM SIGIR Conference on Research and Development in Information Retrieval* (pp. 512-53).

Jeffrey, E. F. (2006). *Mastering Regular Expressions* (3rd ed.). Sebastopol, CA: O'Reilly Media Inc.

Jones, C. B., Alani, H., & Tudhope, D. (2001). Geographical Information Retrieval with Ontologies of Place. In *Proceedings of the International Conference on Spatial Information Theory: Foundations of Geographic Information Science* (LNCS 2205, pp. 322-325).

Jones, C. B., Purves, R., Ruas, A., Sanderson, M., Sester, M., van Kreveld, M., & Weibel, R. (2002). Spatial information retrieval and geographical ontologies: An overview of the SPIRIT project. In *Proceedings of SIGIR-02, the 25th Conference on Research and Development in Information Retrieval* (pp. 387-388).

Kornai, A., & Sundheim, B. (2003). In *Proceedings of the NAACL-HLT Workshop on the Analysis of Geographic References*.

Larson, R. R. (1995). Geographic Information Retrieval and Spatial Browsing. In L. C. Smith & M. Gluck (Eds.), *Geographic Information Systems Patrons Maps and Spatial Information* (pp. 81 123). Champaign-Urbana, IL: University of Illinois.

Larson, R. R. (1996). Geographic information retrieval and spatial browsing. In L. Smith & M. Gluck (Eds.), *GIS and Libraries: Patrons, Maps and Spatial Information* (pp. 681-124). Champaign-Urbana, IL: University of Illinois.

Laurini, R., & Thompson, D. (1992). *Fundamentals of Spatial Information Systems*. New York: Academic Press.

Lee, L., Lee, Y., Lin, C., & Huang, K. (2007). Automatic journey geocoding. In *Proceedings of the Taiwan Geographic Information Conference*.

Lin, K. H.-Y., Hou, W.-J., & Chen, H.-H. (2005) Retrieval of Biomedical Documents by Prioritizing Key Phrases. In *Proceedings of the Fourteenth Text REtrieval Conference (TREC 2005)*, Gaithersburg, MD.

Martins, B., & Silva, M. J. (2005). A graph-ranking algorithm for geo-referencing documents. In *Proceedings of the ICDM-05, the 5th IEEE International Conference on Data Mining* (pp. 741-744).

May, W.-Y., & Chang, K.-J. (2003). Introduction to CKIP Chinese Word Segmentation System for the First International Chinese Word Segmentation Bakeoff. In *Proceedings of the ACL, Second SIGHAN Workshop on Chinese Language Processing* (pp. 168-171).

McCurley, K. S. (2001). Geospatial mapping and navigation of the Web. In *Proceedings of the 10th International World Wide Web (WWW10)* (pp. 221-229).

Mitra, S., & Acharya, T. (2003). *Data Mining: Multimedia, Soft Computing and Bioinformatics*. New York: John Wiley & Sons.

Pérez-Iglesias, J. (2008). *Integrating BM25 & BM25F into Lucene*. Retrieved from http://nlp.uned.es/~jperezi/Lucene-BM25/

Periakaruppan, R., & Nemeth, E. (1999). GTrace – A Graphical Traceroute Tool. In *Proceedings of the 13th Systems Administration Conference (LISA '99)* (pp. 69-78).

Purves, R. R., Sanderson, A., Sester, M. M., Kreveld, M. V., & Weibel, R. (2002). Spatial information retrieval and geographical ontologies an overview of the SPIRIT project. In *Proceedings of the 25th Annual International ACM SIGIR Conference on Research and Development in Information Retrieval* (pp. 387-388).

Reid, J. A. (2003). geoXwalk – A Gazetteer Server and Service for UK Academia. In Koch & Sølvberg (Eds.), *Proceedigns of the Research and Advanced Technology for Digital Libraries: 7th European Conference (ECDL 2003)* (pp. 387-392).

Rijsbergen, V. C. J. (1979). *Information Retrieval* (2nd ed.). London: Butterworth.

Robertson, S. E., Walker, S., Jones, S., Hancock-Beaulieu, M., & Gatford, M. (1994). Okapi at TREC-3. In *Proceedings of the Third Text REtrieval Conference (TREC 1994)* (pp. 109-126).

Salton, G. (1968). *Automatic Information Organization and Retrieval*. New York: McGraw-Hill Inc.

Salton, G., & Buckley, C. (1988). Term-Weighting Approaches in Automatic Text Retrieval. *Information Processing & Management, 24*(5), 513–523. doi:10.1016/0306-4573(88)90021-0

Scharl, A. (2007). Towards the Geospatial Web: Media Platforms for Managing Geotagged Knowledge Repositories. In A. Scharl & K. Tochtermann (Eds.), *The Geospatial Web – How Geo-Browsers, Social Software and the Web 2.0 Shaping the Network Society* (pp. 3-14). London: Springer.

Scharl, A. (2008). Annotating and Visualization Location Data in Geospatial Web Application. In *Proceedings of the First International Workshop on Location and The Web*.

Souza, L., Davis, C. J., Borges, K., Delboni, T., & Laender, A. (2005). The role of gazetteers in geographic knowledge discovery on the web. In *Proceedings of the LA-Web-05, the 3rd Latin American Web Congress* (p. 157).

Tezuka, T., Kurashima, T., & Tanaka, K. (2006). Toward tighter integration of web search with a geographic information system. In *Proceedings of the 15th international conference on World Wide Web* (pp. 277-286).

Tezuka, T., & Tanaka, K. (2005). Landmark Extraction: A Web Mining Approach. In *Proceedings Of the COSIT'2005* (pp. 379-396).

Vestavik, Ø. (2008). *Geographic Information Retrieval: An Overview*. Retrieved August 20, 2008 from http://www.idi.ntnu.no/~oyvindve/article.pdf

Vogel, D., Bickel, S., Haider, P., Schimpfky, R., Siemen, P., Bridges, S., & Scheffer, T. (2005). Classifying search engine queries using the Web as background knowledge. *SIGKDD Explorations Newsletter*, *7*(2), 117–122. doi:10.1145/1117454.1117469

Woodruff, A. G., & Plaunt, C. (1994). GIPSY: Geo-referenced Information Processing System. *Journal of the American Society for Information Science American Society for Information Science*, *45*(9), 645–655. doi:10.1002/(SICI)1097-4571(199410)45:9<645::AID-ASI2>3.0.CO;2-8

This work was previously published in International Journal of Information Technology and Web Engineering, Volume 5, Issue 1, edited by Ghazi I. Alkhatib, pp. 56-75, copyright 2010 by IGI Publishing (an imprint of IGI Global).

Section 4
Web–Based Technologies for Improving QoS

Chapter 13
Quality of Service for Multimedia and Real-Time Services

F. Albalas
Jadara University, Jordan

B. Abu-Alhaija
Middle East University, Jordan

A. W. Awajan
Al-Balqa' Applied University, Jordan

A. A. Awajan
Princess Sumaya University for Technology, Jordan

Khalid Al-Begain
University of Glamorgan, UK

ABSTRACT

New web technologies have encouraged the deployment of various network applications that are rich with multimedia and real-time services. These services demand stringent requirements are defined through Quality of Service (QoS) parameters such as delay, jitter, loss, etc. To guarantee the delivery of these services QoS routing algorithms that deal with multiple metrics are needed. Unfortunately, QoS routing with multiple metrics is considered an NP-complete problem that cannot be solved by a simple algorithm. This paper proposes three source based QoS routing algorithms that find the optimal path from the service provider to the user that best satisfies the QoS requirements for a particular service. The three algorithms use the same filtering technique to prune all the paths that do not meet the requirements which solves the complexity of NP-complete problem. Next, each of the three algorithms integrates a different Multiple Criteria Decision Making method to select one of the paths that have resulted from the route filtering technique. The three decision making methods used are the Analytic Hierarchy Process (AHP), Multi-Attribute Utility Theory (MAUT), and Kepner-Tregoe KT. Results show that the algorithms find a path using multiple constraints with a high ability to handle multimedia and real-time applications.

DOI: 10.4018/978-1-4666-0023-2.ch013

INTRODUCTION

Multimedia and real-time applications are being highly deployed in the Internet world. These applications need Quality of Service (QoS) to assure user satisfaction. The internet sends multimedia through the network as a sequence of IP packets and does not take into account the service characteristics for each type of application being transmitted. This means, that the internet does not provide any quality assurance. Since diverse applications have different network requirements, these requirements should be provided to the network in order to achieve the desired QoS. These requirements are defined through one or more QoS parameters such as bandwidth, delay and jitter (delay variation).

There are three problems that the algorithms proposed aim to solve. The first problem is that there is a variety of diverse multimedia and real-time applications deployed throughout the Internet, where each application has its own QoS requirements to ensure customer satisfaction that differ from the other applications; for example, video conferencing needs higher bandwidth requirements than Voice over Internet Protocol (VoIP) call (Szigeti & Hattingh, 2004). Therefore, the path selected to deliver a certain application to the end user will depend directly on the QoS requirements of the requested service. This problem is solved by creating different service profiles for each type of application where each service profile contains a list of constraints and all necessary information to deliver the service to the client.

The second problem, is that multimedia and real-time applications cannot be delivered using the traditional routing algorithms, where only a single metric is considered; for example 'distance' in Dijkstra's shortest path algorithm (Dijkstra, 1959; Cormen, 2001). Consequently, multimedia and real-time applications are usually associated with multiple constraints such as delay, jitter, loss rate, bandwidth, etc. where QoS routing is applied

(Guo et al., 2009; Xueshun et al., 2009). Unfortunately, routing with multiple constraints has been proven to be an NP-complete problem (Wang & Crowcroft, 1996; Momtazpour & Khadivi, 2009), where the routing problem cannot be solved to find the exact solution in a real time scale. This problem is solved by decreasing the number of paths that the algorithm has to search which in return decreases the complexity problem; this is done by pruning all the links that do not meet any of the metric requirements for a particular service in the search process. Thus, a feasible solution can be found by combining more than one type of metric with any other types of metrics (additive, concave, or multiplicative) (Wang & Crowcroft, 1996; Momtazpour & Khadivi, 2009).

The third problem is selecting the best path that satisfies multiple metrics that have resulted from the filtering technique, where comparing these paths together on more than one metric can be complicated and needs special methods to make the right selection. In this paper, Multiple Criteria Decision Making methods have been used for selecting the optimal path for service delivery.

According to the problems stated above and the solutions proposed the paper aims to develop and implement a QoS routing algorithm that finds the best path from the source node to the destination node that satisfies the QoS requirements for the requested service. This can be achieved by filtering out all the paths that violate the QoS constraints for a particular service. Another aim is to implement multiple criteria decision making methods into the routing process to assist the algorithm choose the best path.

RELATED WORK

Routing can be classified into three types, unicast, multicast and broadcast. Broadcast routing is not a problem because the packets are sent to all receivers, so there is no need to find an optimal path. The routing problem has two major problems, the unicast routing problem and the multicast routing problem. This paper focuses on the unicast routing problem, where the unicast routing problem is to find the best feasible path from the source node to the destination node that satisfies a set of QoS constraints. Unicast routing algorithms can be further divided to single metric, dual-metrics or multiple metrics routing. As this paper aims to provide QoS routing it is necessary to use multiple metrics.

There are many attempts to solve the unicast routing problem with regards to multiple constraints (Leng et al., 2009; Esfahani & Analoui, 2008; Lee et al., 2009) but only a few papers have considered implementing multiple criteria decision making methods to solve this problem, which is one of the main aims in this paper. The closest work related to the work done in this paper is (Alkahtani et al., 2006), where the authors proposed a new approach for path selection based on multiple metrics, known as EBQRMPM (enhanced best effort QoS routing with multiple prioritized metrics). EBQRMPM uses the Analytical Hierarchy Process (Saaty, 1980; Zhou & Huang, 2010), which is a Multi Criteria Decision Making (MCDM) method commonly used in the Operation Research (OR) area, to find the optimal path out of all feasible paths according to multiple metrics. The algorithm performs in two stages; in the first stage, all possible paths between the source and destination are found. In the second stage, the algorithm uses AHP method to choose the best path, according to weights given to the constraints. The similarity between the algorithms in this paper and the algorithm proposed in (Alkahtani et al., 2006) is the use of multiple criteria decision making methods in QoS routing, where one of the proposed algorithms in this paper uses AHP for path selection. The difference between the two algorithms is that EBQRMPM proved to be not scalable because it selects all possible paths and does not consider the QoS constraints for the service being transmitted. The algorithm that uses AHP in this paper (QoS_AHP), includ-

ing the other two, filter out all the paths that do not satisfy the QoS constraints and that is what reduces the complexity problem of routing according to multiple metrics, then applies AHP for path selection.

RESEARCH METHODOLOGY

Algorithm testing and evaluation is very important, but it is impossible to do that on a physical network. Therefore, a network topology is created for simulation purposes. In this paper, the network topology was provided by the easy to use BRITE: Boston University Representative Internet Topology gEnerator (Medina et al., 2001; Wang et al., 2009). The BRITE network generator was chosen because of its ease of use and pertinence to our investigations. BRITE is only used to create a random network and check for connectivity in this project. The link parameters (bandwidth, delay, jitter, reliability and loss in our current investigations) were assigned using uniformly distributed functions to simplify the implementation. BRITE is capable of generating most commonly used network topologies such as Waxman (Waxman, 1988), Barbasi (Barabási & Albert, 1999) and Transit-Stub (Calvert et al., 1997). In this study the random Waxman method (Waxman, 1988) is used. Figure 1 shows an example of a random Waxman network topology generated by BRITE with 10 nodes and 19 edges.

Routing algorithms use metrics to determine the best path from the source to the destination. The most used metrics are hop-count, cost, path-length, bandwidth, delay, jitter (delay variation), security, and loss (unreliability) (Tanenbaum, 2002). Metrics are usually defined for individual links; however, QoS routing concentrates on the end-to-end value of the metric. The end-to-end value for any metric is the combination of all the links that construct the path from the source node to the destination node. Metrics therefore can be classified according to their composition rules

Figure 1. A network with 10 nodes and 19 edges is generated by BRITE

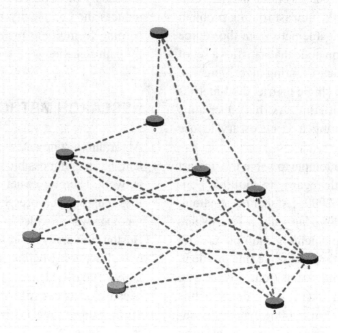

into three classes. If m (n1, n2) is the metric of the link connecting to the two nodes n1, n2 then, for any path P= (n1, n2,…, ni), metrics could be classified in to three types (Wang & Crowcroft, 1996):

1. Additive metrics: where the calculation of the metric for the whole path is calculated as in (1). Delay, delay jitter and cost are considered an additive metrics.

$$m(p) = m(n1,n2) +... + m(ni-1,ni) \qquad (1)$$

2. Multiplicative metrics: where the calculation of the metric for the whole path is calculated as in 2.

$$m(p) = m(n1,n2) \times... \times m(ni-1,ni) \qquad (2)$$

Loss Probability can be indirectly considered a multiplicative metric; because it has to be transformed to an equivalent metric that follows the composition rule.

So if, (1- m(n1,n2)) represents the success ratio over the link n1,n2. Then 3 will represent the loss probability over the whole path.

$$m(p) = 1-[(1- m(n1,n2) \times... \times (1-(m(ni-1,ni))] \qquad (3)$$

3. Concave metrics: where the calculation of the metric for the whole path is calculated as in 4; bandwidth is a concave metric.

$$m(p) = \min[m(n1,n2),..., m(ni-1,ni)] \qquad (4)$$

Different multimedia applications need different requirements to be delivered on the internet, some applications are sensitive on certain metrics more than others, for example video streaming is sensitive on jitter where at the same time it is tolerant to loss (Tanenbaum, 2002) as shown in Table 1.

Table 1 shows examples of real-time applications; each application is sensitive to two or more metrics. Video conferencing is sensitive on delay, bandwidth and jitter. It is also noticeable that all the applications in Table 1 are sensitive to jitter (delay variation).

A Service profile (or QoS) for each appication type is created; the service profile contains all information required to find the best path that satisfies all the constraints needed to deliver the service to the client with the highest quality available; this information includes a list of metrics, the minimum/maximum values, the type of metrics, if it needs to be minimized or maximized, the metric weights that shows the importance of the metric to the service and the curve parameter (R) used by MAUT and will be explained further in the paper. Table 2 shows an example of data stored in a service profile

Table 2 displays the data stored in a QoS profile, each row includes a QoS metric and its related information. The first row in the QoS profile is the bandwidth metric, the table shows that bandwidth is concave, needed to be maximized, should be at least 100 kbps, and its importance is 34% and the value of the R parameter that will be used in the exponential function for MAUT.

On the internet, the paths available between any two nodes can reach up to thousands of paths.

Therefore, to find all the possible paths between the source node and destination node can get very complex. A route filtering method which is an enhanced version of the depth-first search algorithm is implemented which calculates the end-to-end constraints on the paths while they are being constructed. If any of the constraints violate the QoS constraints, while the path is being constructed, then the path and all the nodes beyond that point will be deleted. The filtering mechanism calculates each constraint depending on its type (additive, concave or multiplicative).

IMPLEMENTATION

In this section, a demonstration of how MCDM methods were implemented in QoS routing for communication networks; The MCDM methods used in this paper are the Analytical Hierarchy Process (AHP), Multi-Attribute Utility Theory (MAUT) and Kepner-Tregoe (KT). First a brief discussion of how each method works and then a detailed description of how they are implemented into QoS routing will be introduced.

Table 1. Examples of common applications and the sensitivity of their QoS requirements

Application	Sensitivity				
	Loss	Delay	Bandwidth	Jitter	Security
Audio on demand	Low	Low	Medium	High	Low
Video on demand	Low	Low	High	High	Low
Telephony	Low	High	Low	High	Low
Video conferencing	Low	High	High	High	Low
Confidential Video conferencing	Low	High	High	High	High

Table 2. Information stored in a service profile

Metric	Type	max/min	value	weight	Curve parameter
bandwidth	concave	max	100 kbps	0.34	R= high
Delay	additive	min	150 ms	0.33	R=high
Jitter	additive	min	50 ms	0.33	R= low

Multiple Criteria Decision-Making Methods

When having to deal with everyday simple problems, people tend to use their instinct to take a fast and quick decision, and surprisingly it will produce acceptable results. This is because these kinds of problems often involve a few objectives and only one or two decision makers. But when considering decision making in government departments or big companies, problems are more complex where decisions involve multiple objectives, several decision makers, and are subject to external review. Below is a brief summary of MCDM methods that will be implemented in the proposed algorithms:

The Analytic Hierarchy Process (AHP)

AHP is a mathematical technique developed by Thomas Saaty (Saaty, 1980). AHP is used for complex decision making involving multiple criteria. AHP helps people set priorities and make the best decision when both qualitative and quantitative aspects of a decision need to be considered.

In AHP, a pair-wise comparison method is used to score alternatives for each criterion using a nine-point scale. A value of one in the scale means that the alternatives compared to each other are equally important and a value of nine in the scale means that the first alternative is extremely more important than the second alternative. The comparison of the alternatives will result in a matrix for each criterion; using these matrices, the alternative weights are calculated for each criterion, (see the example in next section). The objective weights are calculated the same way as the alternatives. Then the alternative score is calculated by multiplying the objective weights by the alternative weight for each criterion. The alternative with the highest score is the recommended alternative. An illustrated example can be found in (Saaty, 1980; Winston, 1994).

Multi-Attribute Utility Theory (MAUT)

MUAT is an evaluation method suitable for complex decisions with multiple criteria and many alternatives. MAUT combines different measures of costs, risks, and benefits, along with individual and stakeholder preference into high-level aggregated preferences. MAUT is based on the use of utility functions; utility functions are used to transform different criteria into one universal scale known as the multi-attribute utility, the utility scale is between 0 (worst) and 1 (best). The utility function can be also presented as a graph where a function is created for each criterion (Baker et al., 2001; Bragge et al., 2010). An illustrated example can be found in (Baker et al., 2001).

Utility Function in MAUT

When dealing with multiple metrics it is hard to compare these metrics together because each metric has a different measurement, for example it is hard to compare 256 kbps with 100 ms of delay, therefore, in multiple decision making, objectives(in this case metrics) are converted to one standard value to be compared to each other. In MAUT the scaling function is called a utility function and it converts the objective values to a score between 0 and 1, this score is called the utility score where 0 represents the worst value of the objective and 1 represents the best value. In this study the exponential function taken from (Jensen and Bard, 2003) is used to scale the metric values as below:

$$f(x) = [1-\exp(-(x-a)/R)] / [1-\exp(-(b-a)/R)] \quad (5)$$

Where x is the value of the metric, a is the worst value, b is the best value and R is the curve parameter. The function can be also represented by a graph as in Figure 2.

In Figure 2, the X axis represents the metric value and the Y axis represents its utility score,

Figure 2. A graph representing the exponential utility function

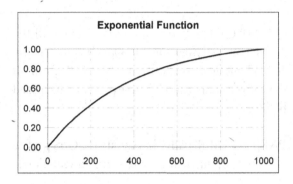

the worst metric value gets a utility score of 0 and the best metric value gets a utility score of 1; between the best metric value and the worse metric value the function is concave if the metric has to be maximized like bandwidth, and convex if the metric has to be minimized like delay. The curve parameter R determines how strong the curve degree between the best and worst metrics value is. Three values for R are defined to be used for investigation depending to the importance of the metric to the service. The values are defined as follows:

- **High R value:** the exponential function has a smooth curve closer to linear approach, this means that the metric is not a very important metric to the service or does not have a very high impact on the service.
- **Medium R value:** the exponential function takes a normal curve, that means that the metric is important to the service and can affect on the service.
- **Low R value:** in this case the curve takes a sharp curve near the turning point, and that indicates that this metric is an important metric and had a high impact on the service provided.

In Figure 3 shows the effect of changing the curve parameter on the exponential function; it is

clear that the increase in the value of R decrease the sharpness of the curve the exponential function will produce.

Kepner-Tregoe (KT)

K-T Decision Analysis is suitable for moderately complex decisions involving a few criteria. The method requires only basic arithmetic. In K-T we first identify the objectives and we call them the WANT objectives. Then each objective is given a score between one (1) and ten (10) depending on its relative importance to the other objectives, these scores are called the criteria weights. The most important objective is identified and given a score of 10, and then all the other objectives are weighted in comparison with the first, from 10 (equally important) down to a possible 1 (not very important) (Kepner & Tregoe, 1981; Bohanec, 2009).

Next, the alternatives are given a score for each objective based on how well they satisfy the WANT objectives. Therefore, the alternative that comes closest to meeting the objective is given a score of 10; and the other alternatives are scored relatively to it. Then, the total score is determined for each alternative by multiplying its score with the criteria weights, and then summing them up for each alternative. The alternative with the highest total score will be the preferred alternative. The

Figure 3. The effect of changing the parameter R on the exponential function

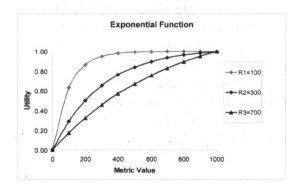

method requires only basic arithmetic. Its main disadvantage is that it may not be clear how much better a score of "10" is than a score of "8", for example. Moreover, total alternative scores may be close together, making a clear choice difficult (Kepner & Tregoe, 1981).

After finding all the paths that satisfy the quality of service constraints for a particular service, they have to be ranked in order to select the best path. It is hard to rank these paths because they have multiple constraints; therefore, the multiple criteria decision making methods are implemented in the algorithm to select the highest ranking path. As there are three methods studied in this paper, there will be three QoS routing algorithms, one for each method. 1) The QoS_AHP algorithm which uses the Analytical Hierarchy Process, 2) the QoS_MAUT algorithm which uses the Multi Attribute Utility Theory and 3) the QoS_KT algorithm which uses the Kepner-Tregoe method. The methods are explained in detail further in this paper (Kepner & Tregoe, 1981).

Implementing the Multiple Criteria Decision Making Methods into QoS Routing

In this section, a detailed description of the three proposed algorithms is illustrated, and how they adapt MCDM methods to select the optimal path. All the proposed algorithms are divided into two parts as shown in Figure 4. The first part finds all possible paths that satisfy the QoS constraints stored in the service profile, and this part is the same for all three algorithms and will be explained once and will be referred to when needed. The second part is to compare all the paths that resulted from the first part and select the best path among them. Each of the three decision making methods described in the previous section are used, which results in three algorithms, QoS_AHP, QoS_MAUT and QoS_KT.

QoS Routing Using MAUT (QoS_MAUT)

QR_MAUT has two main parts, finding all the paths that satisfies the QoS profile and using MAUT method to choose the best path from all the paths found in part1. These parts are discussed below:

PART1: Find all paths that satisfy the QoS profile constraints using route filtering technique

When QR_MAUT receives a request for transmitting a multimedia real-time service, the algorithm loads the service profile for that service, then QR_MAUT starts searching for all possible paths in the network that satisfy the constraints stored the service profile. The search starts from the source node by adding a node each time to the path, each time a node is added to the path the algorithm will calculate the metrics, if at any point the metrics fail to meet the constraints specified in the service profile, that node and all the nodes beyond will be dropped out of the path being built. This step is repeated until all possible paths to the destination node are found; all calculations are made in the source node.

Let's assume that the paths in Table 3 are the paths found to satisfy a certain service profile from the source (3) to the destination (0), we can see that each path has a value for the metrics calculated for all the links in the path.

PART2: Find the best path out of all paths that satisfy the service profile using MAUT

In the previous step six paths have been found, each path has three metrics (length, delay, and bandwidth). Comparing multiple metrics together can get very complex, especially if the number of paths and metrics increase. Therefore, a decision making method has been implemented in the proposed algorithm to break down the complexity

Figure 4. The flowchart for the proposed algorithm

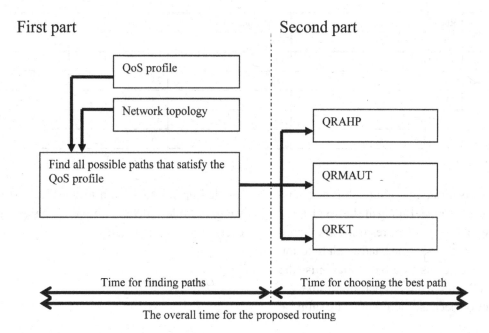

created from multiple metrics. This method is MAUT, and this section of the example MAUT will be used to find the best path from all the paths that were found to satisfy the basic requirements, this is done through the following steps:

1. Import metric weights

In this step weights are given to each metric (bandwidth, delay and jitter), in MAUT the metrics are given scores to reflect the relative importance based on feelings or experience, in the algorithm, the weights should be assigned to the metrics depending on their importance to the service. The sum of the weights are equal to 1, also the metric weights can change depending on the service required. The weights as imported from the service profile are shown in Table 4.

For this simple example the weights were given equal importance, later in this study the metric weights will be prioritized to determine the effect on the performance of the algorithm.

2. Calculate the metric scores for each path

This is done by creating a utility function for each metric, a utility function scales the values of the metric to a score between 0 and 1 and is called the utility weight; the utility weight represents how good the metric scores on a particular path where the best possible value gets a 1 and the worst possible value gets a 0. In this study the exponential function in equation (5) is used.

Using the exponential function a utility graph is created for each metric as shown in Figure 3. As mentioned before, the worst metric value should have a utility value of zero and the best metric value should have a utility of one; in QoS_MAUT, the value of the constraint is considered to be the worst, for example if the bandwidth must be at least 100kbs then that value is given a utility weight zero; whereas, the best metric value is considered to be the best possible obtainable value, for example for bandwidth the best possible value is to get the full capacity from the service provider which in our case is 2Mbps, and this metric value

Table 3. Example of resulting paths and there metric values for a certain service profile

Paths	Route	Bandwidth (kbs)	Delay (ms)	Jitter (ms)
P1	3 4 8 6 9 0	509	76	47
P2	3 4 8 7 9 0	140	67	48
P3	3 4 9 0	181	23	48
P4	3 9 0	829	33	24
P5	3 9 6 8 0	106	65	46
P6	3 9 7 8 0	106	56	47

is given a utility value of 1 although it might be hard to obtain this value. Taking the best and worst metric values into account the resulting graphs are shown in Figure 5 where the bandwidth function is concave because it is required to maximise the value of bandwidth and the other metrics have a convex function because there values are suppose to be minimised.

After the utility function is set, it is used to convert the utility weights for the metric values on each path. For example when the first path metrics (bandwidth=509kbs, delay= 76ms and jitter= 47ms) are converted using the utility functions, the result will be (bandwidth=0.65, delay= 0.77 and jitter= 0.1); all the utility values for the other metric values arc obtained the same way.

3. Find the total score for each path

This step defines the best path by calculating the overall score for each path on all the metrics, the utility weight (value taken from the utility graph) is multiplied by its metric weight (the weight stored in the service profile) as shown in equation (6) where i is the number of paths, N is the number of metrics. Then the results are

Table 4. metric weights

Bandwidth	**Delay**	**Jitter**
0.34	0.33	0.33

totalled up for each path to get the total utility weight. Table 5 details how the final scores for each path are calculated.

$$\sum_{i=1}^{N_{metrics}} Utility_i \cdot Weight_i \qquad (6)$$

In Table 5, each path has a total score where it represents how good that path scores on all the metrics compared to the other paths. QR_MAUT in this example indicates that path 4 with a score of (0.32) is to be the chosen path between all the others.

QOS ROUTING USING KT (QOS_KT)

QoS_KT consists of the following two parts:

PART1: refer to PART 1 of QoS_MAUT
PART2: Find the best path out of all paths that satisfy the QoS profile using KT

From Part 1 many paths can be found to satisfy the service constrains, in Part 2 of the algorithm, K-T will be used to choose one path that best satisfies the constrains in the service profile. This is done by the following three steps:

1. Import metric weights

In this step weights are given to each metric by a special team or an expert according to the

Figure 5. Utility graphs using the exponential function for bandwidth delay and jitter

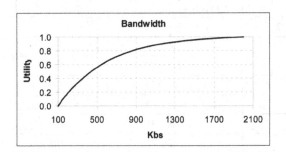

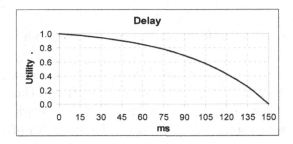

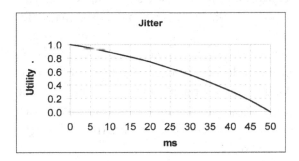

importance of that metric to the service, each metric is given a score between 1 and 10 where the most important metric is given a weight of 10; the other metrics are weighted depending on their relative importance to the first metric, so that means they can be equally important and have a weight of 10 or down to a possible 1 which is not very important compared to the first metric. The weights of the metrics are assumed to be of the same importance, and have been given a score of 10 for all metrics as shown in Table 6.

2.   Calculate metric scores on each path

In order to calculate the score of a path for each metric, the path with the highest metric value is given a score of 10, and the rest of the paths take are given a relative score to the first path, the score for the paths ranges between 1 (very poor compared to the best value) and 10 (equal importance as the best value). These scores are usually compared by a team of experts, but as a modification to KT, a linear equation has been added to replace the human choice, see the equation 7.

$$MetricScore = 1 + \frac{metricvalue - worstvalue}{bestvalue - worstvalue} X9 \qquad (7)$$

In the above equation, the worst metric value is considered the metric constraint stored in the service profile, so for the bandwidth metric its 100kbs; the best metric value is the highest metric value out of all paths available, so for example path 4 has the highest value for bandwidth (829kbs), this value takes a score 10 and then all other values are compared against it. Therefore calculating the bandwidth score on the first path is 1+ (509-100)/ (829-100)*9 = 6, the rest of the metric values are calculated the same way.

3.   Calculate the total score for each path

To get the total metric score for each path, the metric score (calculated in the previous step) is multiplied by its metric weight (the weight stored in the service profile) as shown in equation 8, where *i* is the number of paths, *N* is the number of metrics. Then the results are totalled up for each path to get the total utility weight, this is summarised in Table 7.

$$\sum_{i=1}^{N_{metrics}} Score_i \cdot Weight_i \qquad (8)$$

According to K-T method, (path 4) shows to have the highest score with a total of 300, therefore (path 4) is chosen as the optimal path.

Table 5. Final score results for QoS_MAUT

Path num	Metrics						Total score**
	Bandwidth score		Delay score		Jitter score		$\sum$ metrics
	*Utility	Weight	*Utility	Weight	*Utility	Weight	
Path1	0.65 X 0.34		0.77 X 0.33		0.1 X 0.33		0.51
Path2	0.24 X 0.34		0.81 X 0.33		0.06 X 0.33		0.37
Path3	0.30X 0.34		0.95 X 0.33		0.06 X 0.33		0.44
Path4	0.82 X 0.34		0.93 X 0.33		0.66 X 0.33		0.80
Path5	0.19 X 0.34		0.82 X 0.33		0.13 X 0.33		0.38
Path6	0.19 X 0.34		0.86 X 0.33		0.1 X 0.33		0.38

QOS ROUTING USING AHP (QOS_AHP)

QoS_AHP, consists of the following two parts:

PART1: refer to part1 of QoS_MAUT
PART2: Find the best path out of all paths that satisfy the QoS profile using AHP

From Part 1 many paths can be found to satisfy the service constrains, in Part 2 of the algorithm, AHP will be used to choose the path that best satisfies the constrains in the service profile. This is done by the following three steps:

• Load the metric weights

In AHP, the objective weights (metric weights in our case) are calculated through a series of mathematical calculations. These weights reflect the relative importance based on feelings or experience. In (Alkahtani et al., 2006), one of the modifications on AHP was that the metric weights should be identified by the service provider to reflect the importance of the metric for the delivered service. Initially the metric weights are considered to be of equal importance as shown in Table 8. The metric weights can be prioritized to determine the effect on the performance of the algorithm.

• Calculate the metric scores for each path

This step determines how well each path scores on each metric compared to all the other paths. This is done by first creating a matrix for each metric with a size of $(P \times P)$, where P is the number of paths found in first part of the algorithm. This matrix is called pair-wise comparison matrix for a given metric (pcm (metric)); the main idea of the pcm(metric) is to compare a metric value m, bandwidth for example, of path j to the metric value of all other paths. Let m_i = the metric value for path i and m_j = the metric value for path j. Then the pair-wise comparison matrix (pcm) should be as shown below.

$$pcm(metric(i,j)) = \begin{bmatrix} \dfrac{m_1}{m_1} & \dfrac{m_1}{m_2} & \cdots & \dfrac{m_1}{m_n} \\ \dfrac{m_2}{m_1} & \dfrac{m_2}{m_2} & \cdots & \dfrac{m_2}{m_n} \\ \vdots & \vdots & & \vdots \\ \dfrac{m_n}{m_1} & \dfrac{m_n}{m_2} & \cdots & \dfrac{m_n}{m_n} \end{bmatrix}$$

(9)

Before creating the pcm(metric), it is important to know whether the metric has to be maximised

Table 6. Metric weights

	Bandwidth	Delay	Jitter
Weight in QoS	3.4	3.3	3.3
Equivalent weight	10	10	10

as in bandwidth or minimised as in delay; thus a parameter is identified and stored in the service profile along with the other metric details. The parameter holds two values, min if the metric has to be minimized or max if the metric has to be maximized. According to the type of metric, min or max, the values of the metrics can be insert in the pcm(metric(i,j)) using (10) or (11)

$$pcm(i,j) = mi/mj, \text{ to maximize metric value} \quad (10)$$

$$pcm(i,j) = mj/mi, \text{ to minimize metric value} \quad (11)$$

Where pcm(i,j) is the cell that results from the row i (that represents path i) and column j (that represents path j), m_i= the metric value for path(i), m_j=the metric value for path(j), i, j = (1,2,..,P).

So, for example, to compare path1 with path2 regarding the bandwidth metric, the rule as in (11) is used, since bandwidth should be maximized, pcm(1,2)=(bandwidth for path1/bandwidth for path2), where the total of (mi/mj)) is stored in the first column and the second row and is the called the relative weight.

Obviously, when comparing a path to itself, it will always give a weight value of 1, and comparing path i to path j for a specific metric is reciprocal to comparing path *j* to path *i* for the same metric. In order to save process time, (12) and (13) are used.

$$pcm(metric((i,i)) = 1 \quad (12)$$

$$pcm(metric((j,i)) = \frac{1}{pcm(i,j)} \quad (13)$$

To follow up with the example, a pcm is constructed for each metric (bandwidth, delay, and jitter). Because the same steps apply for all metrics, the example will be shown on one metric and then the final results will be shown for the other metrics. The pcm(bandwidth) is shown in (14).

$$pcm(bandwidth) =$$

$$\begin{bmatrix} 1 & 3.63 & 2.82 & 0.61 & 4.79 & 4.79 \\ 0.28 & 1 & 0.78 & 0.17 & 1.32 & 1.32 \\ 0.35 & 1.29 & 1 & 0.22 & 1.70 & 1.70 \\ 1.63 & 5.90 & 4.59 & 1 & 7.79 & 7.79 \\ 0.21 & 0.76 & 0.59 & 0.13 & 1 & 1 \\ 0.21 & 0.76 & 0.59 & 0.13 & 1 & 1 \end{bmatrix}$$

$$(14)$$

Just for clarification, pcm(bandwidth(1,2) is the bandwidth of path1 (509kps) on the bandwidth of path2 (140kbs), which gives a result of 3.63. All other values are calculated the same way.

At this instant another matrix is generated for each metric. This matrix is called normalized pair wise comparison matrix (npcm (metric)). For each pcm(metric), every element is divided by the sum of the elements in the same column; the npcm can be expressed as in (15).

$$npcm(i, j) = \frac{pcm(i, j)}{\sum column(j)} \quad (15)$$

Where *i, j* = 1,2,..., *P*. Using (15), the npcm(bandwidth) will result in the following:

$$npcm(bandwidth) =$$

$$\begin{bmatrix} 0.27 & 0.27 & 0.27 & 0.27 & 0.27 \\ 0.08 & 0.08 & 0.08 & 0.08 & 0.08 \\ 0.10 & 0.10 & 0.10 & 0.10 & 0.10 \\ 0.44 & 0.44 & 0.44 & 0.44 & 0.44 \\ 0.06 & 0.06 & 0.06 & 0.06 & 0.06 \\ 0.06 & 0.06 & 0.06 & 0.06 & 0.06 \end{bmatrix} \quad (16)$$

Table 7. Final score results using QoS_KT

Path num	Metrics						Total score**
	Bandwidth		Delay		Jitter		∑ metrics
	*Utility	Weight	*Utility	Weight	*Utility	Weight	
Path1	6 X 10		6 X 10		2 X 10		140
Path2	1 X 10		7 X 10		1 X 10		90
Path3	2 X 10		10 X 10		1 X 10		130
Path4	10 X 10		10 X 10		10 X 10		300
Path5	1 X 10		7 X 10		2 X 10		100
Path6	1 X 10		8 X 10		2 X 10		110

It is very clear that by looking at npcm(bandwidth), all the columns in the npcm are identical; and the reason is the relative weights calculated in QoS routing is based on real numerical values and do not depend on human perception. This problem is solved in the algorithm, where only the first column in nppcm is calculated to prevent the overload on process time; this is one of the very important modifications made on AHP because instead of using 2D array for the nppcm, a 1D array is used to store the first column. See modifications details.

Finally, we take the average of each row in the npcm(bandwidth) to generate the average normalized pair-wise comparison matrix (anpcm(bandwidth)). This is illustrated in (17):

$$anpcm(i) = \frac{\sum row(i)}{P} \tag{17}$$

The size of the resulting matrix is [1x P], where P is number of paths. The anpcm values represent the score for each path, so the first value in the matrix is the score for the first path and so on. The values for the anpcm for all metrics are detailed below:

$anpcm(bandwidth) =$
$$(0.27 \quad 0.08 \quad 0.1 \quad 0.44 \quad 0.06 \quad 0.06)$$

$anpcm(delay) =$
$$(0.1 \quad 0.11 \quad 0.32 \quad 0.22 \quad 0.11 \quad 0.13)$$

$anpcm(jitter) =$
$$(0.144 \quad 0.141 \quad 0.141 \quad 0.282 \quad 0.147 \quad 0.144)$$

From looking at anpcm (bandwidth) we can observe that the fourth element (path 4) scores the highest according to bandwidth with a weight of 0.44. The obtained weights determine how well each path scores on each metric.

MODIFICATION DETAILS

As mentioned earlier in this step, that the rows in the npcm (metric) are identical. Thus, when calculating the average of the rows in the anpcm (metric), it will be a waste of time. So by looking at npcm (bandwidth) and at anpcm (bandwidth), it is noticed that the average of the columns in npcm (bandwidth) is obviously the same as the first column. To solve this problem and reduce the

Table 8. metric weights

Bandwidth	Delay	Jitter
0.34	0.33	0.33

computing time and complexity of the algorithm, the first column of the pcm (metric) is calculated only, so calculating the bandwidth score will be as following:

$$pcm(bandwidth) = \begin{pmatrix} 1 \\ 0.28 \\ 0.35 \\ 1.63 \\ 0.21 \\ 0.21 \end{pmatrix}$$

Now the pcm is normalized to generate npcm(bandwidth) the same way as done before but with only one column to result in the following:

$$npcm(bandwidth) = \begin{pmatrix} 0.27 \\ 0.08 \\ 0.1 \\ 0.44 \\ 0.06 \\ 0.06 \end{pmatrix}$$

By having the npcm (bandwidth) we do not need to average the rows because we only have one column; and surprisingly this column is the same as the anpcm(bandwidth) calculated earlier. That means the same result is achieved by only generating the first column in the pcm (metric) then normalizing that column by dividing each element on the sum of the column. This modification reduces the matrix size for all metrics and reduces the complexity and computing time for the algorithm.

• Find the total score for each path

This step defines the best path by calculating the overall score for each path on all metrics. Therefore, for each path the metrics scores are multiplied by the metric weight, and then totalled together to give the overall score of that path. This is repeated for all the paths. The path with the highest total score is the path that best satisfies those metrics.

In Table 9, each path has a total score where it represents how well that path scores on all the metrics compared to the other paths. QoS_AHP in this example indicates that path 4 with a score of (0.32) is to be the chosen path between all the others.

SIMULATION AND RESULTS

To evaluate the proposed algorithms, 100 network scenarios were generated for each network size. For each 100 instances of the network graph, a request for a service was run 1000 times where each request is a simulation cycle. On each request, the network topology was imported, the link parameters were assigned, the source and destination were assigned randomly, the QoS profile was loaded, the three algorithms are applied and the performance measures are recorded. That means for a network of size 10 with 100 instances of a random graph, there will be 1000 simulation requests for each 100 instances, thus giving a total of 100,000 requests for the network of size 10 where it is repeated for all node sizes. The range used for the simulation model is from 10 nodes to 100 nodes increasing by 10 nodes each time.

The reason why the simulation was repeated so many times is because the simulations is based on a lot of random parameters, such as the source and node assignment, link assignment, network topology, etc. Therefore, to assure more accurate values the simulation was repeated many times and then averaged. For instance if a coin was flipped, then the probability of the coin to show any of the sides should be 50%. However, when testing this for a small number of times like 10 times, the percentage for one side might be more or less than 50%. That is why the more the test is repeated on the random values, the more accurate the results will be.

Network Size vs. CPU Running Time

The first result obtained from the simulation is the scalability of the algorithms, where the effect of increasing the size of the network on the CPU running time as shown in Figure 6. It is noticeable from figure 6 that the three algorithms CPU running time increases with the increase of the network size in a linear behaviour. The QoS_KT comes best of the three algorithms and that is because it is the simplest of the three, as it only uses linear equations to select the best path. QoS_AHP comes second best, and that is because it uses a lot of matrices to calculate the best path. Lastly, the QoS_MAUT comes last, as QoS_MAUT uses the exponential functions to calculate the best path and that takes more time than the other two algorithms.

Network Size vs. Percentage of Selecting the Same Path

In this scenario, the selected path of the three algorithms where taken in each simulation to be compared to each other. It has been found that around 60% of the times the three algorithms selected the same path to deliver the service on as shown in Figure 7. It has also been found that

between 30-35% of the times that two of the algorithms out of the three algorithms gave the same path result. In addition, only between 1-10% of the times, all of the algorithms proposed selected a different path. It is noticed that by increasing the number of nodes in the network there is a slight decrease in the percentage where all algorithms gave the same path result. At the same time, there is an increase in the percentage of the paths where two algorithms gave the same path result and an increase in the percentage of the paths where none of the algorithms gave the same path result. This concludes that the increase of the network size, increases the number of paths therefore, making the path scores very close to each other making the selection between the paths very competitive.

The Effect of Changing the Weight of the Constraint on the Selected Path

One of the key variables that play a rule in the selection process is the weight of the constraint which is stored in the QoS profile. The scenario evaluates the effect of changing the metrics weights stored in the QoS profile on the selected path. The scenario was run on a graph with 10 nodes and 20 edges, and the source and destination was set randomly to 2 and 6 respectively. The simulation was run on the three proposed algorithms (QR_AHP, QR_MAUT and QR_KT). There were 15 paths that were found to satisfy the QoS profile requested for this example. The

Table 9. Final calculations using QoS_AHP

Path number	Metrics			
	Bandwidth score	**Delay score**	**Length score**	**Total score**
Path1	0.27 X 0.34	0.10 X 0.33	0.144 X 0.33	0.17
Path2	0.08 X 0.34	0.11 X 0.33	0.141 X 0.33	0.11
Path3	0.10 X 0.34	0.32 X 0.33	0.141 X 0.33	0.19
Path4	0.44 X 0.34	0.22 X 0.33	0.282 X 0.33	0.32
Path5	0.06 X 0.34	0.11 X 0.33	0.147 X 0.33	0.11
Path6	0.06 X 0.34	0.13 X 0.33	0.144 X 0.33	0.11

Figure 6. Network size vs. percentage of similar path selection

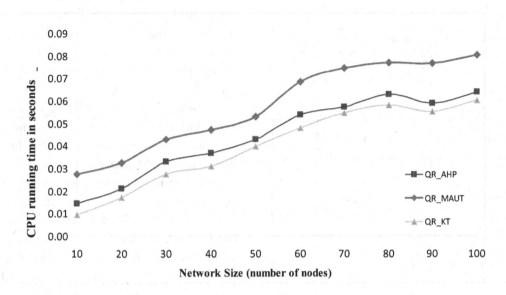

Figure 7. Network size vs. percentage of equal path decision for the proposed algorithms

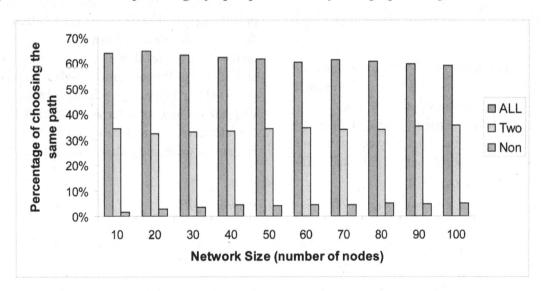

Table 10. QoS profile requested for the simulation process

Metric	Type	max/min	value	weight	Curve parameter
bandwidth	concave	max	256 kbps	0.8	R= high
Delay	additive	min	150 ms	0.1	R=high
Jitter	additive	min	100 ms	0.1	R= low

Table 11. The resulting paths found to satisfy the QoS profile for the scenario in this section

Path Id	Route	Bandwidth (kbps)	Delay (ms)	Jitter (ms)
Path1	2, 7,1, 6	1531	50	79
Path2	2, 7, 6	1290	28	58
Path3	2, 7, 8, 6	850	25	59
Path4	2, 7, 9, 0, 8, 6	1221	76	61
Path5	2, 7, 9, 8, 6	1000	49	63
Path6	2, 9, 0, 8, 6	527	53	32
Path7	2, 9, 0, 8, 7, 6	527	78	73
Path8	2, 9, 3, 4, 8, 6	527	52	77
Path9	2, 9, 4, 8, 6	527	33	66
Path10	2, 9, 5, 6	527	37	54
Path11	2, 9, 7, 1, 6	527	71	56
Path12	2, 9, 7, 6	527	49	35
Path13	2, 9, 7, 8, 6	527	46	36
Path14	2, 9, 8, 6	527	26	34
Path15	2, 9, 8, 7, 6	527	51	75

details of the QoS profile are shown in Table 10, and the resulting paths with the end-to-end value for each metric are shown in Table 11.

Obviously, all the paths in Table 11 guarantee the delivery of the requested service, but the aim is to select the optimal path in this study. The metric weights stored in the QoS profile as shown in Figure 11 is altered each time to study the effect on the selected path. The left side of table 13 shows the metric weights for that simulation and the right side of the table show the selected path for each of the proposed algorithms.

In Table 12, it is noticeable that the proposed algorithms do not always give the same selected path, and that is because each algorithm uses a different MCDM method to compare between the paths. It is also noticeable that by changing the metric weights the selected path changes. The first group in the table gives more importance to bandwidth, according to that, path 1 obtains a high score on bandwidth, whereas when decreasing the value of bandwidth and increasing delay and jitter, path 2 becomes the recommended path by all proposed algorithms.

In the second group, Delay is given a higher priority weight. Therefore, the paths selected by the algorithms are path 2 and path 14 depending on the value of the other two metrics, bandwidth and jitter. For the last group, higher priority is given to jitter. The path selected by all algorithms was path 6 where the jitter weight is high. When the jitter decreases and the delay and bandwidth increase in value, the path selected tends to move

Figure 8. Number of paths vs. Network size for QRAHP and EBQRMM

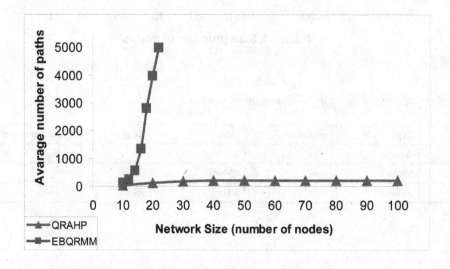

Table 12. The effect of changing the metric weights on the selected optical path for QR_MAUT, QR_AHP and QR_KT

Weights (%)			Selected path		
Band-width	De-lay	Jit-ter	QR_MAUT	QR_AHP	QR_KT
80	10	10	1	1	1
80	15	5	1	1	1
60	20	20	1	1	2
60	30	10	1	2	1
40	30	30	2	2	2
40	40	20	2	2	2
10	80	10	2	2	14
5	80	15	14	14	14
20	60	20	2	2	2
10	60	30	14	14	14
30	40	30	2	2	2
20	40	40	14	14	14
10	10	80	6	14	6
15	5	80	6	6	6
20	20	60	14	14	14
30	10	60	14	14	6
30	30	40	2	2	14
40	20	40	2	2	2

to path 14. Moreover, when the priority for all metrics is nearly equal to each other, the path selected is path 2. It can be recognized that the metric weights effect the path selected but do not affect the number of paths found to satisfy the QoS profile.

Network Size vs. Number of Paths Found to Satisfy QoS Constraints for QRAHP and EBQRMM

As mentioned in the related work, the EBQRMM uses AHP to find the best path, and is the closest algorithm to the work done in this paper. Therefore, QR_AHP and EBQRMM was simulated and compared to each other to study their performance and scalability on different network

sizes. Figure 8 shows the effect of increasing the network size on the number of paths resulting from the search. It is clear that EBQRMM performs poorly by increasing the network size where the number of paths increases dramatically in an exponential behaviour. The number of paths found using EBQRMM rise from around 100 paths in a network with a size of 10 nodes to around 5000 paths in a network with a size of 24 nodes. For the QRAHP, the increase is very slow where the number of paths for a network of size 100 is around 195 paths. Figure 8 clearly shows that QRAHP outperforms EBQRMM in the path discovery process because of the route filtering mechanism implemented in QRAHP.

Network Size vs. CPU Time to Find all Possible Paths that Satisfy the QOS Constraints for QRAHP and EBQRMM

QRAHP and EBQRMM were also evaluated regarding to the CPU running time on different network size. The results are very similar to the previous graph where the CPU running time increases dramatically when increasing the network size in EBQRMM algorithm. For QRAHP the effect of increasing the network size was very low on the CPU running time. QRAHP outperforms EBQRMM in CPU running time as seen in Figure 9. The CPU time is very high in EBQRMM because it searches all possible paths between the client and server whether it satisfies the QoS profile or not.

CONCLUSION AND FUTURE WORK

This paper proposes three algorithms that find the optimal path for delivering multimedia and real-time services that require more the one metric. The algorithms find the path between the source and destination according to the QoS constraints required to deliver a particular service. Although

Figure 9. Number of paths vs. Network size for QRAHP and EBQRMM

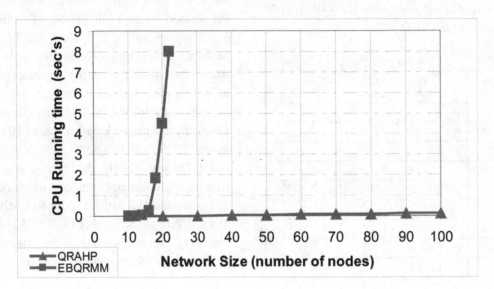

routing with multiple constraints is considered a NP-complete problem, the proposed algorithms have proven to solve this problem by filtering out the paths that do not satisfy the QoS constraints. In addition, the three algorithms, QoS_AHP, QoS_MAUT and QoS_KT each uses a multiple criteria decision making method that compare the paths that satisfy the QoS profile together and chooses the best path among them.

The three algorithms have been tested on Waxman's random network topology in the simulation, also different QoS profiles have been tested in the simulation. The simulation results for the proposed algorithms showed that they are all scalable and does not suffer from the NP-complete problem because of the route filtering used that decreased the complexity of the problem. The algorithms have also proved the ability to handle the delivery of various types of multimedia and real-time services.

The QoS_KT algorithm came first in performance because of its simplicity, then QoS_AHP came in second which used matrices to select the optimal path, and QoS_MAUT was last because of the exponential function used to calculate the utility scores for the metrics. Although QoS_kt came in first, but that does not necessary mean that it is better because QoS_MAUT for example, uses more complicated calculations which represent metrics better than the simple equations used in KT.

There are still much to add to the work presented in this paper, one of the most important is to add mobility to the work where adhoc and sensor networks can be considered. Also implementing a test-bed for physical simulation where results can be more accurate.

REFERENCES

Alkahtani, A. M. S., Woodward, M. E., & Al-Begain, K. (2006). Prioritised best effort routing with four qualities of service metrics applying the concept of the analytic hierarchy process. *Computers & Operations Research, 33*, 559–580. doi:10.1016/j.cor.2004.07.008

Baker, D., Bridges, D., Hunter, R., Johnson, G., Krupa, J., Murphy, J., & Sorenson, K. (2001). Guidebook to Decision-Making Methods. *Department of Energy*, 1-40.

Barabási, A. L., & Albert, R. (1999). Emergence of Scaling in Random Networks. *Science, 286*, 509. doi: 10.1126/science.286.5439.509.

Bohanec, M. (2009). Decision Making: A Computer-Science and Information-Technology Viewpoint. *Interdisciplinary Description of Complex Systems, 7*, 22–37.

Bragge, J., Korhonen, P., Wallenius, H., & Wallenius, J. (2010). Bibliometric Analysis of Multiple Criteria Decision Making/Multiattribute Utility Theory. *Multiple Criteria Decision Making for Sustainable Energy and Transportation Systems*, 259-268.

Calvert, K. I., Doar, M. B., & Zegura, E. W. (1997). Modeling Internet topology. *Communications Magazine, 35*, 160–163. doi:10.1109/35.587723

Dijkstra, E. W. (1959). A note on two problems in connexion with graphs. *Numerische Mathematik, 1*, 269–271. doi:10.1007/BF01386390

Esfahani, A., & Analoui, M. (2008). *Widest K-Shortest Paths Q-Routing: A New QoS Routing Algorithm in Telecommunication Networks*.

Guo, Z., Qiao, J., Lin, S., & Cai, X. (2009). *A distributed parallel QoS routing algorithm with multi-path probing*. Washington, DC: IEEE Press.

Jensen, P. A., & Bard, J. F. (2003). *Operations Research: Models and Methods*. New York: Wiley.

Kepner, C. H., & Tregoe, B. B. (1981). *The New Rational Manager*.

Lee, S., Das, S., Pau, G., & Gerla, M. (2009). *A Hierarchical Multipath Approach to QoS Routing*.

Leng, H., Liang, M., Song, J., Xie, Z., & Zhang, J. (2009). *Routing on Shortest Pair of Disjoint Paths with Bandwidth Guaranteed*. Washington, DC: IEEE.

Medina, A., Lakhina, A., Matta, I., & Byers, J. (2001). BRITE: An Approach to Universal Topology Generation. In *Proceedings of MASCOTS* (Vol. 1).

Momtazpour, M., & Khadivi, P. (2009). New Routing Strategies for RSP Problems with Concave Cost. *Advances in Computer Science and Engineering*, 412-418.

Saaty, T. L. (1980). *The Analytical Hierarchy Process*.

Szigeti, T., & Hattingh, C. (2004). *End-to-End QoS Network Design*. San Jose, CA: Cisco Press.

Tanenbaum, A. S. (2002). *Computer Networks*. Upper Saddle River, NJ: Prentice Hall.

Thomas, H., Cormen, C. E. L., Rivest, R. L., & Stein, C. (2001). *Introduction to Algorithms*. Cambridge, MA: MIT Press.

Wang, B., Zhang, J., Guo, Y., & Zhou, J. (2009). *A Study of Fast Network Self-Healing Mechanism for Distance Vector Routing Protocol*. Washington, DC: IEEE.

Wang, Z., & Crowcroft, J. (1996). Quality-of-service routing for supporting multimedia applications. *Selected Areas in Communications, 14*, 1228–1234. doi:10.1109/49.536364

Waxman, B. M. (1988). Routing of multipoint connections. *Selected Areas in Communications, 6*, 1617–1622. doi:10.1109/49.12889

Winston, W. L. (1994). *Operations Research: Applications and Algorithms*. New York: International Thomson Publishing.

Xueshun, W., Shao-Hua, Y., & Ting, L. (2009). *A Multiple Constraint Quality of Service Routing Algorithm Base on Dominating Tree.*

Zhou, C., & Huang, L. (2010). *Study on the Improvement of Analytic Hierarchy Process under College Course Evaluation System.* Washington, DC: IEEE.

This work was previously published in International Journal of Information Technology and Web Engineering, Volume 5, Issue 4, edited by Ghazi I. Alkhatib, pp. 1-22, copyright 2010 by IGI Publishing (an imprint of IGI Global).

Chapter 14
SPACots:
A Software Tool for Selecting COTS Components

Asmaa Alsumait
Kuwait University, Kuwait

Sami Habib
Kuwait University, Kuwait

ABSTRACT

This paper presents a software tool for integrating a child-friendly computer system based on commercial off-the-shelf (COTS) components. The effective selection of COTS components, which meet a child's requirements and expectations, is a non-trivial and challenging optimization problem. However, many published papers consider the functional requirements while ignoring usability requirements. The functional requirements are concerned with what the computer should be able to do, whereas the usability requirements are concerned with the extent to which the child is able to learn effectively and efficiently throughout the COTS based computer. In this paper, the authors propose an iterative five-task selection and integration of COTS process, including both hardware devices and software modules, to be automated. The core of the automated tool is employing Simulated Annealing (SA) to search the design space to match, select, and integrate COTS components with a maximal satisfaction while neither exceeding a given budget nor violating child and performance constraints. A Monte Carlo simulator was utilized to evaluate the goodness of the COTS based computer design. Computational results based on building a computer for a child handwriting e-learning application show feasibility of SPACots in finding a solution satisfying all constraints while reducing the cost by 58%.

DOI: 10.4018/978-1-4666-0023-2.ch014

INTRODUCTION

There is growing interest in the notion of software development through the planned selection and integration of commercial off-the-shelf (COTS) components. The potential advantages of this integration centric approach are shorter development time and reduced cost. Often a COTS based development process consists of selection, integration, evaluation, adaptation and evolution of components obtained from external vendors. However, most methods focus on system adaptation and integration but many methods neglect the processes of evaluation and selection of COTS with respect to the usability requirements, especially when the intended user is a child. A child has different prospective from an adult, especially when it comes to the usage of computers.

There are several design challenges, which arise during the production of COTS based computer. The first-level is requiring the designers to understand the child's interests and requirements. The second-level is selecting the proper system's (hardware and software) components that can match both the child's physical abilities (e.g., level of eye-hand coordination, keyboard within icons), the child's style of play, and child's cognitive capacity. The third-level is observing and documenting the learning-curve between a child and computer. The fourth-level is reflecting the outcomes of the learning-curve on enhancing the computers to become more user-friendly for a child. The first two-level represents the design prospective, where the last two-level represents the long-term educational effectiveness of COTS based computer.

In this paper, we are presenting a new approach to synthesizing COTS based computer and it is called *selection process approach* (SPA). SPA focuses on the design prospective (first two-level for the production of COTS based computer) by glancing at the COTS selection and integration from the child-computer interaction point-of-view,

where the child's needs are considered early in the requirement phase (Alsumait & Habib, 2009).

SPA comprises of five tasks: defining the user's goals the COTS based computer, defining main software application(s) to be executed on the COTS based computer, formulating and validating the correctness and completeness of all functional and usability requirements, searching for optimal COTS based computer, and reviewing the COTS based computer by the Requirement Engineer (RE) and the stakeholder for final certification. Then, we have developed a software automated tool (SPACots), which assists the (RE) with the challenging tasks of matching and selecting potential hardware and software components, and negotiating changes to the hardware and software components while neither exceeding a given budget nor violating design and performance constraints of the COTS based computer. SPACots focuses on the user-usability requirements, which are ignored by many other COTS selection approaches. Also, SPACots bridges the gap between user-usability requirements, components requirements (hardware and software requirements) and the specifications of COTS products. The contributions of SPACots are:

- Supporting the selection of multiple COTS components in COTS intensive systems.
- Addressing the user-usability requirements during the requirements phase.
- Searching the design space based on a clear formal evaluation method.

The search algorithm utilized within SPACots is based on Simulated Annealing (SA) (Kirkpatrick, 1983), which is a meta-heuristic approach. SA exploits the analogy between the annealing of solid materials and the problem of finding optimum solutions for combinatorial optimization problems. To determine the efficiency of COTS based computer, we have embedded a Monte Carlo Simulator (Habib, 2008) within SPACots to evaluate the performance of COTS based Computer. We have assumed that each software application consist of number of tasks, which can be executed in-order or out-of-order. Also, some of these tasks would be executed more than once. Therefore, our efficiency parameter is to determine how many tasks can be executed over the selected components of the designed computer. This technique is known as the Monte Carlo Simulation, where the order of the tasks is unknown (Metropolis & Ulam, 1949). We have tried several orders and the average value is utilized as the efficiency parameter. Our computational results demonstrate the effectiveness of SPACots in finding good COTS based computers in less than five minutes while satisfying all design and performance constraints and reducing the cost of the final design with respect to the initial design.

The rest of the paper is organized as follows. The next section contains a survey of similar tools for designing COTS based computers. The internal view of the proposed selection process approach (SPA) is described. The transformation of the SPA into working tool as SPACots is shown. The experimental results for an e-learning computer to be used by a child learning how to write the alphabets is presented.

RELATED WORK

The commercial off-the-shelf (COTS) is software, hardware or a technology, which is commercially available for public on lease, license or sale. The use of commercial off-the-shelf (COTS) components is perceived to significantly shorten development time and cost, while improving quality, in developing large, complex software systems (Ncube & Maiden, 1999; Alves & Finkelstein, 2002; Cheng & Atlee, 2007). Ideally, COTS components (such as processors, memory modules, operating systems etc.) offer pre-packaged solutions, which presumably have already gone through the various time consuming and costly phases of requirements specification, design, coding, and testing. Many companies use it as an

economical alternative to in-house development. On the contrary, COTS has disadvantages in terms of difficulty in integration with other software and dependency on COTS suppliers. Because of huge market of COTS software, selection of an efficient and reliable COTS component is also a complex process. This selection process has a high risk of failure due to many limitations, like ill-defined requirements, and incomplete component description. The user requirement engineering and management are a key point of comparison of different approaches used for COTS selection. A structured method has to be followed that may include the activities like domain analysis, building quality model, evaluating COTS components pertaining to the domain and then finally selecting the most suitable COTS. These activities are not just one after the other but they follow an iterative path.

There exist some tools in the market that can be used for the construction of COTS based systems, but we mainly focus on five that are related to our work. OPAL a tool for supporting a COTS selection process (Krystkowiak et al., 2003), eCOTS as a platform for sharing massive information about COTS domains and components (Mielnik et al., 2003), DesCOTS a tool that is based in the use of quality models (Grau et al., 2004), and miniSQUID (Kitchenham et al., 1997) as a tool for defining metrics and quality factors.

OPAL is developed by the Centre de Recherche Public Henri Tudor (CITI) and it focuses on small and medium enterprises (SME) applications (Krystkowiak et al., 2003). It offers and supports the construction of a hierarchical structure to organize requirements. OPAL defines the elements in the hierarchy as quality factors, metrics, requirements or all of them. However, no more than one metric or requirement for quality factor can be defined.

The eCOTS is a platform with varied COTS descriptions (Mielnik et al., 2003). The COTS components are described is detailed description templates, which constitute of a list of criteria. An instance of detailed description template contains data that corresponds to a single COT component. This tool helps in the decision making process but it does not include any direct selection metrics.

DesCOTS is a system for supporting COTS selection processes based on the use of software quality models. DeCOTS supports the definition of selection requirements, and it analyses the stated requirements and the COTS component evaluations in order to inform the selection of a COTS component belonging to a specific domain. Both processes are related to each other and can be applied in an iterative way (Grau et al., 2004). The miniSQUID tool has been developed to store complex software metrics data sets and the metadata that describe them. MiniSQUID uses a *development model* as a framework for the development of the structure in which metrics will be set (Kitchenham et al., 1997).

There are many outstanding questions regarding the use of COTS components including questions related to evaluating COTS components, selecting among COTS components, and the question of which COTS components are more user friendly and easier to use by the end-user. Our proposed tool would lead to more usable systems with a cheaper design cost. Moreover, it would address issues that are not considered by many of the published papers, which are focused on integrating software components only without neither considering user's requirements and needs nor considering the impact of integrating different software and hardware components.

INSIDE-VIEW OF SPA

In spite of the increasing use of COTS hardware and software for systems development, there is little guidance available on how to acquire user usability requirements and component requirements for any system, how to select COTS components compliant with these requirements, or how to interleave requirements acquisition and COTS

component selection to provide the most effective process guidance. In this section, we present the selecting process approach (SPA), which supports the iterative matching, and selection of COTS components, as shown in Figure 1.

The proposed method consists of five tasks, as show in Figure 1. The first task represents the definition of the user's ultimate goal of the computer system. The processes of goals acquisition and specification must be incremental and iterative (Chung & Cooper, 2004; Tsumaki & Tamai, 2005). At the beginning, the high-level goals of the computer system are identified using typical elicitation techniques. From these goals, possible COTS software and hardware candidates are identified in the marketplace, in which new goals and software components may be recognized

The second task represents the definition the main software category or software application. In this task user needs, requirement and main actives to achieve specific goal are identified. The aim is to better understand the end-user capabilities and highlight the characteristics of tasks, which

Figure 1. Overview of the SPA approach

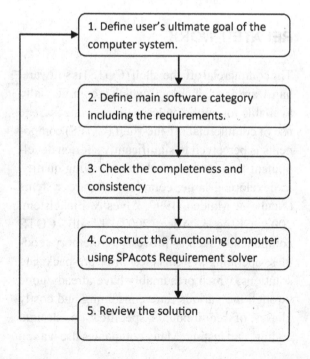

influence usability. The user expectation of the user friendly, efficiency and satisfaction factors should be clearly described.

The third task is checking the completeness and consistency of the requirements. The requirement engineers and stakeholders discuss the software category and main requirements to confirm the completeness and correctness of the second task.

The fourth task automates the construction of a functional COTS based computer by selecting and integrating COTS components. During this task, we ensure that optimization process of the COTS based computer satisfies both the functional and usability requirements. There is an improvement loop, which is iteratively matching suitable hardware and software components that satisfy the selected child's application for the given COTS based computer. Each COTS based computer is candidate solution and the improvement loop would optimize it to build a computer with the lowest possible cost that satisfies the software, hardware, and user usability requirements.

Usability Requirements

The usability is an important factor to be considered in the design of successful and usable systems. It is concerned with the extent to which the users of systems are able to work effectively, efficiently and with satisfaction (ISO/IEC, 1998). There is a lack in the literature of usability metrics that can help to evaluate the hardware and software components during the selection process (Sauro & Kindlund, 2005; Alves et al., 2005; Bertoa, 2006). This was a common problem of most of the proposals and international standards that defined hardware and software component measures. A few papers such as (Sauro & Kindlund, 2005; Alsumait, 2004; Chung et al., 2000) present a set of measures to assess the usability of the components, and describe the method followed to obtain and validate them. In this paper, we have attempted to measure the user-usability from four dimensions: *user friendly*, *look*, *feel*, and *efficiency*. Table 1

Table 1. A Summary of the first three Usability's dimensions

COTS based Computer's Usability	Selecting Rating Range
User-Friendly	1 (very difficult) to 5 (very user-friendly)
Look	1 (bad looking) to 5 (excellent looking)
Feel	1 (bad feeling) to 5 (excellent feeling)

summarizes the first three dimensions of the usability. The user friendly measures how easy a component can be used for a first-time user. We gave it a rating from 1 to 5; where 5 is very easy to use and 1 represents that this components is very difficult to use for the first time. The look and feel is a measure of satisfaction that reflects the conformity and attractiveness of the component with the end-user expectation. They are also given a rating from 1 to 5.

Efficiency Parameter

To determine the efficiency of COTS based computer, we have assumed that each software application consist of number of tasks, which can be executed in-order or out-of-order. Also, some of these tasks would be executed more than once. Therefore, our efficiency parameter is to determine how many tasks can be executed over the selected components of the designed COTS based computer. This technique is known as the Monte Carlo Simulation, where the order of the tasks is unknown (Metropolis & Ulam, 1949). We have tried several orders and the average value is utilized as the efficiency parameter.

SPACOTS TOOL STRUCTURE

The SPACots tool transforms the fourth task of SPA, which is the construction of a functional COTS based computer by selecting and integrating COTS components. It assists the requirement engineers with the challenging tasks of matching and selecting potential hardware and software components. Its main task is to find a list of components that cover the hardware, software, and usability requirements of the COTS based computer. Figure 2 depicts an overview of the SPACots tool, which browses COTS libraries to match the requirements of the first three tasks within SPA.

SPACots gets a software application or a software category as an input. It evaluates all the minimum hardware and software requirements to avoid redundancy and keep track of the minimum accepted requirements. Then, it starts building the first valid solution randomly while maintaining the minimum hardware, prerequisite software requirements and usability requirements. The output of the building process is a functioning computer satisfying the software and hardware requirements of the chosen software category. Figure 3 sketches the output computer in the tool main screen showing the hierarchy of the created computer and it's randomly chosen hardware. SPACots also logs the output computer in an external XML file see Figure 4.

The core of SPACots tool is the gray box, which includes both the search algorithm (Simulated Annealing) and performance evaluation technique (Monte Carlo Simulator). In the next two subsections, we examine in detail the core of SPACots tool.

Search Technique for Finding Feasible Solutions

Simulated Annealing (SA), a meta-heuristic approach, exploits the analogy between the annealing of solid materials and the problem of finding optimum solutions for combinatorial optimization problems (Kirkpatrick et al., 1983). The annealing process involves melting the solid materials being optimized and cooling it by lowering the temperature through a number of slow stages in the vicinity of the freezing point, which continues

until the system freezes into a minimum energy crystalline structure and no further changes occur. The sequences of temperatures and the number of iterations applied to reach equilibrium comprise an annealing schedule. By analogy with the physical process, each step of the SA algorithm replaces the current solution in terms of a random nearby solution, chosen with the probability that depends on the difference between the corresponding function values and on a global parameter T, the temperature. The temperature is gradually decreased during the process, as show in Figure 5. The dependency is such that the current solution changes almost randomly when T is large but by increasingly 'downhill' as T goes to zero. The allowance for 'uphill' moves saves the method from becoming stuck at local minima (Kirkpatrick et al., 1983).

The 'uphill' decision is made by the Metropolis procedure, as shown in Figure 6, where S is the initial solution, T is the temperature and M represents the number of modifications to the solution during a given temperature, and it is also the time until the next parameter update. This procedure allows the system to move consistently towards lower energy states, and helps to 'jump' out of local minima or maxima due to the probabilistic acceptance of some upward moves, using the Boltzmann probability distribution.

Monte Carlo Simulator

Monte Carlo (MC) simulation is based on modeling probabilistic situations, which do not change characteristics over time (Habib, 2008). Figure 7 shows the MC simulator algorithm. The number of tasks executed by the child application is randomly generated. The *for-loop* in lines 5-8 assigns a random request time and calculates processing time and retrieval time for each task's request in the list CR[]. The random number generator, which has a uniform distribution, selects a real number between zero and the duration period. A sort function is called in line 9 to order the list in

Figure 2. An overview of SPACots tool

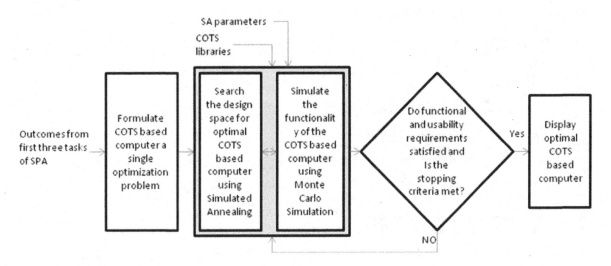

Figure 3. The graphical user interface (GUI) for SPACots

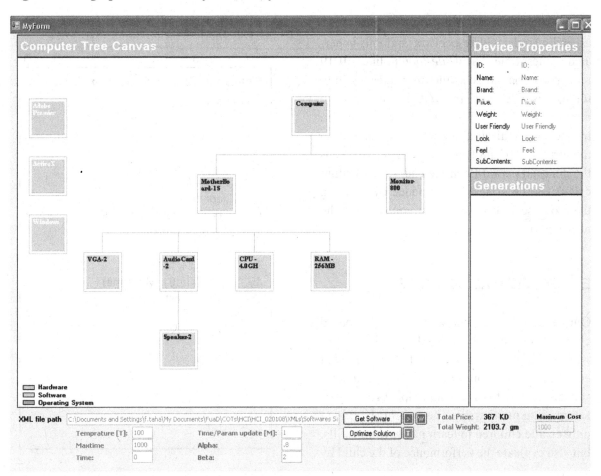

Figure 4. The configured file of the valid solution

```xml
<?xml version="1.0" encoding="utf-8" ?>
<configuration>
  <appSettings>
    <add key="xmlPath" value="C:\Documents and Settings\f.taha\My Documents\FuaD\COTs\HCI\HCI_020108\XM
    <add key="computerArchtecture" value="C:\Documents and Settings\f.taha\My Documents\FuaD\COTs\HCI\F
    <add key="OptimizationOutput" value="C:\Documents and Settings\f.taha\My Documents\FuaD\COTs\HCI\HC
    <add key="SoftwaresDirectory" value="C:\Documents and Settings\f.taha\My Documents\FuaD\COTs\HCI\HC
    <add key="HardwareSatisfactionLook" value="3"/>
    <add key="HardwareSatisfactionFeel" value="3"/>
    <add key="SoftwareSatisfactionLook" value="0"/>
    <add key="SoftwareSatisfactionFeel" value="0"/>
    <add key="HardwareUserFriendly" value="1"/>
    <add key="SoftwareUserFriendly" value="0"/>

    <add key="InternalLoopIndex" value="2"/>

    <!-- Efficiency Values -->
    <add key="Efficiency" value="50"/>
    <add key="MaximumTime" value="70"/>

    <!--C:\Program Files\HCI\HCI_Setup\XMLs\-->
  </appSettings>
</configuration>
```

ascending order with respect to a randomly generated request time. The *while-loop* in lines 10-19 repeatedly tries to schedule all requests within the sort list CR[] until the time_line exceeds the duration period (DP). After that, the algorithm tries to process all child application's tasks within the sort list CR[] until the time_line (TL) exceeds the duration period H. In the end, the algorithm calculates the system performance by dividing the processed requests request_counter over the total number of requests N.

EXPERIMENTAL RESULTS

Our case study is to determine the COTS based computer needed to be utilized for a child handwriting e-learning application. It is a system that enriches the learn-ability of the KG-2 and Grade-1 students to learn to write Arabic alphabetical characters using multimedia. The system should not only enable children to learn handwriting skills, but also evaluate the performance of the child by

Figure 5. Simulated Annealing algorithm

Simulated _Annealing (S$_0$, T$_0$, α, β, M,′ MaxTime)

1. begin
2. T= T$_0$
3. S= S$_0$
4. Time = 0;
5. repeat
6. call Metropolis (S, T, M);
7. Time = Time + M;
8. T = α x T;
9. M = β x M;
10. until (Time ≥ MaxTime)
11. output best solution found;
12. end

Figure 6. An overview of Metropolis procedure

```
Metropolis (S, T, M)

1.  begin
2.      repeat
3.          newS= neighbor (S);
4.          Δh = (Cost (news) – (Cost(S));
5.          If ((Δh <0) or random < e^(- Δh/T))
6.              then S = new S; { accept the
    solution}
7.              M =M-1;
8.      until (M =0)
9.  end
```

Figure 7. Monte Carlo Simulator algorithm

```
Monte_Carlo(N, H, CR[])

1.begin

2. TL = 0;

3. C = 0;

4. request_counter = 0;

5. for i = 0 to N-1

6.  {CR[i].t_out = random_generation;

7.   CR[i].t_in = random_generation;

8.   CR[i].V = random_generation;}

9. Sort (CR[]);

10. while(TL <= H)

11.    begin

12.    if(CR[C].t_out+ CR[C].t_in <= H)

13.        request_counter++;

14.    if(CR[C].t_out > TL)

15.        TL = CR[C].t_out + CR[C].t_in + CR[C].V;
```

measuring the child handwriting of a character to the reference character. In this case study, we have followed the SPA approach, by first defining the goal of the system under development, which is to develop a handwritten recognition e-learning tool that helps children in KG-2 and Grade-1 to learn how to write the Arabic alphabet and simple words.

Then, we have examined the user of the application, which should learn how to recognize the Arabic Alphabet. Plus, the user of the application should recognize the alphabet letters by the sound of the letter. Finally, the user of the application should be able to write each letter neatly and correctly from right to left.

In the next task within SPA, we have defined the main software category or software application including the requirements. In our case we are trying to build a child handwriting e-learning computer system. This COTS based computer should include a handwriting recognition software application, other pre-requisites software such as Adobe Flash Player, Windows as Operating System and required hardware components such as CPU, RAM, Hard Disk, Monitor, Mouse, Speaker, and DrawingPadKit. The computer's requirements for this system could be: hardware/software satisfaction [look /feel] factors should

be at least 3. The hardware/software user friendly factor should be at least 4 to encourage children to use this system. The efficiency should be around 50 and the processes maximum time is at least 70.

Table 2 contains the parameter of Simulated Annealing that the optimization process would start to improve on the initial COTS based computer. Using the initial COTS based computer as a starting point, and the following simulated annealing

Table 3 shows the components that have been changed during the Simulated Annealing process

Table 2. Parameters for the simulated annealing

Simulated Annealing Parameters	Value
Temperature (T)	100
MaxTime	10000
M (number of trials)	1
α (cooling rate)	5%
ß (updating factor for M)	2

while trying to find a cheaper COTS based computer and their effect on the user-usability requirements. Figure 8-10 illustrate the effect of optimizing the solution. In few minutes (under five minutes), SPACots tool has been able to optimize the COTS based computer design by 58% in overall cost, where it reduced the cost from $1254.00 to $529.00 as illustrated in Table 3 and Figure 8. Table 3 shows the optimization process by examining a component (column 1) at each temperature (column 2), where the rest of the columns list the objective function and constraints.

In Figure 8, the plot shows that our SPACots tool follows the classical design cost reduction over time. During the optimization process, the design cost curve went through a number of plateau regions and it was successful to escape from them. This indicates the ability of Simulated Annealing to escape from local minimal. The initial design cost of the computer was $1254.00 and it was optimized to $529.00 while preserving all the constraints, such as user-friendly, satisfaction-look, satisfaction-feel and efficiency at the closed-level.

Figure 9 and Figure 10 illustrate the effects on the four constraints. In Figure 9, the constraints (user-friendly, satisfaction-look and satisfaction-feel) maintain closed-values during the optimization process. The performance of designed computer is illustrated in Figure 10, where its values maintain between 100% to 93% of performance satisfaction of all tasks, except when the hard disk was replaced with a slow transfer rate. Then, the efficiency was dropped from 93.75% to 57.7%. However, the optimization process pulls through the efficiency constraint on the next temperature and it returned to an accepted value above 90% of satisfaction of all tasks.

Table 3. Log file for the SA process

Component	Temperature	Cost	User Friendly	Look	Feel	Efficiency
HardDisk	100	1254	5	4.9	4.9	100
Mouse	95	1256	4.9	4.8	4.9	100
CPU	90.25	1104	4.9	4.8	4.9	93.75
Monitor	85.7375	1084	4.9	4.5	4.6	93.75
HardDisk	81.450625	1064	4.8	4.4	4.5	93.75
Monitor	77.37809375	1079	4.8	4.6	4.7	93.75
RAM	73.50918906	1049	4.8	4.6	4.7	93.75
InputInterface	69.83372961	749	4.8	4.6	4.7	93.75
RAM	66.34204313	769	4.8	4.6	4.7	93.75
DrawingPadKit	63.02494097	629	4.7	4.4	4.5	93.75
MotherBoard	59.87369392	509	4.7	4.4	4.5	93.75
HardDisk	56.88000923	492	4.5	4.1	4.2	57.5
HardDisk	54.03600877	529	4.8	4.5	4.6	93.75
CPU	51.33420833	529	4.8	4.5	4.6	93.75

Figure 8. Optimization curve of the designed computer

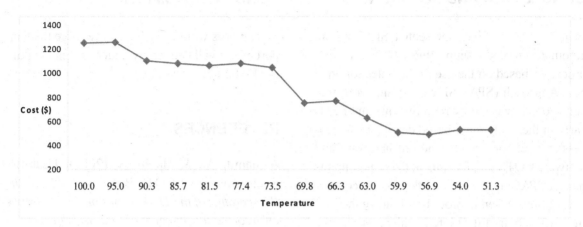

Figure 9. Effect on user-friendly and look-feel satisfaction

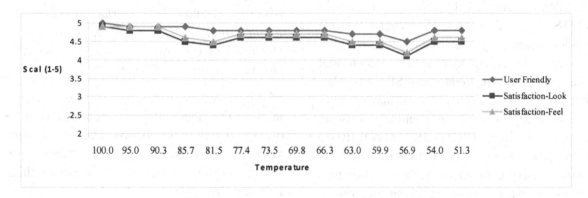

Figure 10. Effect on the performance of designed computer

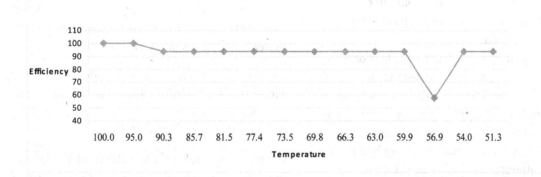

CONCLUSION AND FUTURE WORK

In this paper we have presented SPACots, an automated tool for supporting COTS selection processes based on the use of the Selection Process Approach (SPA). SPA is a framework that considers user usability requirements and needs early in the component-based development process and support the selection of hardware and software COTS components in the requirements phase. SPACots determines the COTS components to be chosen in order to minimize the cost of construction of the system while assuring the satisfaction of hardware, software and user usability requirements.

SPACots supports the selection of multiple COTS components in COTS intensive systems while considering the user-usability requirements. It searches the repository based on a clear formal evaluation method. In addition, SPACots takes into account the dependencies that may exist between the components. For example, a software component could be tied to a set other software components. SPACots will automatically include those required components in the selection process. The experience of designing a computer system for child e-learning applications showed a feasibility of finding a solution satisfying all constrains while minimizing the cost by 58% from the initial COTS based computer design. Moreover, SPACots could be applied to different application domains.

There are still some issues to tackle in order to fully automate the process. For instance we intend to enhance the tool to support the resolution of conflicts between components by assigning priorities to requirements. Moreover, there are some cases where core requirements cannot be entirely satisfied without considerable component adaptation and other cases where these goals must be compromised to match component features. An additional complication is that both goals and component specification might have incompleteness and inconsistencies. Further investigations are required to consider all these matters.

ACKNOWLEDGMENT

The authors would like to acknowledge the support by Kuwait University under a research grant no. EO02/07.

REFERENCES

Alsuamit, A., & Habib, S. (2009). Usability Requirements for COTS Based Systems. In *the Proceedings of the 11th International Conference on Information Integration and Web-Based Application and Services (iiWAS)*, Kuala Lumpur, Malaysia (pp. 393-397).

Alsumait, A. (2004). *User Interface Requirements Engineering: A Scenario-Based Framework*. Unpublished doctoral dissertation, Concordia University, Montreal, Canada.

Alves, C., & Finkelstein, A. (2002). Challenges in COTS decision-making: a goal-driven requirements engineering perspective. In *the Proceedings of the 14th international conference on Software engineering and knowledge engineering*, Ischia, Italy (pp. 789-794).

Alves, C., Franch, X., Carvallo, J., & Finkelstein, A. (2005). Using Goals and Quality Models to Support the Matching Analysis during COTS Selection. In *the Proceedings of International Conference on Composition-Based Software Systems*, Bilbao, Spain (pp. 146-156).

Bertoa, M., Troya, J., & Vallecillo, A. (2006). Measuring the usability of software components. *Journal of Systems and Software, 79*(3), 427–439. doi:10.1016/j.jss.2005.06.026

Cheng, B., & Atlee, J. (2007). Research Directions in Requirements Engineering. In *Proceedings of the IEEE ICSE Future of Software Engineering*, Minneapolis, MN (pp. 285-303).

Chung, L., & Cooper, K. (2004). Matching, Ranking, and Selecting Components: A COTS-Aware Requirements Engineering and Software Architecture Approach. In *Proceedings of the First International Workshop on Models and Processes for the Evaluation of Off-the-Shelf Components*, Edinburgh, Scotland (pp. 41-44).

Chung, L., Nixon, B., Yu, E., & Mylopoulos, J. (2000). *Non-Functional Requirements in Software Engineering*. Dordrecht, The Netherlands: Kluwer Academic Publishing.

Grau, G., Carvallo, J., Franch, X., & Quer, C. (2004). DesCOTS: A Software System for Selecting COTS Components. In *Proceedings of the 30th EUROMICRO Conference*, Rennes, France (pp. 118-126).

Habib, S. (2008). Dynamic Evaluation of Server Placement within Network Design Tool by Embedded Monte Carlo Simulator. *International Journal of Business Data Communications and Networking, 4*(2), 38–57.

ISO/IEC. (1998). *9241-11 Ergonomic requirements for office work with visual display terminals (VDT)--Part 11 Guidance on usability*. ISO/IEC 9241-11.

Kirkpatrick, S., Gelatt, S., & Vecchi, M. (1983). Optimization by Simulated Annealing. *Science, 220*(4598), 671–680. doi:10.1126/science.220.4598.671

Kitchenham, B., Linkman, S., Pasquini, A., & Nanni, V. (1997). The SQUID approach to defining a quality model. *Software Quality Control, 6*(3), 211–233. doi:10.1023/A:1018516103435

Krystkowiak, M., Bucciarelli, B., & Dubois, E. (2003). COTS Selection for SMEs: a report on a case study and on a supporting tool. In *Proceedings of the 1st International Workshop on COTS and Product Software: Why Requirements are so Important (RECOTS)*, Monterey, CA.

Metropolis, N., & Ulam, S. (1949). The Monte Carlo Method. *Journal of the American Statistical Association, 44*(247), 335–341. doi:10.2307/2280232

Mielnik, J., Lang, B., Laurière, S., Schlosser, J., & Bouthors, V. (2003). eCots Platform: An Inter-industrial Initiative for COTS-Related Information Sharing. In *Proceedings of the 2nd International Conference on COTS-Based Software System (ICCBSS)*, Ottawa, Canada (pp. 157-167).

Ncube, C., & Maiden, N. (1999). PORE: Procurement-Oriented Requirements Engineering Method for the Component-Based Systems Engineering Development Paradigm. In *Proceedings of the International Workshop on Component-Based Software Engineering*, Los Angeles, CA (pp. 130-140).

Sauro, J., & Kindlund, E. (2005). A method to standardize usability metrics into a single score. In *Proceedings of the Computer Human Interaction*, Portland, OR (pp. 401-409).

Tsumaki, T., & Tamai, T. (2005). A Framework for Matching Requirements Engineering Techniques to Project Characteristics and Situation Changes. In *Proceedings of the International Workshop on Situational Requirements Engineering Processes - Methods, Techniques and Tools to Support Situation-Specific Requirements Engineering Processes*, Paris (pp. 45-55).

This work was previously published in International Journal of Information Technology and Web Engineering, Volume 5, Issue 3, edited by Ghazi I. Alkhatib, pp. 85-98, copyright 2010 by IGI Publishing (an imprint of IGI Global).

Chapter 15
Quality of Service Management by Third Party in the Evolved Packet System

Ivaylo Atanasov
Technical University of Sofia, Bulgaria

Evelina Pencheva
Technical University of Sofia, Bulgaria

ABSTRACT

The chapter investigates the capabilities for open access to quality of service management in the Evolved Packet System. Based on the analysis of requirements for policy and charging control in the Evolved Packet Core, functions for quality of service (QoS) management and charging, available for third party applications, are identified. The functionality of Open Service Access (OSA) and Parlay X interfaces is evaluated for support of dynamic QoS control on user sessions. An approach to development of OSA-compliant application programming interfaces for QoS management in the Evolved Packet System is presented. The interface's methods are mapped onto the messages of network control protocols. Aspects of interface implementation are discussed, including interface to protocol conversion.

I. INTRODUCTION

The 3rd Generation Partnership Project (3GPP) developed standards for all-IP based architecture referred as Evolved Packet System (EPS) aimed to provide all kinds of multimedia services. EPS enables a feature-rich common packet core which supports high speed radio access networks based on not only 3GPP standards like Long Term Evolution (LTE), but for example WLAN, WiMAX or fixed access. This common core is referred as Evolved Packet Core (EPC).

The requirement to EPC in conjunction with IP connectivity access network (IP-CAN) is to provide quality of service. Quality of service (QoS)

DOI: 10.4018/978-1-4666-0023-2.ch015

is used to differentiate multimedia offering from traditional Internet services which in most cases do not provide QoS. In order to provide a mechanism for service-aware QoS control and coherent charging, 3GPP defines the Policy and Charging Control concept. The Policy and Charging Control (PCC) is a key concept in EPC architecture and it is designed to enable flow-based charging, including, for example, online credit control, as well as policy control, which includes support for service authorization and QoS management (Ouellette, Marchand, Pierre, 2011; Balbas, Rommer, Stenfelt, 2009).

One of the possible IP services layer on the top of EPC is IMS (Gouveia, Wahle, Blum, Megedanz, 2009). IMS stands for IP Multimedia Subsystem and it is an architectural framework for delivering IP multimedia services. The EPC/IMS interworking brings the advantage to integrate broadband access to all voice, messaging and data services. In IMS, the user equipment negotiates with the network the session parameters by means of Session Initiation Protocol (SIP) signaling (Iqbal, Javed, Rehman, & Khanum A., 2010). The interface between IMS and PCC architecture allows transfer of service-related information used to form authorized IP QoS data (e.g. maximum bandwidth and QoS class) and charging rules as well as user plane event reporting (e.g. bearer loss recovery, access network change and out of credit) for any access network (Wang, Liu, & Guo, 2010; Bertenyi, 2011).

To stimulate service provisioning and to allow applications outside of network operator domain to invoke communication functions, an approach to opening the network interfaces is developed (Jain & Prokopi, 2008). The open access to network functions allows third party applications to make use of network functionality and to receive information from the network through application programming interfaces (APIs). Open Service Access (OSA) is defined as service architecture for mobile networks. OSA provides APIs for a palette of network functions such as call control,

data session control, mobility, messaging etc. Currently, no APIs are defined for QoS control on end users' multimedia sessions.

The OSA APIs provide resource programmability and hide the network technology and protocol complexity for application developers but the level of abstraction remains low. Parlay X interfaces are defined to allow open access to network functions for a wider IT community (Yang & Park, 2008). Parlay X Web services are simplified, highly abstracted means for access to network functionality.

The main goal of the chapter is to assess the support of existing APIs for access to PCC functions in EPC and to present an approach to design OSA compliant APIs for QoS management.

Some related works are discussed in section II. The PCC architecture with user data convergence is presented in section III. The requirements for open access to QoS management based on the PCC architectural framework are summarized in section IV. The standardized capabilities for open access to QoS management are evaluated in section V. In section VI, it is described how OSA compliant interfaces can be designed having in mind the identified requirements. The interface implementation requires mapping of interface's methods onto network control protocol messages, and it is described in section VII. The interface usage is illustrated by typical use cases. In section VIII, a model of QoS related data in the user profile is proposed. The interface behavior is modeled by means of state machines in section IX. An approach to formal verification of API to protocol conversion is presented in section X. Finally, section XI concludes the chapter highlighting the benefits of third party QoS management in the EPS.

II. RELATED WORK

Delegating the QoS resource authorization in wireless networks to the service control layer provides a number of benefits such as session and service continuity. Session continuity in EPS provides the ability to maintain continuity of multimedia sessions during terminal mobility events, and facilitates session transfer across user terminals. The policy and charging control allows flexible QoS management in case of changing both the access networks and user devices with different capabilities. The policy and charging control can also contribute to seamless service continuity in case of handover between two wireless networks without user intervention and with minimal service disruptions. The policy and charging control bridges between wireless network services and IMS services.

Good & Ventura (2009) propose a multilayered policy control architecture that extends the general resource management function being standardized; this extended architecture gives application developers greater control over the way their services are treated in the transport layer. Good, Gouveia, Ventura & Magedanz (2010) suggest enhancements to the EPC/PCC framework that extend the end-to-end inter-domain mechanisms to discover the signaling routes at the service control layer, and use this information to determine the paths traversed by the media at the resource control layer. Because the approach operates at these layers, it is compatible with existing transport networks and exploits already existing QoS control mechanisms. Musthaq, Salem, Lohr & Gravey (2008) present an architecture with policy based network management focusing on access network optimization while taking Service Level Agreements (SLAs), business objectives, routing rules, service information, user profiles and platform conditions into account. Kallitsis, Michailidis, & Devetsikiotis (2009) suggest a model for allocating available resources in service-oriented network, with particular focus on delay sensitive services. The proposed policy is dynamic in nature and relies on online measurements of the incoming traffic for adjusting the resource allocations. Extension of the IETF policy-based management framework is given by Haddadou, Ghamri-Doudane, Ghamri-Doudane & Agoulmine, (2006) in order to support dynamic provisioning of short term services (end system signaling) as well as an scalable instantiation scheme. This instantiation scheme is based on the distribution of the provisioning process while keeping centralized only the parts that involve critical resources, i.e. the bandwidth brokerage. Zhao, Jiang & He (2008) present a policy-based radio resources allocation scheme. Different channel allocation algorithms and channel allocation strategies form a series of policies, thus constituting a policy-based channel allocation scheme. Policy-based service provisioning system is proposed by Selvakumar, Xavier & Balamurugan (2008) in order to provide different classes of services.

The necessity of open access to QoS control is substantiated by Elkotob (2008). Santoni & Katchabaw (2007) suggest a solution so that the QoS management functions are delegated to a node monitoring the signaling interchange by the user equipment and the network. Stojanovic, Rakas & Acimovic-Raspopovic (2010) address an open issue of end-to-end service specification and mapping in next generation networks. A centralized approach has been considered, via the third party agent that manages negotiation process in a group of domains. The authors suggest a general structure of the service specification form, which contains technical parameters related with a particular service request. Bormann, Braun, Flake & Tacken (2009) extend the mediation layer between the operators' core network and the charging system with more capabilities for online charging control. The authors present a prototype implementing and extending parts of the 3GPP Policy and Charging Control Architecture by the use of the open source JAIN SLEE framework Mobicents. Akhatar (2009) patents a system and method for providing QoS enablers for third party applications. In one embodiment, the method comprises user equipment establishing a communications session with a third party application server hosting a selected third party application and receiving from the third party application server QoS information comprising at least one of a plurality of QoS attributes and configuring a QoS of a radio access network in accordance with the received QoS information. The method further comprises activating the radio access network QoS for the selected application; and establishing an application session with the third party application server via the radio access network. Koutsopoulou, Kaloxylos, Alonistioti & Merakos (2007) present a platform that extends the existing charging collection information mechanisms and billing systems to provide for advanced and flexible charging mechanisms, and pricing policies. An approach to per flow charging with increased scalability of QoS support charging is suggested by Duan (2007).

OSA APIs, standardized by 3GPP, provide access to network functions upon which application developers can rely when designing new services or enhancements (versions) of already existing ones. The OSA API for QoS management is called "Connectivity Manager" and it is defined in (3GPP TS 23.198-10 v9.0.0, 2009). The API is used for negotiation and management of QoS and service level agreements in IP networks. The OSA "Policy Management", API defined in (3GPP TS 29.198-13 v9.0.0, 2009), is used for definition of information including service level agreements, evaluating policies, and subscription for and notification of policy events.

The Parlay X "Application-Driven Quality of Service" (ADQ), defined in (3GPP TS 29.199-17 v9.0.0 (2009) allows applications to control the QoS available on user connection.

The OSA "Charging" API, defined in (3GPP TS 29.198-10 v9.0.0, 2009), can be used by applications to debit or credit amounts and/or units towards a user, to create and extend the lifetime of a reservation, and to get information about what is left of the reservation. The Parlay X "Payment" Web Service defined in (3GPP TS

29.199-6 v8.1.0, 2009), supports both pre-paid and post-paid models.

The OSA API and Parlay X interfaces are defined before the standardization of EPC/IMS PCC. The analysis on PCC functions shows that these interfaces do not support all QoS management functions that network operator can expose.

III. ARCHITECTURE FOR OPEN ACCESS TO POLICY AND CHARGING CONTROL

The deployment of open service access in PCC architecture is shown in Figure 1.

Policy and Charging Control architecture is defined in 3GPP TS 23.203 specifications. The Policy and Charging Rule Function (PCRF) encompasses policy control decision and flow based charging control functionalities. The Policy and Charging Enforcement Function (PCEF) includes service data flow detection, policy enforcement and flow based charging functions. It is located at the media gateway. The Online Charging System (OCS) performs online credit control functions. It is responsible for interacting in real time with the user's account and for controlling or monitoring the charges related to service usage. Offline Charging System (OFCS) is responsible for charging process where charging information is mainly collected after the session and it does not affect in real time the service being used.

The Home Subscriber Server (HSS) contains all subscription related information needed for PCC rules. If the PCC architecture supports User Data Convergence (UDC) defined in 3GPP TS 23.335 then the User Data Repository (UDR) acts as a single logical repository for user data. The user data may for example contain information about default QoS parameters which have to be applied each time the user creates a session. Functional entities such as HSS and Application Servers keep the application logic, but they do not locally store user data permanently.

Figure 1. Open service access in PCC architecture

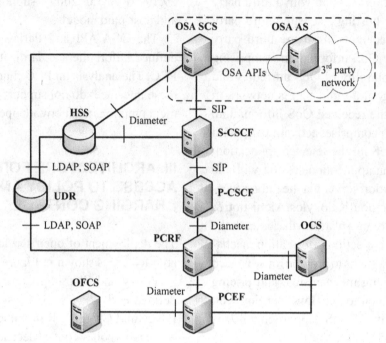

Call Session Control Functions (CSCFs) include functions that are common for all services. The Proxy CSCF (P-CSCF) is the first point of contact for user equipment. It deals with SIP compression, secured routing of SIP messages and monitoring SIP sessions. Serving CSCF (S-CSCF) is responsible for user registration and session management.

Application Servers (AS) run 3rd party applications which are outside the network operator domain. OSA Service Capability Server (SCS) is a special type of AS that provides OSA APIs for 3rd party applications and supports control protocols toward the network.

Diameter is the control protocol in interfaces where authentication authorization and accounting functions are required. The control protocol in interfaces where session management is performed is Session Initiation Protocol (SIP). Lightweight Data Access Protocol (LDAP) and Simple Object Access Protocol (SOAP) are the control protocols used to create, read, modify and delete user data

in the UDR, and to subscribe for and receive notifications about user data changes.

Note that not all charging related interfaces and policy control functions are shown in Figure 1 for the sake of simplicity.

In the next section, we study the functionalities of PCC and UDC in order to determine the requirements for open access to QoS management.

IV. REQUIREMENTS FOR OPEN ACCESS TO POLICY AND CHARGING CONTROL

The PCC includes mechanisms for controlling the bearer traffic by using IP policies.

A. Gating and QoS Control

During the multimedia session establishment and modification, the user equipment negotiates a set of media characteristics. If the network operator applies policy control then the P-CSCF sends

Figure 2. Open service access to QoS control

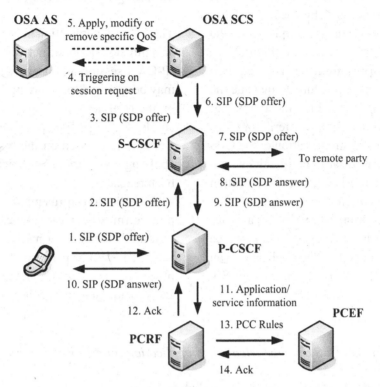

the relevant session description information to the PCRF in order to form IP QoS authorization data. The 3rd party application can be involved in the process of QoS authorization by requesting specific QoS parameters to be applied, modified or removed. Figure 2 illustrates the application control on QoS resource authorization for given SIP session.

Functional requirement 1: During the SIP session establishment, 3rd party application may require to apply or to modify temporary specific QoS features on user session(s). The required functions include: apply temporary QoS parameters, modify temporary QoS parameters and remove QoS parameters for a predefined duration (e.g. for session duration). The application logic is activated in case of session initiation, modification or termination.

In EPS, it is primary the network which decides what kind of bearer user equipment needs when communicating. Having application/service information and based on subscription information and policies, PCRF provides its decision in a form of PCC rules which are used by the PCEF for gating control.

Any QoS events, such as indication of bearer release or bearer loss/recovery, are reported by the PCEF to the PCRF and P-CSCF. Using the policy control capabilities, the P-CSCF is able to track status of the IMS signaling and user plane bearers which the user equipment currently uses, and to receive notifications when some or all service data flows are deactivated. To receive notifications about QoS events the 3rd party application needs to manage its subscriptions for notifications. By using information about bearer and signaling path status, the 3rd party application can improve service execution.

For example, the application can initiate session release on behalf of the user when gets indication that all service flows assigned to the ongoing session are released, but the P-CSCF has not received session termination request from the user itself. The scenario is shown in Figure 3.

Functional requirement 2: The required functions for 3rd party application to manage the QoS event subscription include the following: create notifications and set the criteria for QoS; change notifications by modification of the QoS event criteria; enable/disable notifications, and query for the event criteria set; report notification upon QoS event occurrence.

Functional requirement 3: The 3rd party application must be able to request QoS resource release. Using this function, the application can prevent unauthorized bearer resources after SIP session termination.

B. Usage Monitoring

The 3rd party application may be interested in the accumulated usage of network resources on per IP-CAN session and user basis. This capability may be required for applying QoS control based on the total network usage in real-time. For example, the 3rd party application may change the charging rate based on the resource usage (e.g. applying low tariff after a specified volume have been reached).

Functional requirement 4: The 3rd party application must be able to set and send the applicable thresholds for monitoring. Usage monitoring, if activated, shall be performed for a particular application, a group of applications or all detected traffic belonging to a specific multimedia session.

Figure 3. Notification of QoS resource release and application- initiated session release

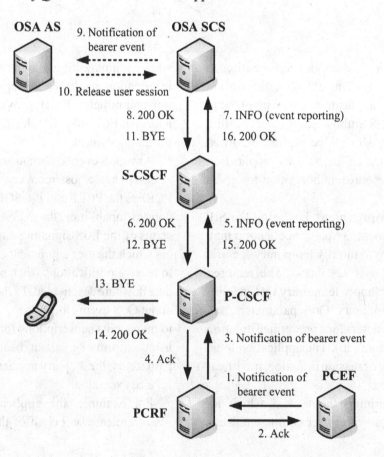

C. Application Detection and Control

The PCC architecture allows application detection and control. This feature comprises the request to detect specific application traffic, to report to the PCRF on initiation or termination of application traffic and eventually to apply the specified enforcement actions. The 3rd party application may be interested in application detection and control for example for statistics or detection of application misuse.

Functional requirement 5: The 3rd party application must be able to instruct the PCRF, on which applications to detect and report to the 3rd party application. The 3rd party application must be able to instruct the PCRF for detection of applications. Additionally, instructions may be given to the PCRF regarding appropriate actions that have to be taken upon application traffic detection.

D. User Data Access

The 3rd party application may need to retrieve QoS-related user data which are stored in the UDR. Query function may be used by the 3rd party application to obtain the QoS related data from the user profile or its specific components. List function may be used by the 3rd party application to browse the existing QoS related data in user profiles in the various UDRs. Create function may be used by the 3rd party application to add new QoS-related data in the user profile. Delete function may be used by the 3rd party application to remove specific QoS-related data from the repository. Modify function may be used by the 3rd party application to change the QoS related data in the profile components. It is the responsibility of the Service Provider to define which QoS-related data may be modified or deleted by application providers.

The application access to QoS related data, stored in the user profile, is shown in Figure 4.

Functional requirement 6: The required functions for access to QoS related user data include the following: query QoS data in order to retrieve the QoS parameters applied to user sessions by default; create QoS data in order to add new QoS parameters in user profile; modify QoS data in order to set new default QoS parameters; and delete QoS in order to erase the QoS parameters from the user profile.

Subscription/notification procedures allow the OSA SCS to get notified when particular QoS data for specific user are updated in the UDR. Using functions for access to QoS related user data, the 3rd party application can receive up to date

Figure 4. Open access to QoS related user data

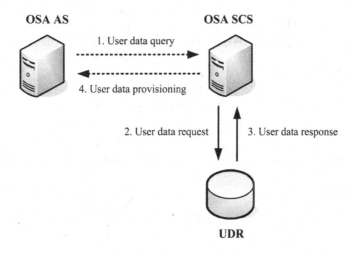

information. Subscribe function may be used by the 3rd application to request notifications about changes in QoS related data in the user profile as shown in Figure 5. Unsubscribe function may be used by the 3rd party application to cancel one or several existing subscriptions. Notify function is required when the data identified in subscription are changed or when the invoked subscribe func-

tion requests retrieval of all initial values of the referenced data as shown in Figure 6.

Functional requirement 7: To be aware about user's data changes, the 3rd party application needs functions for subscription management and means for notifications when such QoS related events occur.

Figure 5. Subscription to QoS related user data change

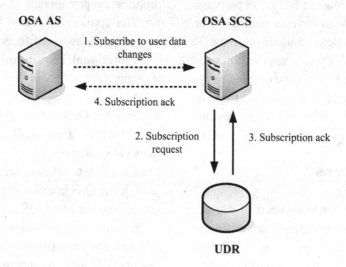

Figure 6. Notifications of user data changes

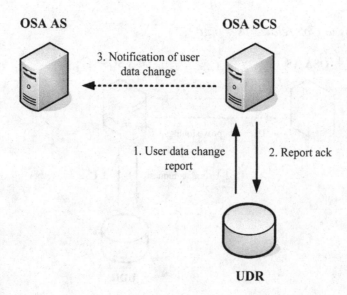

E. Charging Control

The charging function in PCC supports the following charging models: volume based charging, time based charging, time and volume based charging, event based charging and no charging. It is possible to apply different rates and charging models (e.g. depending on the user location). The charging system selects the applicable rate based on QoS provided for the service, time of day, etc. In case of online charging, the charging actions are taken upon PCEF events (e.g. re-authorization upon QoS change).

Functional requirement 8: In addition to functions for online and offline charging control, notification function is also required. To provide QoS-based charging and flow-based charging, the 3rd party application needs to get notifications when some service data flows (e.g. video stream) or all service data flows (i.e. media streams of particular SIP session) have been deactivated, when the session has been terminated or access network has been changed. When the 3rd party application provided usage monitoring thresholds to the PCRF, the PCRF shall report accumulated usage to the 3rd party application, after the PCRF detects that the usage threshold provided by the 3rd party has been reached.

The event types that can be reported to the 3rd party application involved in QoS management are summarized in Table 1. These event types can affect the QoS resource authorization and charging.

V. EVALUATION OF STANDARDIZED CAPABILITIES FOR OPEN ACCESS TO POLICY AND CHARGING CONTROL

A. Open Service Access Interfaces

The OSA "Connectivity Manager" API provides configuration and control of the attributes of IP connectivity within and between IP domains. The APIs can be used to configure QoS parameters in a virtual private network (VPN) supported by IP networks. The VPN is provisioned using virtual leased line concept termed as virtual provisioned pipe (VPrP). Elements that can be specified for a VPrP include attributes such as packet delay and packet loss. Characteristics of traffic that enters the VPrP at its access point to the provider network can be also specified by attributes such as maximum rate and burst rate. The APIs allow QoS attributes to be set as default but do not provide means for assessment of the provisioned QoS. The APIs are more of operational control and cannot be used for dynamic QoS control on multimedia sessions. Lee, Leaney, O'Neill & Hunter (2005) suggest an improvement of OSA Connectivity

Table 1. QoS-related event types

Event type	Description
Loss/Release of bearer	Loss of bearer that can result in QoS degradation (e.g. the service data flows are deactivated as a consequence). If all the bearers are lost, the application can request QoS resource release.
Recovery/establishment of bearer	Recovery or establishment of a new bearer.
IP-CAN change	The access network providing IP connectivity is changed which can result in applying specific charging.
Out of credit	The user credit limit is reached.
Session termination	The session terminates normally.
Usage report	Reports that the usage threshold provided by the 3rd party application has been reached.

Manager but it does not provide real time control on QoS parameters.

The OSA "Policy Management" APIs can be used to define policy rules, concerning provisioned QoS, and to evaluate the IP policy. An IP policy is ordered combination of rules defining how to administer and control the access to QoS resources. The combination of conditions and actions in a rule determines how and when the rule has to be applied in policy evaluation. The APIs can be used to evaluate the provisioned QoS, but not for dynamic QoS control.

The OSA "Charging" APIs support the functions for online and offline charging allowing volume and/or time based charging, but not for QoS based charging.

Currently, there are no OSA APIs for dynamic QoS control and access to QoS related user data.

B. Parlay X Web Services

The "Application-Driven Quality of Service" (ADQ) is a Parlay X Web Service that allows applications to control the QoS available on user connection. Configurable service attributes are upstream rate, downstream rate and other QoS parameters specified by the service provider. Changes in QoS may be applied either for defined time interval, or each time user connects to the network. The "ADQ" Web Service enables applications to register with the service for notifications about network events that affect QoS, temporary configured on the user's connection. The current Parlay X standardized interfaces do not support functions for usage monitoring, application detection and control, and resources release.

As to (3GPP TS 29.214 v11.1.0, 2011), there are indications reported over the Rx reference point by the PCRF to the P-CSCF such as recovery of bearer, establishment of bearer, IP-CAN change, out of credit and usage report. These indications can not be provided to the 3rd party application by the existing definition of the enumerated type QoSEvent.

The Parlay X "ADQ" Web service provides operations which allow retrieval of the current status of user sessions, including history list of all QoS transactions previously requested against a user session. As far as the getQoSStatus operation of the ApplicationQoS interface is used by the 3rd party application to access the currently available QoS features on a user session, it is impossible for 3rd party application to retrieve the configured QoS features stored in the user profile. Further, if the QoS related data in the user profile have been changed by administrative means, the 3rd party application can not be notified.

The Parlay X "Payment" Web Service supports payment reservations, pre-paid payments, and post-paid payments. It supports charging of both volume and currency amounts, a conversion function and a settlement function in case of a financially resolved dispute. When combined with "ADQ" Web Service, the "Payment" may be used for charging based on the negotiated QoS. The features for QoS based charging are restricted to temporary configured QoS parameters but can not reflect the dynamic QoS change during the session. Flow-based charging is also impossible, as far as the Parlay X "Call notification" Web Service, defined in (3GPP TS 29.199-3 v8.1.0, 2009), does not provide notifications about media addition or deletion for a particular session.

Web services are currently a good solution to integrate existing heterogeneous applications and a way to access to the Web, but they still have some open problems. Except the above mentioned limitations, there are no warranties about the adequacy of service repositories and the performance is low with current synchronous protocol invocations. Further, providing high level abstraction of network functions, Parlay X interfaces are not as flexible in expressing communication details as OSA interfaces.

In the next section, an approach to design of APIs for Application-managed QoS is presented. The API design follows the approach adopted by OSA and it is based on the identified requirements for open access to PCC functions in IMS. The idea is to substantiate the necessity for definition of OSA API for dynamic QoS control.

VI. AN APPROACH TO DESIGN INTERFACES FOR QOS MANAGEMENT

The Application-managed Quality of Service consists of the following Service Capability Features:

- QoS Resources is a service that allows 3rd party applications to access functions for QoS management including dynamic control of QoS parameters available on user sessions.
- QoS-related User Data is a service that allows access to and permanent changes in QoS data in user profile.

To provide compliance of the designed services to OSA API, two interface packages have to be defined. One of the packages contains description of the Application-managed Quality of Service interfaces on the network side, and the other one contains description of the application's interfaces. Figure 7 presents the QoS Resources interface diagram. Figure 8 presents the QoS-related User Data.

A. IpQoSManager Interface

The IpQoSManager inherits from OSA IpService and it is manager of the Application-managed Quality of Service. The interface provides functions for subscription management over QoS events. The IpQoSManager interface methods include the following:

- enableQoSNotification() enables notifications about QoS events. It is used by 3rd party application to indicate its interest in receiving notifications about QoS events related to a particular user or a group of users. The parameter eventCriteria describes the event type. Possible exceptions that can rise are invalid QoS event type and criteria;
- disableQoSNotification() disables QoS event notifications. The method is used by the 3rd party application to stop the QoS event notifications;
- changeQoSNotification() changes the criteria for generating notifications. The method is used by 3rd party application to modify the existing criteria defined with enableQoSNotification();
- getQoSCriteria() queries about criteria for sending notifications. It is used by 3rd party application to query about criteria, defined by enableQoSNotification() and changeQoSNotification();
- getQoSHistoryEvents() traces the dynamic changes related to QoS parameters. It is used by 3rd party application to receive information about applied QoS features on user sessions;
- createQoSResources() creates an object representing authorized QoS resources for given user session.

B. IpAppQoSManager Interface

The IpAppQoSManager interface inherits from OSA IpInterface. It provides methods for QoS event notifications. The interface methods include the following:

- notifyQoSEvent() notifies the application about occurrences of QoS events. The parameter eventInfo specifies event related data;
- notificationQoSInterrupted() indicates to the application that QoS event notifications are temporary interrupted, for example due to detected faults;

Figure 7. Class diagram of QoS resources interfaces

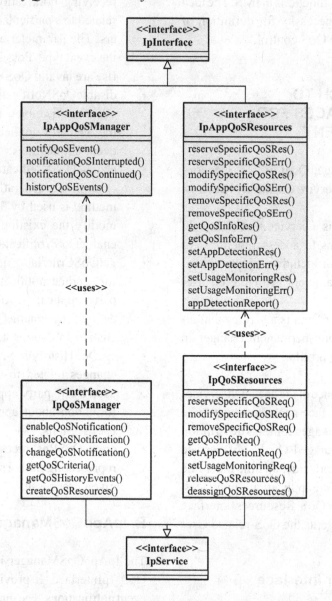

notificationQoSContinued() indicates to the application that QoS event notifications are continued;

historyQoSEvents() provides the application with a list of QoS changes.

C. IpQoSResources Interface

The IpQoSResources interface inherits from OSA IpService and provides methods for control on QoS resource authorization. The IpQoSResources interface methods include the following:

reserveSpecificQoSReq() requests authorization and reservation of QoS resources. It

Figure 8. Class diagram of QoS-related user data interfaces

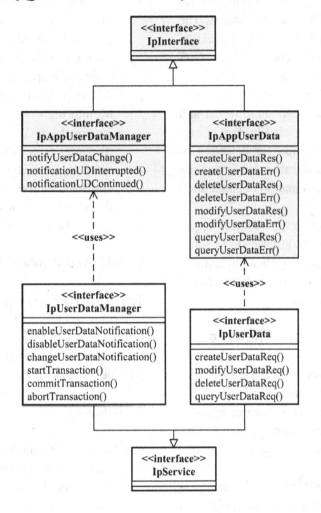

is used by the 3rd party application to request temporary reservation of resources with specific QoS. The method parameters include traffic class, uplink and downlink bit rate, duration for the reservation;

- modifySpecificQoSReq() initiates modification of the temporary established QoS. It is used by 3rd party application to modify the QoS features set by reserveSpecificQoSReq();
- removeSpecificQoSReq() removes the temporary established QoS. It is used by the 3rd party application to remove the QoS

features set by reserveSpecificQoSReq() or modifySpecificQoSReq();

- releaseQoSResources() releases the QoS resources reserved for the user session. It is used by the 3rd party application to release the authorized QoS resources (e.g. on receiving notification that all bearers assigned to user session are lost);
- deassignQoSResources() requests the relationship between the 3rd party application and the QoS resources and associated objects to be destroyed. The 3rd party application has no further control on QoS resources;

- getQoSInfoReq() requires information related to the temporary QoS features. The method is used by the 3rd party application to query about the currently available QoS features set on user sessions;
- setAppDetectionReq() instructs the network on which applications to detect and report. The method is used by the 3rd party application to request for detection of specified application traffic and for reports on initiation and termination of application traffic;
- setUsageMonitoringReq() requests applicable thresholds for monitoring to be set. It is used by the 3rd party application to apply usage monitoring for the accumulated usage of network resources.

D. IpAppQoSResources Interface

The IpAppQoSResources interface is implemented by the application and inherits from OSA IpInterface. The interface methods are designed to receive results of the application required actions in the network as follow:

- reserveSpecificQoSRes() and reserveSpecificQoSErr() indicate to the application the result of QoS resource authorization and reservation;
- modifySpecificQoSRes() acknowledges the modification of the requested QoS features while modifySpecificQoSErr() reports that the requested QoS features can not be modified because of errors;
- removeSpecificQoSRes() acknowledges the cancellation of previously defined QoS features while removeSpecificQoSErr() reports that the requested QoS features can not be removed because of errors;
- getQoSInfoRes() and getQoSInfoErr() report information related to currently available QoS or errors;

- setAppDetectionRes() and setAppDetectionErr() report that the application detection and control feature on specified application traffic is activated or can not be activated respectively;
- setUsageMonitoringRes() and setUsageMonitoringErr() report that usage monitoring for the accumulated usage of network resources is activated or can not be activated respectively;
- appDetectionReport() is used to report to the 3rd party application about initiation or termination of specific application traffic.

E. IpUserDataManager Interface

The IpUserDataManager inherits from OSA IpService and it is the manager of user data. The interface provides functions for management of subscriptions for changes of user's data. It also provides means for transactions handling. The IpUserDataManager interface methods include the following:

- enableUserDataNotification() allows notifications about changes in the user profile. The method is used by the application to indicate its interest in receiving notifications about user data changes;
- disableUserDataNotification() removes the subscription about user data changes. The method is used by the application to indicate that it is no longer interested in receiving notifications about user data changes;
- changeUserDataNotification() is used to change the criteria for notifications set by enableUserDataNotification();
- startTransaction() opens a transaction. All modifications to the user data up to the call to either commitTransaction() or abortTransaction() will be treated as part of this transaction. The transaction brackets consisting of startTransaction() and commitTransaction() are generally used

to perform changes in an atomic way, i.e. to ensure that either all changes are made persistent or all changes are undone in case of failure of even a single action. Any other clients reading data modified by this transaction will see the existing data until commitTransaction() is called;

- commitTransaction() commits a transaction. All modifications to the data in the user profile made since the last call to startTransaction() will be committed;
- abortTransaction() aborts a transaction. All modifications to the data in the user profile made since the last call to startTransaction() will be discarded.

F. IpAppUserDataManager Interface

The IpAppUserDataManager interface inherits from OSA IpInterface. It provides methods for notifications about changes of user's data. The interface methods include the following:

- notifyUserDataChange() notifies the application about occurrence of user data change;
- notificationUDInterrupted() indicates to the application that the notifications about user data changes are temporary interrupted, for example due to congestion situation;
- notificationUDContinued() indicates to the application that the notifications about user data changes are continued.

G. IpUserData Interface

The IpUserData interface inherits from OSA IpService and provides methods for access to QoS data stored in the user profile. The interface includes the following methods:

- createUserDataReq() inserts a new user data record into the UDR. It is used by

the 3rd party application to add new QoS-related user data for an existing user;
- deleteUserDataReq() deletes a user data record stored in the UDR. The method is used by the 3rd party application to delete QoS-related user data for an existing user;
- modifyUserDataReq requires permanent changes in user profile related to the QoS features. It is used by the application in order to change the parameters of agreements for the service provisioned. The method parameters include the traffic class, uplink and downlink rates etc.;
- queryUserDataReq() requests to retrieve user data from UDR. It is used by the 3rd party application to query about QoS related data stored in the user profile;

H. IpAppUserData Interface

The IpAppUserData is an application interface and inherits from OSA IpInterface. The interface methods are designed to receive results of the actions in the network required by the application as follow:

- createUserDataRes() indicates that the new user data record is inserted in the UDR while cretesUserDataErr() reports that a new user data record can not be inserted due to errors;
- modifyUserDataRes() and modifyUserDataErr() indicate to the 3rd party application that the requested change is done or can not be done because of errors respectively;
- deleteUserDataRes() and deleteUserDataErr() indicate to the 3rd party application that the QoS data is deleted or can not be deleted because of errors respectively
- queryUserDataRes() and queryUserDataErr() report QoS data stored in the user profile or errors.

VII. MAPPING OF APPLICATION-MANAGED QUALITY OF SERVICE INTERFACES ONTO NETWORK PROTOCOLS

In order to make an adequate the implementation of the open access to QoS management functions in the network, the interfaces methods have to be mapped onto messages of network control protocol.

A. SIP Based Interfaces

The interfaces between the application server (OSA SCS) and S-CSCF, and between S-CSCF and P-CSCF are SIP-based. SIP session information (including QoS parameters) is described by means of Session Description Protocol (SDP) and is transferred within the SIP message body. The initial request is sent as SIP INVITE message. The SIP re-INVITE message is used for modification of established session. QoS related information about SIP session is transferred by INFO message. The management of the subscription to QoS related events and notifications about QoS related events are provided by means of SIP SUBSCRIBE/NOTIFY mechanism. The initial filter criteria for application triggering are stored as a part of user data stored and are downloaded to the S-CSCF on user registration.

Table 2 show the mapping of QoS Resource interfaces onto SIP signaling.

The methods getQoSCriteria(), getQoSHistoryEvents(), historyQoSEvents(), createQoSResources(), notifyQoSInterrupted(), and notifyQoSContinued() do not require any signaling in the network, and only some actions in the OSA SCS.

B. LAPD and SOAP Based Interfaces

All procedures related to query or deletion data from the UDR and to creation or update data within the UDR are controlled by LDAP as specified in

Table 2. Mapping overview of QoSResources interfaces onto SIP

QoS Resources interface method	SIP message
enableQoSNotification()	SUBSCRIBE/200[SUBSCRIBE]
disableQoSNotification()	SUBSCRIBE/200[SUBSCRIBE]
changeQoSNotification()	SUBSCRIBE/200[SUBSCRIBE]
notifyQoSEvent()	NOTIFY/200[NOTIFY]
reserveSpecificQoSReq()	INVITE
modifySpecificQoSReq()	re-INVITE
removeSpecificQoSReq()	re-INVITE
setUsageMonitoringReq()	INFO
setAppDetectionReq()	INFO
appDetectionReport()	INFO
releaseQoSResources()	BYE, 200 [BYE]
reserveSpecificQoSRes()	200 [INVITE]
reserveSpecificQoSErr()	4xx, 5xx, 6xx
modifySpecificQoSRes()	200 [re-INVITE]
modifySpecificQoSErr()	4xx, 5xx, 6xx
removeSpecificQoSRes()	200 [re-INVITE]
removeSpecificQoSErr()	4xx, 5xx, 6xx
setUsageMonitoringRes()	200 [INFO]
setUsageMonitoringErr()	4xx, 5xx, 6xx
setAppDetectionRes()	200 [INFO]
setAppDetectionErr()	4xx, 5xx, 6xx

3GPP TS 29.335. The subscription/notification messages related to changes in user data stored within the UDR are transferred by HTTP in SOAP envelopes.

The method startTransaction() opens a transaction and all modifications to the user database up to the call to either method commitTransaction() or abortTransaction() will be treated as part of this transaction. To initiate an LDAP session, the OSA SCS first establishes a transport connection with the UDR and then initiates an LDAP session by sending a BindRequest message. In order to allow the application to relate a number of operations and to have them performed in one unit of interaction a transaction is used as shown in Figure 9.

Figure 9. Opening a link for LDAP session and transaction for access to user data

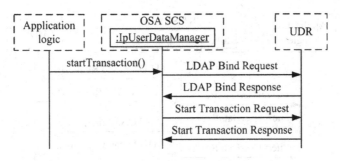

Figure 10. Closing a transaction

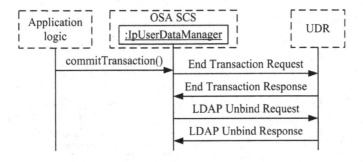

The commitTransaction() method makes all modifications to user data base made since the last call to startTransaction() to be committed. The OSA SCS will end the transaction within the UDR and if it is the last transaction a termination of the LDAP session will be done as shown in Figure 10.

3rd party applications that wish to participate in user data management process have to use the methods of the IpUserData interface to access a user data of interest. The OSA SCS translates the method invocations in Query, Create, Delete, Modify and Update messages for UDR. Figure 11 shows an example protocol mapping.

Using the IpUserDataManager interface an authorized application may request subscription for and notifications of events related to user data change. On receiving subscription request the OSA SCS makes use of the HTTP Post method contained in a SOAP message envelope. The information flow for the subscription procedure is shown in Figure 12.

The notifyUserDataChange() method is invoked on the application upon occurrence of the specified event. The UDR notifies the OSA SCS about any changes in user data using Notify request messages sent by the HTTP Post method in a SOAP message envelope. Notify response messages are coded as HTTP response message and contain a SOAP message envelope as shown in Figure 13.

C. Diameter Based Interfaces

When User Data Convergence is not supported the OSA SCS is connected to the HSS. The protocol between the OSA SCS and HSS is Diameter and

Figure 11. User data creation and listing

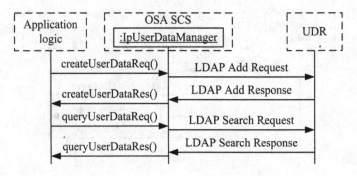

Figure 12. Subscription to user data changes

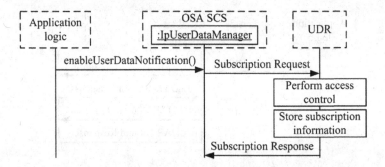

the 3rd party application access to user data is through Diameter commands.

The startTransaction() method opens a Diameter dialogue and the commitTransaction() method closes the identified Diameter dialogue.

All application initiated updates in user data are reflected in the HSS through the Diameter commands Profile-Update-Request/Answer (PUR/PUA). The access to user data is provided by the Diameter commands User-Data-Request/Answer (UDR/UDA).

The OSA SCS subscribes to receive notifications on behalf of the 3rd party application using Diameter commands Subscribe-Notifications-Request/Answer (SNR/SNA). Push-Notification-Request/Answer (PNR/PNA) commands are used to notify the OSA SCS about events of interest.

The Rx reference point is between the P-CSCF and the PCRF. It is used for policy and charging

control. In the context of PCC, the Diameter Authentication-Authorization-Request/Answer (AAR/AAA) commands are used to deliver SIP session information. The Re-Authorization-Request/Answer (RAR/RAA) commands report events related to QoS. The Session-Termination-Request/Answer (STR/STA) commands are used to release the resources, authorized earlier for a SIP session. The Abort-Session-Request/Answer (ASR/ASA) commands are used to provide information that all bearer resources, allocated to SIP session, are released.

Figure 14 shows an example usage of Application-managed Quality of Service interfaces for charging, based on the provided QoS on user session. The 3rd party application uses also the OSA Multiparty Call Control API (3GPP TS 29.198-4-3, 2009) and Charging API (3GPP TS 29.198-12, 2009).

Figure 13. Notification about user data change

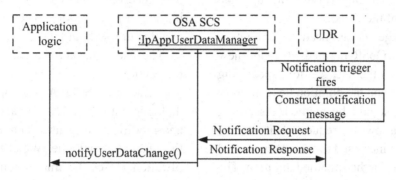

Figure 14. Quality of service based charging

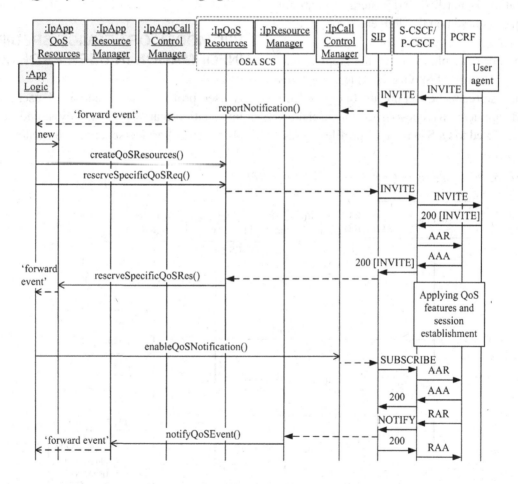

In the example, a mobile user is at the stadium enjoying a football match. The user decides to share the emotion with his friend who is away. The user wants to send to him a video of the football match. However the current service offering does not support the requested rate and hence it is required a temporary bit rate upgrade for the duration of the video.

The QoS management application then invokes the reserveSpecificQoSReq() method to apply new QoS parameters to the user session, specifying the higher bit rate and the duration he temporary QoS parameters should be applied. Assuming that the network allows the requested bit rate, the user's rate will be increased to the rate requested by the application for the specified duration. The application subscribes to notifications of events related to QoS available on user session. During the multimedia session the QoS goes down, the application is notified and generates charging information based on the delivered QoS, not the requested one.

Another example of usage the Application-managed Quality of Service interfaces concerns QoS data stored in the user profile. In Figure 15, a 3rd party application requests a permanent change of data related to QoS in the user profile, e.g. the

user receives a bonus of increased maximum bit rates as a promotion. The example considers architecture without UDC.

The user data change takes place within a transaction. When the 3rd party application starts a transaction, the OSA SCS opens a Diameter dialogue with the HSS. The access to user data and modification of user data are performed in the Diameter dialogue. When the 3rd party application closes the transaction, the OSA SCS closes the Diameter dialogue.

A model of QoS related data that may be stored in the user profile is suggested in the next section.

VIII. MODEL OF SERVICE SPECIFIC INFORMATION IN USER PROFILE

A user profile is a collection of user-specific information that is permanently stored in UDR or in HSS. Service-specific information about

Figure 15. Open access to QoS data in the user profile

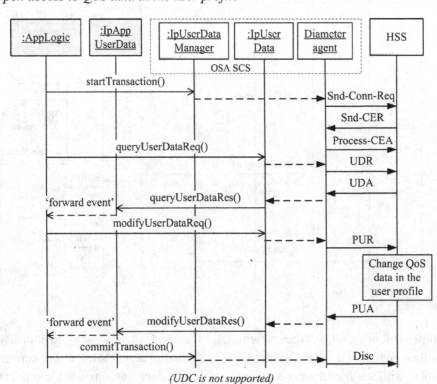

(UDC is not supported)

the user is defined as a part of the Generic User Profile which is owned by the user or value added service provider as to 3GPP TS 29.328. The data are specific to individual services (standardized or non-standardized). These could be service customisation data of the user, and/or service authentication- and authorization data (for "single sign on") like keys, certificates, passwords. These data are transparent i.e. the data are understood syntactically but not semantically by the UDR. The data format is not standardized and may be defined by the user or 3rd party service provider.

A possible content of the authorized and subscribed service information for EPS may define QoS related parameters for the user. Monitoring of QoS parameters defined in user profile and provisioned on user connections is a way to improve the network throughput (Gochev, Poulkov & Iliev, 2010).

Figure 16 shows the suggested model of the class RepositoryData, representing the QoS parameters that may be used for policy control. The ServiceIndication together with user identity and data reference identifies the service-related transparent data. The ServiceNumber is a unique service

identifier. The ServiceData contain information about the access network that provides Internet Protocol connectivity (IP-CAN) and information related to charging policy. Examples of IP-CANs are LTE and WiMax.

The QoS parameters that may be used in policy-based decisions for authorization of bearer resources intended for IMS traffic include the following:

- **Traffic class:** four different traffic classes defined for UMTS are conversational, streaming, interactive and background. By including the traffic class, network can make assumptions about the traffic source and optimize the transport for that traffic type.
- **Guaranteed Bit Rate (GBR):** describes the bit rate the UMTS bearer service will guarantee to the user or application.
- **Maximum Bit Rate (MBR):** describes the upper limit a user or application can accept or provide. This allows different rates to be used for operation (e.g., between GBR and MBR).

Figure 16. Model of the RepositoryData hierarchy

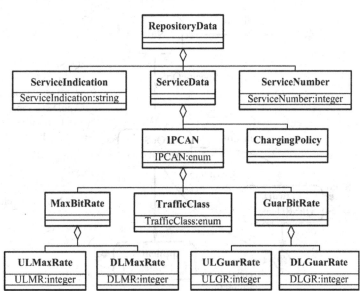

MBR and GBR are defined separately for the uplink and the downlink. Further elaboration of the model may include conditions on values related to MBRs.

IX. STATE MACHINES RELATED TO APPLICATION-MANAGED QUALITY OF SERVICE

An OSA SCS that provides Application-managed Quality of Service interfaces toward applications and control protocols toward the network has to maintain state machines for the QoSManager, QoSResources, UserDataManager and UserData objects. The application view of QoS Manager object and QoSResources object has to be synchonized with the corresponding SIP session state machine. The application view of UserDataManager and UserData objects has to be synchonized with the corresponding Diameter peer state machine.

The application view on the states of QoSResources object has to be synchronized with the SIP session states. From the application point of

view, the QoSResources object state is one of the states, as shown in Figure 17.

In Null state, the QoSResources object is created but no QoS resources are authorized for the session. In QoSAuthorized state, the rights for establishment of sessions with specific QoS features are verified and QoS resources are allocated. Any QoS modifications requested by the application require re-authorization of QoS resources. In QoSAuthorized state, the 3rd party application may request detection and reporting of specific application traffic as well as usage monitoring of network resources. QoS event reports might be result of temporary QoS degradation or release. For example, if all media flows are deactivated but BYE message is not received, the application can request session release. Notifications about access network change can affect charging, if access specific services are provided. In ApplicationQoSReleased state, the application has requested to release the QoSResources object, the media gateway only collects the related QoS information. In case the application has not requested additional QoS related information, the QoSResources object is destroyed immediately.

Figure 17. Application view on the states of QoSResources object

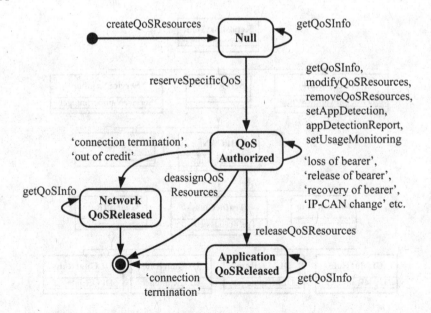

In NetworkQoSReleased state, the authorized QoS resources are released and the media gateway can return QoS information to the application if requested. In this state, the application can release the QoSResources object by invoking the deassignQoSResources method. Note that the application has to release the object itself as good practice requires that when an object was created on behalf of a certain entity, the entity is also responsible for destroying it when the object is no longer needed. In Figure 17, all QoS events causing transitions can be reported by the notifyQoSEvent method.

The application view of the states of UserDataManager object is shown in Figure 18.

During signaling of service agreement, the IpUserDataManager interface is created by the OSA SCS. To access the user data the 3rd party application starts a transaction which activates the UserDataManager object. When the 3rd party application invokes enableUserDataNotification method, it indicates its interest in receiving notifications about user data changes. The UserData-

Manager object enters Active state. The Active state can be further decomposed into Normal state and Interrupted state. Being in Normal state, the UserDataManager can inform the application about user data change event occurrences by invoking the method notifyUserDataChange. When the UserDataManager object is in Interrupted state, events requested with enableUserDataNotification will not be forwarded to the application. There exist multiple reasons for this: for instance it is possible that the 3rd party application receives more notifications from the network than defined in the Service Level Agreement. Another example is that the Service has detected that it receives no notifications from the network due to e.g. link failure. In Active state, the 3rd party application is allowed to change the criteria for notifications by invoking changeUserDataNotification. In Active state, the 3rd party application can also indicate that it is no longer interested in user data events by calling disableUserDataNotification().

The 3rd party application view of the states of IpUserData object is shown in Figure 19.

The IpUserData interface is created after service agreement sign. When the 3rd party application wants to get access to or to change user data, an UserData object is created and provides access to the user data. When the user data are provided to the 3rd party application or changed as required

Figure 18. Application view of the states of User-DataManager object

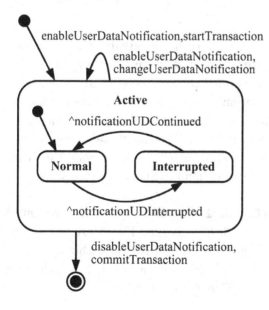

Figure 19. Application view of UserData object state machine

by the 3rd party application, the UserData object is terminated.

X. AN APPROACH TO VERIFICATION OF BEHAVIOR EQUIVALENCE

Any implementation of OSA SCS in the network requires single point of contact. Instead of coordinating the connections in circuit switched and packed switched networks, the OSA SCS governs the interface to the underlying network. Bearer and media connections rely on connections made by the session signaling, e.g. SIP signaling. Interfacing between OSA API requests and responses, and SIP signaling, the OSA SCS needs to care for both state machines – one for the application view of the states of QoS resources and another for the SIP session states. Both state machines have to expose behavioral equivalence, i.e. their observable behavior should be equivalent.

In this section an approach to formal verification of the conformance of the application view of QoS resources to SIP session states is proposed. The approach is useful in design and implementation of OSA SCS. The same approach may be applied in verification of functional behavior of OSA SCS regarding UserData interfaces translation into Diameter protocol. The Diameter peer state machines are defined in (RFC 3588, 2003). These state machines have to be synchronized with the 3rd party application view on the UserDataManager object and UserData object i.e. the corresponding state machines have to expose equivalent behavior also.

Because of OSA SCS complexity, model-based testing techniques assist in its systematization. By starting with a formal model, test cases can be derived automatically in order to prove the conformance of implementations with respect to their specifications. Automation of some parts of the testing activity, using models and formal methods is a way to improve the quality and to reduce the design cost.

To prove behavioral equivalence between state machines formally, the notion of labeled transition systems is used (Panangaden, 2009).

Definition 1: A Labeled Transition System (LTS) is a quadruple $(S, Act, \rightarrow, s_0)$, where S is countable set of states, Act is a countable set of elementary actions, $\rightarrow \subseteq S \times Act \times S$ is a set of transitions, and $s_0 \in S$ is the set of initial states.

A labeled transition system $(S, Act, \rightarrow, s_0)$ is called finitely branching, if for any state $s \in S$, set $\{(a, s'), a \in Act \mid (s, a, s') \in \rightarrow\}$ is finite. The labeled transition systems we consider in the paper are finitely branching.

We will use the following notations:

- $s \xrightarrow{a} s'$ stands for the transition (s, a, s');

- $s \xrightarrow{a}$ means that $\exists s': s \xrightarrow{a} s'$;

- $s \Rightarrow s_n$, where $\mu = a_1, a_2,..., a_n$: $\exists s_1, s_2, ..., s_n$, such that $s \xrightarrow{a_1} s_1 ... \xrightarrow{a_n} s_n$;

- $s \overset{\mu}{\Rightarrow}$ means that $\exists s'$, such as $s \overset{\mu}{\Rightarrow} s'$;

- $\overset{\hat{\mu}}{\Rightarrow}$ means $\Rightarrow$ if $\mu \equiv \tau$ or $\overset{\mu}{\Rightarrow}$ otherwise,

where τ is one or more internal (invisible) actions.

The concept of bisimulation is used to prove that two labeled transition systems expose equivalent behavior. The strong bisimulation possesses strong conditions for equivalence which are not always required. For example, there may be internal activities that are not observable. The strong bisimulation ignores the internal transitions.

Definition 2: Two labeled transition systems $T = (S, A, \rightarrow, s_0)$ and $T' = (S', A, \rightarrow', s_0')$ are weekly bisimilar if there is a binary relation $U \subseteq S \times S'$ such that if $s_1 \, U \, t_1$: $s_1 \subseteq S$ and $t_1 \subseteq S'$ then $\forall a \in Act$:

○ $s_1 \overset{a}{\Rightarrow} s_2$ implies $\exists \ t_2: t_1 \overset{\hat{a}}{\Rightarrow}' t_2$ and $s_2 \ U$ t_2;

○ $t_1 \overset{a}{\Rightarrow}' t_2$ implies $\exists \ s_2: s_1 \overset{\hat{a}}{\Rightarrow} s_2$ and $s_2 \ U$ t_2.

The SIP session state machine and the QoS resources state model as seen by OSA application are formally described by labeled transition systems which are intentionally simplified.

By $T_{SIPQ} = (S_{SIPQ}, Act_{SIPQ}, \rightarrow_{SIPQ}, s_0)$ it is denoted a labeled transition system which represents a simplified SIP session state model (see Exhibit 1).

In the 3rd party application view of QoS resources, there are no authorized QoS resources for the SIP session in the Null, NetworkQoSReleased, and ApplicationQoSReleased states. Hence, from session control point of view, these states share a common abstraction and can be labeled by single label.

By $T_{RM} = (S_{RM}, Act_{RM}, \rightarrow_{RM}, s_0')$ it is denoted a labeled transition system representing the application view of the states of QoS resources authorized for a SIP session (see Exhibit 2).

The transitions triggered by notifyQoSEvent() method report QoS events enumerated in Table 1

and shown as informal text in Figure 16. The getQoSInfo method does not affect the session state and QoS resource state, so it is excluded when proving behavior equivalence.

Proposition: The labeled transition systems T_{RM} and T_{SIPQ} are weakly bisimilar.

Proof: To prove the bisimulation relation between two labeled transition systems, it has to be proved that there is a bisimulation relation between their states. With U it is denoted a relation between the states of T_{RM} and T_{SIPQ} where U = {(Null, $Idle_{SIPQ}$), (QoSAuthorized, Established)}. Table 3 presents the bisimulation relation between the states of T_{RM} and T_{SIPQ}.

The mapping between the Quality of Service Management interface methods and SIP messages in Table 2, shows the action similarity. Based on the bisimulation relation between the states of T_{RM} and T_{SIPQ} it is proved that both systems expose equivalent behavior.

The identified bisimulation relations between QoS resources finite state machine and SIP session finite state machine shown in Table 3 might be used when testing the labeled transition systems.

Exhibit 1. SIP session state model

$S_{SIPQ} = \{$ $Idle_{SIPQ}$, SessionProgress, WaitPRack, PRacked, WaitUpdate, Updated, Wait180, Alerting, WaitAck, Established, $Wait200_{INFO}$, $Wait200_{INVITE}$, $Wait200_{BYE}$ $\}$;

$Act_{SIPQ} = \{$INVITE, BYE, 183, PRACK, 200_{PRACK}, UPDATE, INFO, 180, 200_{INVITE}, 200_{INFO}, 200_{UPDATE}, 200_{BYE}, ACK$\}$;

$\rightarrow_{SIPQ} = \{Idle_{SIP}$ INVITE SessionProgress,

       SessionProgress 183 WaitPRack,

       WaitPRack PRACK PRacked,

       PRacked 200_{PRACK} WaitUpdate,

       WaitUpdate UPDATE Updated,

       Updated 200_{UPDATE} Wait180,

       Wait180 180 Alerting,

       Alerting 200_{INVITE} WaitAck,

       WaitAck ACK Established,

       Established INFO $Wait200_{INFO}$,

       $Wait200_{INFO}$ 200_{INFO} Established,

       Established INVITE $Wait200_{INVITE}$,

       $Wait200_{INVITE}$ 200_{INVITE} Established,

       Established BYE $Wait200_{BYE}$,

       $Wait200_{BYE}$ 200_{BYE} $Idle_{SIP}\}$;

    $s_0 = \{Idle_{SIP}\}$.

Exhibit 2. QoS resources authorized for a SIP session

S_{RM} ={ Null, QoSAuthorized};
Act_{RM} = {reserveSpecificQoS, modifySpecificQoS, removeSpecificQoS, qosApproval, setAppDetection, setUsageMonitoring, appDetectionReport, releaseQoSResources, notifyQoSEvent};
$\rightarrow_{RM}$ = {Null reserveSpecificQoS QoSAuthorized,
    QoSAuthorized setAppDetection QoSAuthorized,
QoSAuthorized appDetectionReport QoSAuthorized,
    QoSAuthorized modifySpecificQoS QoSAuthorized,
    QoSAuthorized removeSpecificQoS QoSAuthorized,
    QoSAuthorized qosApproval QoSAuthorized,
    QoSAuthorized setUsageMonitoring QoSAuthorized,
    QoSAuthorized notifyQoSEvent QoSAuthorized,
    QoSAuthorized notifyQoSEvent Null,
    QoSAuthorized releaseQoS Null};
$s_{0'}$ = {Null}.

Table 3. Bisimulation relation between QoS resources and SIP session state machines

Transitions in T_{RM}	Transitions in T_{SIPQ}
Null reserveSpecificQoS QoSAuthorized	$Idle_{SIPQ}$ INVITE SessionProgress SessionProgress 183 WaitPRack, WaitPRack PRACK PRacked, PRacked 200_{PRACK} WaitUpdate, WaitUpdate UPDATE Updated, Updated 200_{UPDATE} Wait180, Wait180 180 Alerting, Alerting 200_{INVITE} WaitAck, WaitAck ACK Established
QoSAuthorized modifySpecificQoS QoSAuthorized,	Established INVITE $Wait200_{INVITE}$, $Wait200_{INVITE}$ 200_{INVITE} Established
QoSAuthorized removeSpecificQoS QoSAuthorized	Established INVITE $Wait200_{INVITE}$, $Wait200_{INVITE}$ 200_{INVITE} Established
QoSAuthorized setAppDetection QoSAuthorized	Established INFO $Wait200_{INFO}$ $Wait200_{INFO}$ 200_{INFO} Established
QoSAuthorized appDetectionReport QoSAuthorized	Established INFO $Wait200_{INFO}$ $Wait200_{INFO}$ 200_{INFO} Established
QoSAuthorized notifyQoSEvent QoSAuthorized	Established INFO $Wait200_{INFO}$, $Wait200_{INFO}$ 200_{INFO} Established
QoSAuthorized notifyQoSEvent Null	Established INFO $Wait200_{INFO}$, $Wait200_{INFO}$ 200_{INFO} $Idle_{SIPQ}$
QoSAuthorized releaseQoS Null	Established BYE Wait200 Wait200 200_{BYE} $Idle_{SIPQ}$

Implementation relations, formalizing the notion of correctness with respect to labeled transition system specifications, are defined analogous to the theories of testing equivalence and preorder, and refusal testing.

XI. CONCLUSION

The open access to QoS management functions allows for 3rd party applications dynamic control on QoS available on user sessions. The required functionality for open access to QoS management might be derived from the functional architecture of policy and charging control in the Evolved Packet System. The access to QoS control, gating control, flow-based charging and user data management provides 3rd party applications with flexibility in QoS management.

The standardized application programming interfaces do not support the entire policy and charging control functionality that network operator can expose. The research on existing functions for QoS management accessed by 3rd party applications substantiates the need of definition of abstraction of management functions in order to provide flexibility in expressing communication details.

If the Open Service Access approach is adopted in interface definition, besides the access to dynamic QoS control, 3rd party applications can benefit from other OSA APIs that expose a palette of network functions. Implementation issues of the APIs concern the interfaces toward the network which are not standardized. In EPS, the functionality of QoS management interfaces has to be mapped onto control protocols SIP, Diameter, LDAP and SOAP. The OSA SCS has to maintain the state machines representing the 3rd party application view of interface objects and the control protocol state machines. These state machines have to be synchronized and to expose equivalent behavior.

The open access to QoS control provides more flexibility in resource management as far as the QoS provisioning is one of the main requirements to the EPS. Possible stakeholders that may benefit from Application-managed Quality of Service include Value Added Service providers for QoS management and 3rd party provided services that run on application servers outside the network on behalf of particular users.

REFERENCES

Akhatar, H. (2009). *System and method for providing quality of service enablers for third party applications*. (Patent application number: 20090154397). Retrieved from http://www.faqs.org/patents/app/20090154397

Balbas, J., Rommer, S., & Stenfelt, J. (2009). Policy and charging control in the evolved packet system. *IEEE Communications Magazine, 47*(2), 68–74. doi:10.1109/MCOM.2009.4785382

Bertenyi, B. (2011). *Key drivers for LTE success: Services evolution.* 3GPP Seminar, Moscow.

Bormann, F., Braun, A., Flake, S., & Tacken, J. (2009). Towards a policy and charging control architecture for online charging. *International Conference on Advanced Information Networking and Applications Workshops*, (pp. 524-530).

Calhoun, P., Guttman, E., Zorn, G., & Arkko, J. (2003). *RFC 3588: Diameter base protocol.*

Chang, Y. (2010). Consolidation of EPC and heterogeneous home network. *TamKang Journal of Science and Engineering, 13*(1), 21–28.

Duan, X. (2007). *Method for establishing diameter session for packet flow based charging*. Retrieved from http://www.freshpatents.com/Method-for-establishing-diameter-session-for-packet-flow-based-charging-dt20070816 ptan 20070189297.php

Elkotob, M. (2008). *Autonomic resource management in IEEE 802.11 open access networks*. Dissertation, Lules University of Technology, Sweden. Retrieved from http://epubl.ltu.se/1402-1757/2008/38/LTU-LIC-0838-SE.pdf

Gochev, H., Poulkov, V., & Iliev, G. (2010). Uplink power control for LTE improving cell edge throughput. *International Conference on Telecommunications and Signal Processing, TSP'2010*, (pp. 465-467).

Good, R., Gouveia, F., Ventura, N., & Magedanz, T. (2010). Session-based end-to-end policy control in 3GPP evolved packet system. *International Journal of Communication Systems, Special Issue: Part 1: Next Generation Networks, 23*(6-7), 861–883.

Good, R., & Ventura, N. (2009). Application-driven policy-based resource management for IP multimedia subsystems. *International Conference on Testbeds and Research Infrastructures for the Development of Networks & Communities, TridentCom'2009*, (pp. 1-9).

Gouveia, F., Wahle, S., Blum, N., & Megedanz, T. (2009). *Cloud computing and EPC / IMS integration: New value-added services on demand.* 5th International ICST Mobile Multimedia Communications Conference'2009.

3GPP. (2009a). *TS 23.198-10 v9.0.0: Open service access (OSA)- Connectivity manager service capability feature* (SCF), (release 9).

3GPP. (2009b). *TS 24.229 v9.2.0: IP multimedia call control protocol based on session initiation protocol (SIP) and Session description protocol* (SDP), (Release 9).

3GPP. (2009c). *TS 29.198-4-3 v9.0.0: Open service access (OSA)- Multiparty call control, subpaer 3, service capability feature* (SCF), (Release 9).

3GPP. (2009d). *TS 29.198-12 v9.0.0: Open service access (OSA)- Charging service capability feature* (SCF), (Release 9).

3GPP. (2009e). *TS 29.198-13 v9.0.0: Open service access (OSA)- Policy management service capability feature* (SCF), (Release 9).

3GPP. (2009f). *TS 29.199-6 v8.1.0: Open service access (OSA)- Parlay X Web Services, part 6: Payment*, (Release 9).

3GPP. (2009g). *TS 29.199-17 v9.0.0: Open service access (OSA)- Parlay X Web services, part 17: Application-driven quality of service* (QoS), (Release 9).

3GPP. (2011a). *TS 23.203 v11.2.0: Policy and charging control architecture*, (Release 9).

3GPP. (2011b). *TS 29.214 v11.1.0: Policy and charging control over Rx reference point*, (Release 9).

Haddadou, K., Ghamri-Doudane, S., Ghamri-Doudane, Y., & Agoulmine, N. (2006). Designing scalable on-demand policy-based resource allocation in IP networks. *IEEE Communications Magazine, 44*(3), 142–149. doi:10.1109/MCOM.2006.1607878

Iqbal, U., Javed, Y., Rehman, S., & Khanum, A. (2010). SIP-based QoS management framework for IMS multimedia services. *International Journal of Computer Science and Network Security, 10*(5), 181–188.

Jain, M., & Prokopi, M. (2008). The IMS 2.0 service architecture. *Second International Conference on Next Generation Mobile Applications, Services and Technologies, NGMAST '08*, (pp. 3-9).

Kallitsis, M., Michailidis, G., & Devetsikiotis, M. (2009). Measurement-based optimal resource allocation for network services with pricing differentiation. *Performance Evaluation, 66*(9-10), 505–523. doi:10.1016/j.peva.2009.03.003

Koutsopoulou, M., Kaloxylos, A., Alonistioti, A., & Merakos, L. (2007). A platform for charging, billing, and accounting in future mobile networks. *Computer Communications, 30*, 516–526. doi:10.1016/j.comcom.2005.11.022

Lee, S., Leaney, J., O'Neill, T., & Hunter, M. (2005). Open service access for QoS control in next generation networks – Improving the OSA/ Parlay connectivity manager. *Journal of Operations and Management in IP-Based Networks*, *3751*, 29–38. doi:10.1007/11567486_4

Musthaq, S., Salem, O., Lohr, C., & Gravey, A. (2008). Policy-based QoS management for multimedia communication. Retrieved from https://portail.telecom-bretagne.eu/publi/public/download.jsp?id...542.6

Ouellette, S., Marchand, L., & Pierre, S. (2011). A potential evolution of the policy and charging control/QoS architecture for the 3GPP IETF-based evolved packet core. *IEEE Communications Magazine*, *49*(5), 231–239. doi:10.1109/MCOM.2011.5762822

Panangaden, P. (2009). *Notes on labelled transition systems and bisimulation.* Retrieved from http://www.cs.mcgill.ca/~prakash/Courses/comp330/Notes/lts09.pdf.

Santoni, D., & Katchabaw, M. (2007). Resource matching in a peer-to-peer computational framework. *International Conference on Internet Computing 2007*, (pp. 89-95).

Selvakumar, S., Xavier, S., & Balamurugan, V. (2009). Policy based service provisioning system for WiMAX network: An approach. *ICSCN International Conference on Signal Processing, Communications and Networking,* (vol. 4-6, pp 177-181).

Stojanovic, M., Rakas, S., & Acimovic-Raspopovic, V. (2010). End-to-end quality of service specification and mapping: The third party approach. *Computer Communications*, *1*, 1354–1368. doi:10.1016/j.comcom.2010.03.024

Wang, Y., Liu, W., & Guo, W. (2010). Architecture of IMS over WIMAX PCC and the QoS mechanism. *IET 3rd International Conference on Wireless, Mobile and Multimedia Networks, ICWMNN'10*, (pp. 159- 162).

Yang, J., & Park, H. (2008). A design of open service access gateway for converged Web service. *10th International Conference on Advanced Communication Technology 2008*, (pp.1807- 1810).

Zhao, F., Jiang, L., & He, C. (2008). Policy-based radio resource allocation for wireless mobile networks. *Proceedings on International Conference on Neural Networks and Signal Processing,* (pp. 476-481).

Chapter 16
The Role of Physical Affordances in Multifunctional Mobile Device Design

Sorin Adam Matei
Purdue University, USA

Anthony Faiola
Indiana University, USA

David J. Wheatley
Motorola Applied Research, USA

Tim Altom
Indiana University, USA

ABSTRACT

As designers of mobile/media-rich devices continue to incorporate more features/functionality, the evolution of interfaces will become more complex. Meanwhile, users cognitive models must be aligned with new device capabilities and corresponding physical affordances. In this paper, the authors argue that based on HCI design theory, users approach objects by building mental models starting with physical appearance. Findings suggest that users who embrace a device's multifunctionality are prevented from taking full advantage of an array of features due to an apparent cognitive constraint caused by a lack of physical controls. The authors submit that this problem stems from established mental models and past associated behaviors of both mobile and non-mobile interactive devices. In conclusion, users expressed a preference for immediate access and use of certain physical device controls within a multi-tasking environment, suggesting that as mobile computing becomes more prevalent, physical affordances in multifunctional devices may remain or increase in importance.

DOI: 10.4018/978-1-4666-0023-2.ch016

INTRODUCTION

The cell phone and other small mobile devices are rapidly becoming the preferred access points to, and storage repositories for, personal messages and media, such as music, photos, and video. Such devices are transforming person-to-person mobile communication into a convergence of voice and media sharing communication, i.e., devices with multifunctional capabilities (Heo, Ham, Park, Song, & Yoon, 2009; Monk, Fellas, & Ley, 2004). The functionality of these devices is further enhanced by the possibility of transferring media content to a fixed interface display, such as a personal computer (PC), TV, or stereo system. The emergence of such multimedia-enabled mobile devices creates a number of physical and conceptual design challenges that revolve around two issues.

The first issue is related to the fact that commercial devices are relatively shrinking –even the iPad is smaller than a typical laptop– yet they continue to incorporate more features and functionality. Consequently, controls and interfaces have either become more crowded or have been buried in complex hierarchically structured graphical user interfaces (GUIs) (Vivrou & Kabbasi, 2002). The second issue is that the ever-increasing functionality offered by these novel technologies is limited by the socio-cultural maps and cognitive models that users and designers carry in their minds about mobile and non-mobile device capabilities and their corresponding physical affordances (Faiola, & MacDorman, 2008; Gibson 1977, Hartson, 2003, Norman, 2002).

The concept of affordance designates the capacity of a device to suggest a particular kind of use by virtue of some physical attribute. For example, a cell phone's most significant affordance is related to voice calling, as an object made to be grasped with one hand and positioned between the ear and mouth. This affordance is reinforced both by the physical design of the device and its controls (softkeys, menu options, hardwired buttons, etc.), as well as the way in which people approach it cognitively.

Physical affordances are extremely effective when they are incorporated into simple/unifunctional devices with limited functions (Norman, 1998). However, problems often emerge when these multimedia devices additionally serve as gateways and transfer devices for video, photos, text, and various other types of information and media content between different types of platforms. For example, how should the device make its non-talking affordances visible and immediately understandable to the user who associates the device with a more basic cell phone or home phone that only makes calls? The process of "unveiling" the functional potential of the multimedia device relies heavily on creating features and interface and interaction design solutions that suggest the idea that the device is not just a cell phone, but also a vehicle for content capture, storage, and transfer between various platforms.

The study described here suggests that users approach such multifunctional devices with cognitive models derived from their prior experiences of using phones, cameras, camcorders, PDAs, and PCs. A directive principle, should be that industrial and interaction designers must consider the users' prior experience with media devices, in order to avoid conflicts with existing cognitive or mental models associated with the use of these single appliance interactive devices. Forlizzi (2007) specifically recommends thorough examination of priori subjective experiences with mobile products, which can lead to generalizable knowledge for design activities. At the same time, the design process should not be dogmatic. Ecological, participatory design is preferable. Prototypes that reflect a combination of existing and novel features and practices need to be continuously tested and "winning" features sorted out in the process of actual usage. Patters of use need to be monitored and the conclusions of such monitoring fed back into participatory design activities (Forlizzi, 2007).

This paper, which is theoretically informed by affordance theory (Norman, 2002), addresses the role that particular affordances and cognitive models play in facilitating or hindering the use of the cell phone as a gateway for multimedia content generation, transfer, and visualization. More specifically, we reconsider the role of the touch screen interface and the physical device affordances in such multifunctional mobile devices.

MULTIFUNCTIONAL DEVICES AND AFFORDANCES

Discussions about how new audio and video technologies and information management should be embedded in the mobile communication environment and assimilated by users go back to the 1980s and to Xerox's Palo Alto Research Center (Abowd & Mynatt, 2000; Dourish, 2001). Nevertheless, the practical integration of multimedia functionality within a mobile device is an evolving concept (Goldstein, Nyberg, & Anneroth, 2003). As third and four generation wireless technologies (3/4 G) become more prevalent, users are finding themselves dealing with the escalating complexities of multifunctional devices. As a result, they will need interface affordances that are capable of bridging the gap between the device's functional capabilities and the user's outdated mental models (Agre, 2001; Dourish, 2001).

The importance of physical affordances for mobile devices

True affordance means that each function and its corresponding method of operation are apparent from the device's visual appearance, haptics, or other direct sensory indicators. Simply put, affordances should provide strong clues about the operation of the device (Norman, 2002) and make its function intuitively obvious, or using a consecrated usability research term, immediately learnable (see below for details about this and other core usability concepts). Norman distinguishes between real and perceived affordances. Real affordances include the physical characteristics of a device or interface that facilitate its operation (Hartson, 2003). In other words, physical affordances help users with their physical actions (Hartson, 2003).

Perceived affordances are another class of device features (2002), which offer users not only clues related to the proper operation of a device, but help with their cognitive actions. Both of these fundamental types of affordances should be deployed harmoniously to "show how good design can make the appropriate use of a device clear and obvious to a user" (Dourish, 2001, p. 119).

Developers of multipurpose mobile devices have traditionally assumed that affordances that leverage what users already know about interfaces and social connectivity will result in greater user recognition in the context of new computing experiences (Agre, 2001; Klopfer, Squire, & Jenkins, 2004). Yet, such assumptions are often optimistic, both in terms of appropriateness of such conventions for the hand-held mobile experience and also in terms of the ability of first time users to transfer past experience to the new interface context. In an early study on the interface design challenges that surround multimedia mobile devices, research findings suggested that the efficacy of such devices decreases dramatically with their functional complexity (Goldstein et al., 2003). The study notes that elimination of traditional camera control in a multifunctional device that could take pictures, prevented many users from using this particular functionality. Goldstein et al. drew this broader conclusion:

The porting of the stationary computer metaphor to a mobile multipurpose device may prove ineffective if it is in conflict with previously acquired efficient source metaphors when using information appliances tailored to accomplish a single task. It is mandatory that efficient source metaphors are given proper attention. Omitting well known affordances under the assumption that the intelligent user is capable of accomplishing the task efficiently anyhow is a bold assumption. All applications included in a multipurpose device must be designed with the respective information appliance in mind (Goldstein et al., 2003, p. 373).

To address this problem, Goldstein et al. suggest that physical affordances be generously applied to device design to ensure effective utilization of device capabilities. Weiser's (1993) paper on ubiquitous computing essentially agrees, emphasizing that the marriage between mobility and multimedia should be marked by a dramatic shift from the desktop paradigm and its ancillary metaphors to devices and interfaces that are already well known (cameras, MP3 players, etc). The reason is that such devices and the conventions they are associated with are easier to incorporate into the expected flow of human behaviors and actions, activity that we might term "natural." One of Abowd and Mynatt's (2000) observations in their review of ubiquitous computing is also pertinent here:

We desire natural interfaces that facilitate a richer variety of communications capabilities between humans and computation. It is the goal of these natural interfaces to support common forms of human expression and leverage more of our implicit actions in the world. Previous efforts have focused on speech input and pen input, but these interfaces still do not robustly handle the errors that naturally occur with these systems; also these interfaces are too difficult to build (Abowd & Mynatt, 2000, p. 42).

These requirements are in tune with the emerging research agenda that emphasizes the physical or "embodied" aspect of technological design and use (Dourish, 2001). Brereton and McGarry (2000) have drawn our attention to the fact that people habitually think in terms of objects, even when they design devices that do not require direct physical manipulation to affect the device's functions. This idea is congruent with what Dourish (2001) de-

scribes as the "embodied" nature of technologies. This concept addresses the fact that technologies exist and are interacted with in the physical world in specific contexts and with specific expectations for immediate and facile manipulation. In this process, users prefer tangible (physically manipulable) controls and rely on intuitive use, primed by context and by the task at hand, with "intuitive" here meaning "according to the manner in which we understand affordance theory to suggest that humans use reality and the objects within it." As already mentioned above and will be detailed below, more intuitive affordances are also easy to learn and degree of "intuitiveness" translates into enhanced "learnability" (Nilesen & Mack, 1994).

From this perspective, the future belongs to small information appliances that maximize, not minimize, external affordances, says Dourish, who envisions adaptive interfaces that rely directly on the physical appearance of the objects. For example, picking up a multimedia mobile device with both hands, and holding it lengthwise makes it ready for taking a picture, while grasping it with one hand and taking it to the ear triggers the cell phone functionality. It is true that users ultimately learn new paradigms of manipulation and control, but it is also true that such learning is difficult and not optimal when existing affordances insufficiently developed.

Although Dourish's research stretches into the future, his emphasis on the need to turn back to physical controls and physically intuitive devices is a fruitful change of focus for understanding and designing more usable multipurpose mobile devices. In this new light, issues of scale, input/output methods, autonomy, and connectivity as they are relevant to the new generation of portable platforms, represent a universe that is yet to be fully explored. Furthermore, Dourish's work makes it easier to understand why the new generation of mobile multimedia devices, with their new combinations of on-screen and direct physical affordances can be challenging even for expert users.

Getting beyond calls for simplicity: Finding the right balance between physical and on-screen controls

Typical problems and ideal parameters for designing multipurpose multimedia computing devices have been discussed in the literature by Norman (2002) or Oulasvirta and Saariluoma (2004). Norman argued that the human mind has a restricted working memory, not being able to handle more than four or five items at a time. Hence, external mnemonic aids are critical to support cognition and to prevent human error. Furthermore, features in complex systems should be made as visible as possible to reduce information overload, while functionality and feedback should be clearly visible in the interface. Or, even better, as Raskin (2000) suggested, features should be "detectable," i.e., the user should be able to pick out the features that he or she needs from the multitude of functionalities available at any particular time. A delicate balancing act is required to achieve this detectability. Emphasizing some features at the expense of others might make certain features unavailable for unsophisticated users. At the same time, an excess of visibility makes the device "gadget-ridden and feature-laden," which can intimidate even experienced users" (Norman, 2002).

This tendency of devices to appear "busy" is particularly acute in the context of multipurpose devices, which combine the functionality and versatility of many individual technologies, each involving its own complexities. For example, a multipurpose device tasked to capture, enrich, and transfer moving and still pictures would combine some of the more basic features from current digital video and photo cameras, PDAs and desktop computers. Collectively, these devices include over 20 physical affordance points. For example, a typical digital camera has seven contact points located in two areas of the device, a camcorder can have as many as eight contact points located in three areas, a PDA may have six contact points, located in two areas, and the PC may have four major contact points, in three areas. Even if this collection of physical affordances could be reduced or consolidated, it would be difficult to squeeze them within the confines of smaller mobile device. Confronted by this challenge, designers may prefer to hide and then flexibly make functionalities available through a range of GUI solutions. However, for many users, these solutions may prove to be suboptimal, since the main way they recognize a specific functionality is through its corresponding physical affordance (Figure 1).

The multipurpose device studied in this paper is a good illustration of such dilemmas. Capable of taking pictures and videos, group voice calling, and device-to-device media transfer, the device relies on screen interfaces that borrow heavily from the feel and iconic vocabulary of the personal computing and PDA environment. This puts the device squarely in the situation described above by Goldstein et al. (2003). Manipulating a camera phone through an on-screen interface is in latent conflict with the familiar operation and the mental model of clicking a shutter button and with holding the camera with both hands. Goldstein et al suggest that although on-screen controls seem to be more efficient from a usability perspective, they can in fact make devices more confusing and difficult to use. In what follows, after we will position our paper in the broader field of mobile device usability research, we will discuss the physical-affordance-related design issues that emerged in the prototyping process of a mobile, multifunctional device.

APPROACH AND PREVIOUS RELATED WORK

The present paper reports results of a paper-prototyping and early usability study, informed by methodological principles recommended by

Figure 1. A multipurpose device that can capture, store, and transfer images would combine about 25 physical affordances currently employed by digital cameras, camcorders, PDAs, and PCs.

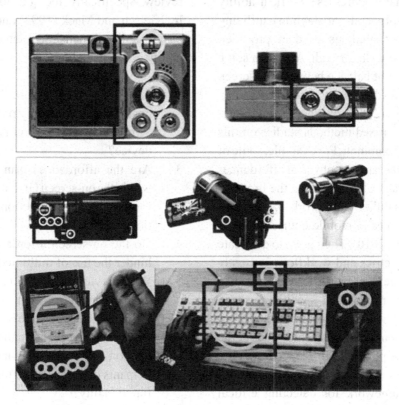

Nielsen and Mack (1994), Norman (1998; 2002), and in broad theoretical terms by Dourish (2001). The paper builds upon or is related to a class of empirical studies that aim to explain how physical interfaces can impact adoption, use and the evolution of multifunctional devices from stand-alone tools into gateways for a web-based world of multimedia content and services (Brereton & McGarry, 2000; Goldstein et al., 2003; Kangas & Kinnunen, 2005; Jin & Ji, 2010; Heo, Ham, Park, Song, & Yoon, 2009; Marti & Schmandt, 2005; Vivrou & Kabbasi, 2002).

The empirical/mixed methods study reported here aims to determine to what degree multi-functional mobile devices need or not physical affordances and how our findings might change future design decisions. Our study borrows a number of principles and common metrics from usability research. Of the five quality metrics of usability proposed by Nielsen, we focus on three:

learnability (How easy is it for users to accomplish basic tasks the first time they encounter the design?), error or failability (How many errors do users make? How severe are these errors?), and satisfaction (How satisfied are the users with the design?). As we will explain below, in our study learnability and failability are operationalized as questions about how "intuitive" the interface is (easy to learn and subsequently operate), while satisfaction is dealt with separately. Because our study evaluates a paper prototype whose functionality was simulated and was limited in time to a single research activity, the other quality attributes of usability research recommended by Nielsen, assessment of efficiency (How quickly can users perform tasks?) and memorability (When users return to the device after a period of not using it, how easily can they reestablish proficiency?), were not included in our study.

Our study extends current research on physical affordances in that it addresses specific usability issues (learnability and satisfaction) utilizing in-depth thematic analysis of data provided through cognitive walk through. Although some research on the right balance between on-screen and physical affordances has been conducted, notably Wheatly (2007) or Faiola and Matei (2010), in depth, mixed methods studies on this topic are not very common. For example, some of the most recent research on physical affordances are still at the stage of determining the broadest outlines of a usability framework for judging physical affordances in mobile devices (Heo et al., 2009; Jin & Ji, 2010). Their goal, to delineate some common metrics and workflow, is mostly orientative and general-heuristic. Compared to them, our study proposes a specific research framework with measurable parameters and actionable recommendations. Our study adds to the research space of physical affordance a specific thematic-analysis methodology, which includes an empirical framework for detecting critical issues. This and the specific methods utilized in our work are described below.

RESEARCH QUESTIONS

We use an X[1] mobile multimedia device study to investigate the relative value of on-screen and physical affordances. The study was designed and conducted in 2004 in collaboration with the device manufacturer. Its most immediate aim was to support the design process. The broader, more theoretical goal of the study is to better illuminate how physical affordances in multifunctional mobile devices may remain or increase in importance in the design and use of such devices. Specifically, we examined how GUI controls, as opposed to hardwired buttons and input mechanisms, facilitate or hinder user understanding of the multifunctional nature of the device. The study was driven by a set of research questions, which were designed to

tap into the major issues discussed in the literature review. Specifically, the questions are informed by Nielsen and Mack (1994) and aim to explore basic usability dimensions related to device learnability and satisfaction:

1.   Is the concept of a multifunctional device positively evaluated by participants?
2.   Would participants actually use such a device?
3.   Are the affordances primarily embedded within the on-screen GUI interface sufficient for suggesting the functionality and utilization of the device?
4.   Do the on-screen affordances facilitate or hinder the use of multimedia functions?
5.   What previous conventions and associations did participants use in integrating the device into their use repertoire?
6.   Does the level of technological sophistication affect the manner in which participants discovered and understood device functionality?

Method

A user study was carried out to explore reactions to an early prototype of an X cell phone capable of media capture, streaming, and transfer. The primary method of delivering the treatment was through one-on-one interview sessions, in which the seven task scenarios described below were carried out.

Participants

There were 23 participants between the ages of 18 and 28 (12 female and 11 male, average age 20.7 yrs). They were recruited through announcements distributed through the student class registration and grade notification accounts. Students saw the announcements when they checked their grades or registered the classes. Subjects were recruited from a pool of 71 individuals, of which 23 made and came to a research appointment. They represented a variety of majors (Engineering, Liberal Arts, Management, Hospitality). One was a graduate student, four were juniors, eight sophomore, and ten seniors. Each student received financial compensation for her/his time. Although not a fully random sample, the respondents reflected the group of young, relatively technically literate individuals, who could become one of the target user bases for this device.

Treatment

To illustrate the functions of this future application in a way that participants could understand and relate to, a Microsoft PowerPoint storyboard scenario was used for the study, which hypothesized the context of use and included specific functional tasks. The scenario was animated and presented using a laptop computer. It contained seven discrete sequential user tasks, which were carried out by the participants using three paper prototypes of device interfaces.[2] It should be noted that to reduce the possibility of "priming" the subject, the scenarios were not device-specific, but mentioned only a generic type of task that

is generally performed with the type of device being investigated here. The scenarios referred to core device capabilities, especially its multifunctional features. They aimed to capture basic user reactions to core features, such as capturing and moving media across platforms and sharing between users. These were considered core and novel capabilities, which involved a new approach to control and interface design.

The prototypes[3] consisted of a basic foam-core block, representing the approximate size and format of the device, onto which interchangeable screens could be placed in response to the participants' operations and menu selections on previous screens (Figure 2). The paper prototypes were based on an X cellular handset, which used touch screen and stylus operation.

Procedure

Participants were asked to perform the tasks by tapping on various screens or buttons. The tasks were not device specific, but activity specific. People were shown life scenarios, such as the need to call two other people simultaneously. For each action they were presented with different versions

Figure 2. Example of paper prototype screens

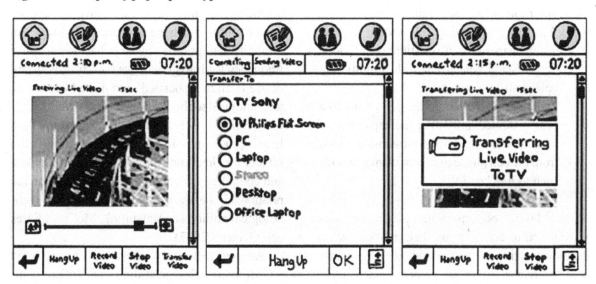

of paper screen interfaces. Paper screen prototypes were specifically chosen so that participants would perceive them as being very early in the development process and would therefore be more willing to provide both positive and negative feedback to influence the development. They were also hand drawn, rather than computer generated, to help reinforce this early stage, developmental feel. If the prototype and screens were perceived as being more finished, participants could feel that their feedback would be less influential in the development of the application and consequently they might be more reluctant to be critical.

The task scenarios specifically probed user values associated with the following functional capabilities, for which physical device affordances were not necessarily obvious:

1. Group voice calling, providing the capability to simultaneously initiate a multi-way voice call between three or more people.
2. Group presence information, providing visual information relating to the availability and geographic location of call recipients prior to the initiation of a voice call.
3. Wirelessly sending and receiving media files concurrent with a group voice call, providing the capability to share still images and short video clips within a multi-way call without interrupting the voice call.
4. Wirelessly streaming live video, providing the capability of real-time (live) video transmission concurrent with a multi-way voice call.
5. Transfer of media from the cell phone to a nearby device, providing the capability of transferring a picture or streaming video from the cell phone to a fixed display device such as a PC or TV.

In addition to performing the tasks, participants were asked to comment on their actions during execution, using a "think-aloud" technique, and also to answer a number of open-ended ques-

tions about their understanding of the tasks, the prototype features, and additional contexts or situations in which these functions might be used. Furthermore, if a screen was not understood, or if one was felt to be missing, the subject was asked to sketch what they felt would work better for them, or to sketch out the missing screen. Participants were also asked to answer a series of open-ended questions related to their subjective evaluation of the group voice calling and multimedia capabilities of the device. The questions were open-ended to reduce bias. In order to determine that the participants were a representative sample, including highly technical and less technical users and both early and later adopters of technology, they completed short questionnaires that assessed their level of technological sophistication and speed of adoption.

The questionnaire for technical ability included two sets of questions. The first set referred to frequency of using personal computers, mobile devices (cell phone, iPod, digital camera, or PDA), and mobile/digital communication services (instant messaging, text messaging, or email). The actual question was: "How often have you used the following devices or services in the last month?" The response alternatives were: more than once a week, once a week, once or twice a month, and never in the last month. The answers allowed quantitative separation into groups on the basis of interest in using common technology. An implicit assumption we make is that increased use frequency leads to greater use facility. Frequency of use is often associated with digital literacy or technological sophistication; the more technologically capable users are also frequent users. Van Braak (2003) has found strong and significant effects for time and frequency of computer use and computer use competence. (For more on the relationship between frequency of use and technological ability, see Ballantine, McCourt Lares and Oyelere, 2007).

Answers to this set of questions were weighted on different scales, according to degree of complexity and relevance of the particular device to this study. Thus, frequency of use of most common devices and services (personal computer, cell phone, or email) was evaluated on a 0-1 scale (0 never, 0.25 once-twice a month, .5 every week, 1 more than once a week), while more sophisticated devices or services (cameras, instant messaging, or text messaging) were rated on a 1-3 scale (0 never, 1 once-twice a month, 2 every week, 3 more than once a week). Use of iPods and PDA was rated on a 0-5 scale (0 never, 3 once-twice a month, 4 every week, 5 more than once a week). This higher rating reflected a lower frequency of occurrence of these devices among participants, though for different reasons. The iPod was relatively new and expensive and not pervasive among the student population at the time of the study, while the PDA was becoming somewhat redundant due to the prevalence of PDA-like functions on other mobile devices, including cell phones.

Frequency of use offers only a very crude measure of technical ability. A second set of questions more directly captured technological ability, adding a qualitative dimension to the quantitative dimension of frequency of use. This second set of questions collected information on participants' proficiency (speed) at adopting and using new devices. The literature on user adoption is extensive, and we do not propose to summarize all current models here, but it is well accepted that users fall into distinct groups of adoption patterns.

Four self-reported items were formulated as affirmative statements with which respondents were asked to agree, disagree, or be neutral. The statements were:

1. I am one of the first people among my friends or relatives to buy a new electronic device.
2. I like to customize the settings on my computer or electronic devices.
3. I rarely use the more advanced features on my cell phone, computer or electronic devices.
4. I learn faster than most of my peers how to use a new technological gadget.

The three answer categories were weighted using a score of 5 for agreement, 0 for neutral, and -5 (negative value) for disagreement.

A final index was constructed through item summation. *M*=20.9, *SD*=11.5, *Median*=22, *Range*= 1 to 42. To facilitate further analysis, the sample was split at the median. Two groups, almost equal in size, were created, labeled as "high" (*N*=11) and "low" (*N*=12) technological sophistication respondents.

ANALYSIS

Video recordings of the "think-aloud" narrative and responses to open-ended questions were transcribed from all subject sessions. Transcriptions for each subject were unitized into discrete items, one for each task (T), question (Q), or subject. EZAnswer, a qualitative analysis software developed by CDC (Centers for Disease Control) was used for managing and analyzing the data.

In addition to unitizing the data by task, question and subject a series of codes was generated for identifying discrete user reactions to the tasks and questions raised by the scenarios. The coding procedure relied on the Applied Hermeneutic Methodology (AHM) developed by Ross and Wallace, which was successfully applied in analysis of large-scale qualitative databases (Wallace, Ross, & Davies, 2003). The procedure emphasizes, in line with the hermeneutic tradition of qualitative analysis, the fact that inter-rater reliability is a social consensus process. AHM requires that the coding taxonomy be agreed upon by all the coders. When a suitable taxonomy has been agreed upon, a process of reading based on the 'hermeneutic circle' is formalized. The text is broken down into units and re-read a number of times by all the coders who then decide on the proper unitization and coding as a group. When a consensus has emerged and a common interpretation of the codes is achieved, the coding can continue either in a group setting or individually.

Eighty-four unique codes were generated through this iterative procedure. They reflected the most important reactions to the prototype identified in the data. Codes included generic and specific issues. They focused on actual use, immediate or intended, and on positive and negative reactions. For example, when evaluating user actions and reactions during the task that involved transferring media content to a nearby display, we used 4 codes, 2 for evaluation: "Respondent positively (negatively) evaluates content transfer to nearby device" and 2 for intention to use: "Respondent would (not) transfer content to nearby device."

We applied the codes to a text unit, which comprised the outcomes of specific tasks performed and the answers to questions asked during each users' interaction with the prototype. To each segment we allocated one or more codes by a panel of three coders, based on consensus voting and iterative reading of the session transcriptions as described above.

The present paper analyzes only the reactions to the tasks and the answers to the questions that directly referred to the multimedia/multipurpose capabilities of the device. These were as follows:

T3: Subject sends live video from a remote location.
Q6: Would the subject point camera at self or at surroundings while capturing video at a remote location?
T4: Subject transfers live video received on a cell phone to a nearby TV.
Q9: Would subject transfer media content to a nearby device if they had a device like the one evaluated?
T5: Subject sends photo to a group concurrent with a voice call.
Q13: What current multimedia sharing practices are used by the subject?
T6: Subject sends photo to a group listed in address book.

T7: Subject transfers received photo to a PC followed by a call back.
Q19: Is the device a good idea?
Q20: What do you think about the device?
Q21: What did you like about it?
Q22: What did you dislike about it?
Q23: What did you find most difficult?
Q24: What would you change in the prototype?
Q25: Open ended comments.

Lists of code frequencies for each question/group of questions or tasks were generated during the analysis and each list presented in descending order the number of participants that offered answers or comments that fit a specific code. The position of each code (issue) on the list was assumed to indicate its relevance and importance.

For example, Table 1 presents one of the "importance" lists generated during the analysis. It lists the codes applied while analyzing the question: "What did you like about the device?" (Q21). The figure indicates that the preference for the "all in one" device is the most important issue found in the answers to this question, with 43% of participants indicating this preference, followed by preference for media sharing, media transfer, etc. Each task was similarly analyzed and relevant code rankings were generated.

RESULTS

Subjective Evaluation of the Multifunctional Approach

The first research question was aimed at determining whether the idea of a multifunctional device was evaluated positively by participants. The results indicate that the theoretical concept of a multifunctional device capable of media capture, storage, transfer, and voice calling was generally well received by participants. Analysis of all the answers to the questions and of the comments made while the tasks were performed indicates

Table 1. Example of "Importance" list: "What did you like about the device?"

Code	Issue	Number of respondents who mention the issue at least once	Percent of respondents who mention issue at least once*
WFMULAY	Likes all in one	10	43.48%
WFMSAY	Likes media sharing	6	26.09%
WFMXAY	Likes media transfer	4	17.39%
MQUALAY	Video quality concerns	2	8.70%
PHYBOAY	Keyboard/hard-button related	2	8.70%
ICONHAY	Understands icon meaning	2	8.70%
PHYHAAY	Physical deployment related	1	4.35%
CORDSAY	Expects cords and cables	1	4.35%
SIMPLAN	Screens, menus sufficient	1	4.35%
Note=* percentages within each category; figures in parentheses represent absolute values.			

that 70% of participants positively evaluated the ability to exchange media content during an ongoing voice call (7th most salient issue)[4]. In addition, when asked to make open-ended comments about the device (Q19, Q20, and Q25), 66% positively appreciated its ability to share content during an ongoing voice call. Furthermore, 96% of the respondents positively evaluated the idea of transferring media content to a PC (2nd most salient issue).

Expected Usage of the Multifunctional Device

The data indicated that participants would use such a multifunctional device, but that they would not use its entire array of functionality equally. Detailed analysis indicated that participants were much more likely to transfer a media file to a nearby device than to stream media in real-time. While all participants would use the file transfer feature, only 21% would send live video to other people or stream media. These reactions indicate that, although multimedia capabilities seemed like a good idea to the participants, these functions might not be immediately utilized when participants interacted with a cell phone. Moreover,

from some of the verbal comments, it appeared that the participants' inclination toward the most commonly used multimedia feature, i.e., media transfer to a nearby display device, was driven, at least in part, by obstacles which existed in transferring camera-phone media content to other devices. Many of the participants repeatedly asked, before performing the transfer procedure, if the transfer was to be done using wired connections. When told that this was not necessary, participants generally expressed relief and amazement that the procedure could be so easy to perform, implying that the current procedures were not as facile as those presented in the study scenario.

Intuitiveness of GUI-Based Affordances

The GUI-based affordances related to image capture and content transfer/sharing that were common at the time of this study (and were integrated into the experimental prototype) were not adequate to suggest the multifunctional possibilities (voice + media sharing/transfer) of the device. Although the GUI symbols used to indicate the various multimedia features of the device were quite conventional and straightforward—mostly icons and metaphors borrowed from desktop and web interfaces (e.g., camera, address book, handset, home, etc.)—40% of participants affirmed that they did not understand their meaning in the current context (while 40% indicated that they recognized the icons and 20% did not express an opinion). Thus they were not easy to learn and the interface partially failed the test of "learnability" (Nielsen and Mack, 1994).

This finding becomes more understandable in light of the fact that many participants expected to control the multimedia features of the device through hardwired buttons and controls instead of by selecting icons on a touch screen. Sixty-five percent of the participants raised various issues about the physical deployment of the device, especially its lack of clear function affordances. Their specific verbal comments frequently indicated that they expected a button for taking a picture or video, just like still image and video cameras. Other comments suggested a preference for a clam-shell design, which would permit a larger usable area to accommodate both a larger LCD screen and a greater number of hardwired controls. Suggestions also included having hardwired buttons not only for video recording and image capture, but also for sending video, and using a physical cradle for synching/sending media with/to PCs or TVs.

An interesting comment, which indicated that current use patterns might impede the full and advantageous adoption of device multifunctionality, was that cell phones are often used in multitasking situations (e.g., shopping, walking, doing house chores, etc.). In such contexts, embedded multimedia or other controls in a touch screen display, which require the user to look at the screen, would make utilization of those functions very difficult. For many respondents, a multimedia, multifunctional device made sense only insofar as it allowed hardwired controls that facilitated multitasking. They indicated a preference for controls that can be found and operated by touch, using one hand, and without having to look at the screen.

These comments are best summarized by what is probably the most important finding of our study. Sixty-one percent of participants indicated that they wanted a hardwired button for controlling the multimedia capabilities of the device. Moreover, when specifically asked to indicate which feature they disliked the most, 52% indicated physical deployment issues, such as those related to buttons/control layout, number of screens or menus that users had to go through to accomplish a specific task, the size of the prototype, and the lack of hardwired buttons.

Intuitiveness of Multifunctional Affordances

As mentioned in the literature review, devices with too few affordances are insufficiently informative, while devices with too many are confusing. Therefore, it was important in this study to determine whether the multifunctional affordances led to confusion when it came to conceptualizing and using the multimedia functions. Or, using a term consecrated in usability research (Nielsen & Mack, 1994), the study aimed to find out if the multifunctional affordances are "learnable."

Streaming live video wirelessly from a remote location was one of the central and most novel purposes of the device. Although this functionality was explicitly described to the participants, and the "live" nature of the sharing was also indicated in the on-screen menus, when asked to send live video, a majority of participants (87%) behaved as if they were sending a pre-recorded file (i.e., recording a video clip, then saving it, and finally sending it as a digital file). The fact that the task required a live feed, or that the GUI included functions that allowed live broadcasting, did not seem to be sufficient to define the live streaming capability. Not surprisingly, while carrying out the task a third of the respondents indicated that they were confused by the GUI.

The failure of many users to engage this feature clearly highlighted how a function can be ignored or misunderstood if lacking an obvious physical affordance. Many respondents indicated during debriefing that what was missing was a sense of "YOU ARE BROADCASTING LIVE NOW." To make this functionality apparent, they asked for a prominent and clearly labeled "GO LIVE" on-screen button or, better still, to include an actual physical button for triggering the video capture and/or video transmission, just like the ones on video-cameras. The fact that the entire operation took place in a desktop-like GUI environment might have interfered with the mental model of live broadcasting. Users rarely capture streaming video on a computer, generally preferring to download pre-recorded video clips, and the "download first" stereotype appears to have been transferred to this mobile device.

Embedded Conventions and Mental Models

The data also provided some insights into how previous conventions and mental models were used by participants for integrating the multifunctional device into their repertoire. Throughout the study, participants seemed to rely heavily on their previous mobile experience for making sense of the new device. Many indicated that they approached the device as a cell phone, the multimedia functions being seen as optional/added services. They understood the functionality of the device in terms of the cell phone sets they already possessed. Some of the participants even used their personal cell phones to show the interviewer what the prototype functionalities should look like. In addition, when confronted by the new capabilities, the participants often referenced other mental models: "this is like a camera-phone, right?" or "like an instant messaging program," or "like a palm-top." This strongly indicates that participants called upon, or referred to, multiple pre-existing mental models when trying to understand and integrate the device functionality into their use-repertoire.

Effect of Technological Sophistication

Data was collected to assess the level of technological sophistication of the study participants so as to determine whether it had any impact on the manner in which they discovered and understood the device functionality and took advantage of its advanced utilities. Utilization and understanding differences arising from level of technological sophistication were assessed by cross-tabulations. We used the dichotomized level of technological sophistication (low vs. high) as the independent variable, and presence/absence of mentions of key study issues by each subject (coded according to the system described in the data analysis section) as the dependent variables.

This analysis produced two relevant findings. Highlighting the importance of physical affordances in the context of multipurpose devices, more technologically aware respondents indicated in higher proportion than those with a lower level of technological awareness that they would prefer a physical button over an on-screen menu item for accessing certain device functions. More specifically (Table 2), one third (4) of the highly sophisticated, versus none of the low sophisticated, respondents expressed a preference for physical buttons and controls (*Chi-squared*=5.3, *p*<.05). In addition, 90% (10) of the highly sophisticated, versus half (6) of the low sophistication respondents, indicated that they would prefer to have seen a clearly labeled "media" place or icon incorporated in the controls (*Chi-squared*=4.5, *p*<.05).

These findings suggest that physical affordances and clear (on-screen or otherwise) indicators of multimedia device functionality become more, not less relevant, to users who are more technologically sophisticated. One possible explanation for this is that this category of users has a greater breadth of experience with a wide range of media and communications devices. Consequently, they may have developed, and called upon, a wider range of mental models or func-

Table 2. Affordance related issues: Overall and specific (by technological sophistication) analysis.

User Issue	Low sophistication N=12	High sophistication N=11	All subjects N=23
Respondents would prefer a physical vs. a screen control	0%* (0)	36%* (4)	14% (4)
Respondents would prefer a "Media Home" button or place among the device controls	50%* (6)	91%* (10)	70% (16)
Concurrent media sharing – positive evaluation			70% (16)
Media transfer – positive evaluation			96% (22)
Icons not fully understood			40% (9)
Usability issues; physical deployment			65% (15)
Respondents want hardwired controls			61% (14)
Note=* percentages within each category; figures in parentheses represent absolute values.			

tional expectations, which, rather than being advantageous, could lead to greater confusion when learning a new device, particularly one that does not fit into one of these pre-existing mental models or device categories.

DISCUSSION

HCI researchers are increasingly in need of methodologies that can assist them in creating mobile devices that take into consideration the limitations of human cognition and the context of use. As Raskin (2000) holds, to truly be "responsive to human needs and considerate of human frailties" (p.7) design must conform to the natural limitations of human physical and cognitive abilities. The role human-computer interaction design plays in the design of complex multifunctional media devices will depend heavily on our understanding of the challenges associated with human-centered design and the tangible issues surrounding physical affordances (Faiola, 2006). Consequently, affordances and direct device controls are essential for fitting new capabilities and experiences into users' old mental maps.

Findings from this study indicate that users who theoretically embrace a device's multifunctionality are prevented from actually using its

capabilities by an apparent cognitive constraint. Participants in the study preferred the multifunctional device, but could not always take advantage of its full range of features. Moreover, participants seemed to fall back on known metaphors especially related to file sharing, saving, and sending and mental models, sometimes from a desktop computing environment, which did not serve their purposes well in the context of the study. A recurring theme throughout the study was the need for more physical controls and hardwired buttons. This response emerged not only because the users had formed particular mental models and associated behaviors from their past use of other devices. They also indicated their positive preference for immediate access and use of the controls within a multi-tasking environment, i.e., one-handed operation with little or no visual demand. Ultimately, the participants clearly preferred a device that had more direct entry points to some basic functionality, which would make it easier to handle.

In part, our findings are validated by the development of devices that use controls, which involve physical gestures that mimic or rely on hardwired controls. In this respect, devices such as the Apple iPhone do not contradict, but rather validate our conclusions. The iPhone is considered by many HCI researchers a "convergent device," because its multi-touch on-screen interaction style converges with affordances associated with physical controls (Anderson, 2008). The multi-touch screen, and the gestures that are associated with it, require users to physically interact with the device by pressing buttons on the display. These actions build on and involve gesture repertoires and conventions derived from physical affordances.

It is also important to note that while the iPhone has eliminated physical buttons (except for the one primary and second volume control), it has limited usage. The lack of a physical keyboard is a clear disadvantage when it comes to business applications. The Blackberry mobile device, with its keyboard and plethora of physical buttons, is preferred for sending and receiving email (Visiongain, 2009). Moreover, the iPhone has limited multimedia capabilities. At least presently, it does not support live video streaming. Hence, based on these existing limitations to current mobile devices, our findings need to be qualified. That is to say, when designing a new generation of multifunctional devices it is imperative to utilize gestures and tools that leverage to our embodied experience and that employ real or simulated physical affordances. That is, the specific choice for physical or onscreen controls should be decided based upon the nature of the task.

CONCLUSION, LIMITATIONS, AND FUTURE WORK

In summary, our findings suggest that the increasing portability of devices with greater computing power and multifunctionality will force interaction designers to rethink the current emphasis on abstract, screen based controls, derived from desktop computing. Although on-screen controls are necessary, due to the increasing capabilities and shrinking physical size of mobile devices, they can become cumbersome and obscure in a media rich device. Natural interfaces, which take advantage of existing behaviors, physical capabilities, and

human-centered abilities, are more appropriate in this context. At the same time, on-screen controls should emphasize simplicity and should utilize intuitive icons.

These recommendations need, however, to be considered in the context of the limitations implicit in design of our study. Our study reports 6 years old data. A new generation of devices that use only on screen controls, especially the iPhone, seem to have put to rest the issue of physical affordances. Yet, as it is immediately obvious for any iPhone or iPad user, these devices do not eliminate physical affordances. They in fact turn the entire touch screen into a powerful touch, sweep, or pinch interface that acts like a physical affordance. The iPhone repertoire of gestures and icons associated with multimedia (e.g., camera controls) are far closer associated with physical devices than with computer screen metaphors, as it was the case for Windows Mobile devices of the pre-Windows Mobile 7 generation. Moreover, the staying power of physical affordances is suggested by the resilience of Blackberry and of a variety of Android phones that use various physical controls, including physical keyboards. In fact, the fastest growing smartphone brands are those that utilize multiple physical affordances, such as the Motorola Droid (Wireless and Mobile News, 2010). Our findings and interpretations are thus still relevant and can add a possible explanation for these developments.

Another potential limitation of our study is the small sample. In size and range of user experiences it offers a rather small segment of possible reactions and experiences. The size of the sample was dictated, however, by the in-depth research protocol, which involved about 10 hours of data collection and processing per individual. Paper prototype cognitive walk-through activities are powerful heuristic tools, yet they have inherent limitations, notably their inability to correctly represent system status and action feedback. These limits might have triggered exaggerated negative reactions to on screen features that could not be fully represented in a paper prototype and artificially stimulated our respondents' declared need for physical affordances. However, the findings are too consistent and emerge in even straightforward use situations to completely warrant this alternative explanation. Another possible limitation of our study is the fact that the sample did not cover a very wide range of technological abilities. The finer differences we have proposed for groups of respondents that vary in terms of technological ability need to be interpreted as orientative indicators. The limited nature of this data also prevented us from deeper analyses in this direction. This and other issues need to be more and better tested on actual physical devices in a study that utilizes a larger and more diverse sample.

In the future, designers of multi-functional devices should investigate easy-to-use physical affordances, while optimizing their size, location, and interrelationship with groups of functions. Another key issue to be explored is users' ability to configure new or revised mental models that allow them to switch between various purposes and functions of multiple physical affordances. We recommend that a strong balance be struck between intelligent interfaces and hardwired controls to assist users in the adoption of multifunctional devices.

As Dourish (2001), Norman (2002), and other (Suchman, Randall, & Blomberg, 2002) hold, people approach objects by building mental maps of their functions, starting with their physical and outward appearance. Hence, next generation designers of mobile multifunctional media-rich devices should be cognizant of the emerging findings related to physical affordances in HCI and human behavioral studies. In this way, the complex issues surrounding content creation and delivery, context of use, accessibility of information, and universal access can be adequately addressed in the design and implementation of future mobile devices.

REFERENCES

Abowd, G. D., & Mynatt, E. D. (2000). Charting past, present, and future research in ubiquitous computing. *ACM Transactions on Computer-Human Interaction*, 7(1), 29–58. doi:10.1145/344949.344988

Agre, P. E. (2001). Changing places: Contexts of awareness in computing. *Human-Computer Interaction*, 16(2-4), 177–192. doi:10.1207/S15327051HCI16234_04

Anderson, G. (2008). Let's get physical. *Interaction*, 15(5), 68–72. doi:10.1145/1390085.1390101

Ballantine, J. A., Larres, P. M., & Oyelere, P. (2007). Computer usage and the validity of self-assessed computer competence among first-year business students. *Computers & Education*, 49(4), 976–990. doi:10.1016/j.compedu.2005.12.001

Brereton, M., & McGarry, B. (2000). An observational study of how objects support engineering design thinking and communication: implications for the design of tangible media. In *Proceedings of the ACM Computer Human Interaction Conference*, The Hague, The Netherlands (pp. 217-224).

Dourish, P. (2001). *Where the action is: The foundations of embodied interaction*. Cambridge, MA: MIT Press.

Faiola, A. (2006). Designing humane technologies: A potential framework for human-computer interaction design. *The International Journal of the Humanities*, 2(3), 1877–1886.

Faiola, A., & MacDorman, K. (2008). Exploring the influence of web designer cognitive style on information design: A cross-cultural comparison of a holistic and analytical perspective. *Information Communication and Society*, 11(3), 348–374. doi:10.1080/13691180802025418

Faiola, A., & Matei, S. (2010). Enhancing human–computer interaction design education: teaching affordance design for emerging mobile devices. *International Journal of Technology and Design Education*, *20*, 239–254. doi:10.1007/s10798-008-9082-4

Forlizzi, J. (2008, April 24). The Product Ecology: Understanding Social Product Use and Supporting Design Culture. *International Journal of Design*, *2*(1). Retrieved from http://www.ijdesign.org/ojs/index.php/IJDesign/article/view/220/143.

Gibson, J. J. (1977). *The theory of affordances*. Hillsdale, NJ: Erlbaum Associates.

Goldstein, M., Nyberg, M., & Anneroth, M. (2003). Providing proper affordances when transferring source metaphors from information appliances to a 3G mobile multipurpose handset. *Personal and Ubiquitous Computing*, *7*(6), 372–380. doi:10.1007/s00779-003-0252-9

Hartson, H. R. (2003). Cognitive, physical, sensory, and functional affordances in interaction design. *Behaviour & Information Technology*, *22*(5), 315–338. doi:10.1080/01449290310001592587

Heo, J., Ham, D., Park, S., Song, C., & Yoon, W. C. (2009). A framework for evaluating the usability of mobile phones based on multi-level, hierarchical model of usability factors. *Interacting with Computers*, *21*(4), 263–275. doi:10.1016/j.intcom.2009.05.006

Jin, B. S., & Ji, Y. G. (2010). Usability risk level evaluation for physical user interface of mobile phone. *Computers in Industry*, *61*(4), 350–363. doi:10.1016/j.compind.2009.12.006

Kangas, E., & Kinnunen, T. (2005). Applying User-Centered Design to Mobile Application Development. *Communications of the ACM*, *48*(7), 55–59. doi:10.1145/1070838.1070866

Klopfer, E., Squire, K., & Jenkins, H. (2004). Environmental detectives: PDAs as a window into a virtual world. In Kerres, M., Kalz, M., Stratmann, J., & De Witt, C. (Eds.), *Didactik der notebook-universitat*. Munster, Germany: Waxmann Verlag.

Marti, S., & Schmandt, C. (2005). Physical embodiments for mobile communication agents. In *Proceedings of the 18th annual ACM symposium on User interface software and technology*, Seattle, WA (pp. 231-240). New York: ACM.

Monk, A., Fellas, E., & Ley, E. (2004). Hearing only one side of normal and mobile phone conversations. *Behaviour & Information Technology*, *23*(5), 301–305. doi:10.1080/0144929041000171 2744

Nielsen, J., & Mack, R. L. (1994). *Usability Inspection Methods* (1st ed.). New York: John Wiley & Sons, Inc.

Norman, D. (1998). *The invisible computer: why good products can fail, the personal computer is so complex, and information appliances are the solution*. Cambridge, MA: MIT Press.

Norman, D. A. (2002). *The design of everyday things*. New York: Basic Books.

Oulasvirta, A., & Saariluoma, P. (2004). Long-term memory and interrupting messages in human-computer interaction. *Behaviour & Information Technology*, *23*(1), 53–64. doi:10.1080/0144929 0310001644859

Raskin, J. (2000). *The humane interface: New directions for designing interactive systems*. Reading, MA: Addison-Wesley.

Snyder, C. (2003). *Paper prototyping*. San Francisco: Elsevier Science.

Suchaman, L., Randall, T., & Blomberg, J. (2002). Working artefacts: Ethnomethods of the prototype. *The British Journal of Sociology*, *53*(2), 163–179. doi:10.1080/00071310220133287 van Braak, J. P. (2004). Domains and determinants of university students' self-perceived computer competence. *Computers & Education*, *43*(3), 299–312. doi:10.1016/j.compedu.2003.09.006

Visiongain. (2009). *Mobile email 2009: Challenging Blackberry and succeeding in the consumer market*. Retrieved December 10, 2009, from http://www.visiongain.com/Report.aspx?rid=377

Vivrou, M., & Kabbasi, K. (2002). Reasoning about users' actions in a graphical user interface. *Human-Computer Interaction*, *17*(4), 369–398. doi:10.1207/S15327051HCI1704_2

Wallace, B., Ross, A. J., & Davies, J. B. (2003). Applied hermeneutics and qualitative safety data: The CIRAS project. *Human Relations*, *56*(5), 587–607. doi:10.1177/0018726703056005004

Weiser, M. (1993). Some Computer Science Issues in Ubiquitous Computing. *Communications of the ACM*, *36*(7), 75–84. doi:10.1145/159544.159617

Wheatley, D. (2007). User-Centered Design and Evaluation of a Concurrent Voice Communication and Media Sharing Application. In *Human-Computer Interaction. HCI Intelligent Multimodal Interaction Environments* (LNCS 4552, pp. 990-999). Berlin: Springer.

Wireless and Mobile News. (n.d.). *Android Top Users Fastest Growing Smartphone Makers, IPhone Declines, Says iSuppli*. Retrieved October 21, 2010, from http://www.wirelessandmobile-news.com/2010/10/android-top-users-fastest-growing-smartphone-makers-iphone-declines-says-isuppli.html

ENDNOTES

[1] X stands for a cell phone brand, which was anonymized for the review process.

[2] Though the examination of device affordances was not one of the primary research motivations for this study, the acquired data suggested an inquiry into these issues. By utilizing basic paper prototyping rather than a tangible physical handset platform, participants were nonetheless able to conceptualize and predict some of the physical usage issues which would arise with a real physical device. At the same time, we agree that not all physical usage issues may be replicated in a paper prototype. Hence, we submit that it has sufficient physicality to test the applicability of affordances, more so than doing so virtually would offer.

[3] Paper prototyping has long since proved its worth in both design and research applications. Prototypes have four distinct dimensions: breadth, depth, interaction, and look (Snyder, 2003). Breadth is the span of functionality presented, depth is the degree to which the real world is replicated in detail, interaction is the faithfulness with which interactivity is replicated, and look is the appearance. Of these, we agree with Snyder that breadth is the least important. We also agree that paper prototypes offer the best balance of depth and look when simply constructed and versatile. In addition, paper prototypes can reproduce affordances with reasonable fidelity, a key benefit for the present study.

[4] Salience in this context signifies the relative ranking of the code among all the 84 codes generated through the analysis. Specifically, the code was the 7th most mentioned issue among the questions studied.

This work was previously published in International Journal of Information Technology and Web Engineering, Volume 5, Issue 4, edited by Ghazi I. Alkhatib, pp. 40-57, copyright 2010 by IGI Publishing (an imprint of IGI Global).

Compilation of References

3GPP. (2009a). *TS 23.198-10 v9.0.0: Open service access (OSA)- Connectivity manager service capability feature* (SCF), (release 9).

3GPP. (2009b). *TS 24.229 v9.2.0: IP multimedia call control protocol based on session initiation protocol (SIP) and Session description protocol* (SDP), (Release 9).

3GPP. (2009c). *TS 29.198-4-3 v9.0.0: Open service access (OSA)- Multiparty call control, subpaer 3, service capability feature* (SCF), (Release 9).

3GPP. (2009d). *TS 29.198-12 v9.0.0: Open service access (OSA)- Charging service capability feature* (SCF), (Release 9).

3GPP. (2009e). *TS 29.198-13 v9.0.0: Open service access (OSA)- Policy management service capability feature* (SCF), (Release 9).

3GPP. (2009f). *TS 29.199-6 v8.1.0: Open service access (OSA)- Parlay X Web Services, part 6: Payment*, (Release 9).

3GPP. (2009g). *TS 29.199-17 v9.0.0: Open service access (OSA)- Parlay X Web services, part 17: Application-driven quality of service* (QoS), (Release 9).

3GPP. (2011a). *TS 23.203 v11.2.0: Policy and charging control architecture*, (Release 9).

3GPP. (2011b). *TS 29.214 v11.1.0: Policy and charging control over Rx reference point*, (Release 9).

Abowd, G. D., Atkeson, C. G., Hong, J., Long, S., Kooper, R., & Pinkerton, M. (1997). Cyberguide: A mobile context-aware tour guide. *ACM Wireless Networks*, *3*(5), 421–433. doi:10.1023/A:1019194325861

Abowd, G. D., & Mynatt, E. D. (2000). Charting past, present, and future research in ubiquitous computing. *ACM Transactions on Computer-Human Interaction*, *7*(1), 29–58. doi:10.1145/344949.344988

Ackerman, M., & Carft, R. (2002). Telemedicine technology. *Telemedicine Journal and e-Health*, *8*(1). doi:10.1089/15305620252933419

ActiveCollab. (2011). *Project collaboration software*. Retrieved September 8, 2011, from http://www.active-collab.com

Adda, M., Valtchev, P., Missaoui, R., & Djeraba, C. (2007). Toward Recommendation Based on Ontology-Powered Web-Usage Mining. *IEEE Internet Computing Journal*, 45-52.

Agrawal, R., & Srikant, R. (1994). Fast algorithms for mining association rules. *VLDB '94, Proceedings of 20th International Conference on Very Large Data Bases* (pp. 487-499). Santiago de Chile, Chile: Morgan Kaufmann.

Agrawal, R., & Srikant, R. (1995). Mining Sequential Patterns. *Eleventh International Conference on Data Engineering, IEEE Computer Society* (pp. 3-14). Taipei, Taiwan: IEEE.

Agre, P. E. (2001). Changing places: Contexts of awareness in computing. *Human-Computer Interaction*, *16*(2-4), 177–192. doi:10.1207/S15327051HCI16234_04

Akhatar, H. (2009). *System and method for providing quality of service enablers for third party applications*. (Patent application number: 20090154397). Retrieved from http://www.faqs.org/patents/app/20090154397

Akilandeswari, J., & Gopalan, N. P. (2008). An Architectural Framework of a Crawler for Locating Deep Web Repositories Using Learning Multi-agent Systems. In *Proceedings of the 2008 Third International Conference on Internet and Web Applications and Services* (pp.558-562).

Alfresco Software. (2011). *Alfresco*. Retrieved September 8, 2011, from http://www.alfresco.com

Ali, A. A., Hamidah, I., & Nur Izura, U. (2009). Improved integrity constraints checking in distributed database by exploiting local checking. *Journal of Computer Science and Technology*, 24(4), 665–674.. doi:10.1007/s11390-009-9261-0

Alkahtani, A. M. S., Woodward, M. E., & Al-Begain, K. (2006). Prioritised best effort routing with four qualities of service metrics applying the concept of the analytic hierarchy process. *Computers & Operations Research*, 33, 559–580. doi:10.1016/j.cor.2004.07.008

Allen, D. (2003). *Getting things done: The art of stress-free productivity*. Harlow, UK: Penguin.

Alsuamit, A., & Habib, S. (2009). Usability Requirements for COTS Based Systems. In *the Proceedings of the 11th International Conference on Information Integration and Web-Based Application and Services (iiWAS)*, Kuala Lumpur, Malaysia (pp. 393-397).

Alsumait, A. (2004). *User Interface Requirements Engineering: A Scenario-Based Framework*. Unpublished doctoral dissertation, Concordia University, Montreal, Canada.

Alves, C., & Finkelstein, A. (2002). Challenges in COTS decision-making: a goal-driven requirements engineering perspective. In *the Proceedings of the 14th international conference on Software engineering and knowledge engineering*, Ischia, Italy (pp. 789-794).

Alves, C., Franch, X., Carvallo, J., & Finkelstein, A. (2005). Using Goals and Quality Models to Support the Matching Analysis during COTS Selection. In *the Proceedings of International Conference on Composition-Based Software Systems*, Bilbao, Spain (pp. 146-156).

Amitay, E. Har'E, N., Sivan, R., & Soffer, A. (2004). Web-a-Where: Geotagging Web content. In *Proceedings of SIGIR-04, the 27th Conference on Research and development in information retrieval* (pp. 273-280).

Anand, P., Herrington, A., & Agostinho, S. (2008). Constructivist-based learning using location-aware mobile technology: an exploratory study. In *Proceedings of World Conference on Educational Multimedia, Hypermedia and Telecommunications 2008* (pp. 2312-2316). Chesapeake, VA: AACE.

Anderson, R., Manifavas, C., & Sutherland, C. (1996). Netcard - A practical electronic cash system. In M. Lomas (Ed.), *Proceedings of 1996 International Workshop on Security Protocols, Berlin, Germany, Lecture Notes in Computer Science, 1189*, (pp. 49-57). Berlin, Germany: Springer Verlag.

Anderson, G. (2008). Let's get physical. *Interaction*, 15(5), 68–72. doi:10.1145/1390085.1390101

Andrade, L., & Silva, M. (2006) *Relvance Ranking for Geographic IR*. Paper presented at the Workshop on Geographical Information Retrieval (SIGIR'06).

ANSI/NISO Z39. 50. (2003). *Information Retrieval: Application Service Definition and Protocol Specification*. Retrieved from http://www.niso.org/standards/standard_detail.cfm?std_id=465.

Ardissono, L., & Bosio, G. (2011). Context-dependent awareness support in open collaboration environments. *User Modeling and User-Adapted Interaction - The Journal of Personalization Research*, in press.

Ardissono, L., Bosio, G., & Segnan, M. (2011). An activity awareness visualization approach supporting context resumption in collaboration environments. In A. Paramythis, L. Lau, S. Demetriadis, M. Tzagarakis, & S. Kleanthous (Eds.), *International Workshop on Adaptive Support Team Collaboration 2011* (pp. 15-25). Aachen, Germany: CEUR.

Ardissono, L., Bosio, G., Goy, A., & Petrone, G. (2009a). Context-aware notification management in an integrated collaborative environment. In A. Dattolo, C. Tasso, R. Farzan, S. Kleanthous, D. Bueno Vallejo, & J. Vassileva (Eds.), *International Workshop on Adaptation and Personalization for Web 2.0* (pp. 21–30). Aachen, Germany: CEUR.

Ardissono, L., Bosio, G., Goy, A., Petrone, G., & Segnan, M. (2009b). Managing context-dependent workspace awareness in an e-collaboration environment. In S. Yamada & T. Murata (Eds.), *International Workshop on Intelligent Web Interaction* (pp. 42–45). New York, NY: IEEE Press.

Ardissono, L., Goy, A., Petrone, G., & Segnan, M. (2009c). SynCFr: Synchronization collaboration framework. In M. Perry, H. Sasaki, M. Ehmann, G. Otiz Bellot, & O. Dini (Eds.), *5th Conference on Internet and Web Applications and Services* (pp. 18–23). New York, NY: IEEE Press.

Ardissono, L., Bosio, G., Goy, A., Petrone, G., Segnan, M., & Torretta, F. (2010). Collaborative service clouds. *International Journal of Information Technology and Web Engineering*, *5*(4), 23–39. doi:10.4018/jitwe.2010100102

Arlotta, L., Crescenzi, V., Mecca, G., et al. (2003). Automatic annotation of data extracted from large Web sites. In *Proceedings of the 6th International Workshop on Web and Databases*, San Diego, CA (pp. 7-12).

Armstrong, M. P., & Bennett, D. A. (2005). A manifesto on mobile computing in geographic education. *The Professional Geographer*, *57*(4), 506–515. doi:10.1111/j.1467-9272.2005.00495.x

Aschoff, F. R., Schmalhofer, F., & Elst, L. (2004). Knowledge mediation: A procedure for the cooperative construction of domain ontologies. *Proceedings of the ECAI-2004 Workshop on Agent-Mediated Knowledge Management* (AMKM-2004), (pp. 29-38).

Baba, N. (2007). *Ripple Effect in Web Applications*. Unpublished master's thesis, Lebanese American University, Beirut, Lebanon.

Bajwa, I. S. (2010). Virtual telemedicine using natural language processing. *International Journal of Information Technology and Web Engineering*, *5*(1), 43–55. doi:10.4018/jitwe.2010010103

Baker, D., Bridges, D., Hunter, R., Johnson, G., Krupa, J., Murphy, J., & Sorenson, K. (2001). Guidebook to Decision-Making Methods. *Department of Energy*, 1-40.

Baker, J., Cunei, A., Kalibera, T., Pizlo, F., & Vitek, J. (2009). Accurate garbage collection in uncooperative environments revisited. *Concurrency and Computation*, *21*(12), 1572–1606. doi:10.1002/cpe.1391

Baker, S. (2010). Web services and CORBA. [Berlin, Germany: Springer.]. *Lecture Notes in Computer Science*, *2519*, 618–632. doi:10.1007/3-540-36124-3_42

Balbas, J., Rommer, S., & Stenfelt, J. (2009). Policy and charging control in the evolved packet system. *IEEE Communications Magazine*, *47*(2), 68–74. doi:10.1109/MCOM.2009.4785382

Ballantine, J. A., Larres, P. M., & Oyelere, P. (2007). Computer usage and the validity of self-assessed computer competence among first-year business students. *Computers & Education*, *49*(4), 976–990. doi:10.1016/j.compedu.2005.12.001

Bandyopadhyay, S., Maulik, U., Holder, L.-B., & Cook, D.-J. (2005). Advanced Methods for Knowledge Discovery from Complex Data. 95-121. Springer Berlin Heidelberg.

Barabási, A. L., & Albert, R. (1999). Emergence of Scaling in Random Networks. *Science, 286*, 509. doi:10.1126/science.286.5439.509

Barai, M., & Caselli, V. (2008). *Service oriented architecture with Java*. Packt Publishing Ltd.

Barbosa, J., Hahn, R., Rabello, S., & Barbosa, D. (2008). Local: a model geared towards ubiquitous learning. In *Proceedings of the 39th SIGCSE technical symposium on Computer science education* (pp. 432-436).

Beck, K. (2000). *Extreme programming explained*. Reading, MA: Addison-Wesley.

Bell, E. (2002). *Fundamentals of Web applications using. Net*. Upper Saddle River, NJ: Prentice Hall.

Berner, E. S. (2007). *Clinical decision support systems*. New York, NY: Springer. doi:10.1007/978-0-387-38319-4

Bertenyi, B. (2011). *Key drivers for LTE success: Services evolution*. 3GPP Seminar, Moscow.

Bertoa, M., Troya, J., & Vallecillo, A. (2006). Measuring the usability of software components. *Journal of Systems and Software*, *79*(3), 427–439. doi:10.1016/j.jss.2005.06.026

Bhatia, K. K., & Sharma, A. K. (2008). A Framework for Domain Specific Interface Mapper (DSIM). *International Journal of Computer Science and Network Security*, *8*, 12.

Black, S. (2001). Computing ripple effect for software maintenance. *Software Maintenance. Research and Practice, 13*(4), 263–278.

Boguraev, B., & Neff, M. S. (2000). Discourse segmentation in aid of document summarization. In *Proceedings of the Hawaii International Conference on System Sciences (HICSS)* (pp. 10-17).

Bohanec, M. (2009). Decision Making: A Computer-Science and Information-Technology Viewpoint. *Interdisciplinary Description of Complex Systems, 7*, 22–37.

Boley, H., & Chang, E. (2007). Digital ecosystems: Principles and semantics. In *Proceedings of the Inaugural IEEE International Conference on Digital Ecosystems and Technologies* (pp. 398-403). Cairns, Australia.

Bormann, F., Braun, A., Flake, S., & Tacken, J. (2009). Towards a policy and charging control architecture for online charging. *International Conference on Advanced Information Networking and Applications Workshops*, (pp. 524-530).

Borriello, G., Chalmers, M., Lamarca, A., & Nixon, P. (2005). Delivering real-world ubiquitous location systems. *Communications of the ACM, 48*(3), 36–41. doi:10.1145/1047671.1047701

Bouquet, P., Serafini, L., & Zanobini, S. (2003). Semantic coordination: A new approach and an application. In *International Semantic Web Conference*, (pp. 130-143).

Bragge, J., Korhonen, P., Wallenius, H., & Wallenius, J. (2010). Bibliometric Analysis of Multiple Criteria Decision Making/Multiattribute Utility Theory. *Multiple Criteria Decision Making for Sustainable Energy and Transportation Systems*, 259-268.

Bratman, M. E. (1999). *Intention, plans, and practical reason.* Center for the Study of Language and Informatics.

Breitman, K., Casanova, M. A., & Truszkowski, W. (2007). *Semantic Web: Concepts, technologies and applications.* London, UK: Springer-Verlag.

Brereton, M., & McGarry, B. (2000). An observational study of how objects support engineering design thinking and communication: implications for the design of tangible media. In *Proceedings of the ACM Computer Human Interaction Conference*, The Hague, The Netherlands (pp. 217-224).

Briand, L., Labiche, Y., & Soccar, G. (2002). *Automating Impact Analysis and Regression test Selection Based on UML Designs (Tech. Rep. No. SCE- 02-04).* Carleton.

Brigham, F. J., Scruggs, T. E., & Mastropieri, M. A. (1995). Elaborative maps for enhanced learning of historical information: Uniting spatial, verbal, and imaginal information. *The Journal of Special Education, 28*(3), 440–460. doi:10.1177/002246699502800404

BrightPlanet.com LLC. (2000, July). *White Paper: The Deep Web: Surfacing Hidden Value.*

Bryce, C., Oriol, M., & Vitek, J. (1999). *A coordination model for agents based on secure spaces* (LNCS 1594, pp. 4-20). Berlin: Springer Verlag.

Budinsky, F., DeCandio, G., Earle, R., Francis, T., Jones, J., & Li, J. (2004). WebSphere Studio overview. *IBM Systems Journal, 43*(2), 384–419. doi:10.1147/sj.432.0384

Buyukkokten, O., Cho, J., Garcia-molina, H., Gravano, L., & Shivakumar, N. (1999). Exploiting geographical location information of web pages. In *Proceedings of the ACM SIGMOD Workshop on the Web and Databases (WebDB'99)* (pp. 91-96).

Byrd, R., & Ravin, Y. (1999). Identifying and extracting relations in text. In *Proceedings of the Applications of Natural Language to Information Systems (NLDB)* (pp. 149-154).

Cai, G. (2002). GeoVSM: An integrated Retrieval model for geographic information. In M. J. Egenhofer & D. M. Marks (Eds.), *GIScience 2002* (LNCS 2478, pp. 65-79).

Calhoun, P., Guttman, E., Zorn, G., & Arkko, J. (2003). *RFC 3588: Diameter base protocol.*

Calvert, K. I., Doar, M. B., & Zegura, E. W. (1997). Modeling Internet topology. *Communications Magazine, 35*, 160–163. doi:10.1109/35.587723

Campi, A., Ceri, S., Duvall, E., Guinea, S., Massart, D., & Ternier, S. (2008, January). Interoperability for searching Learning Object Repositories: The ProLearn Query Language. *D-LIB Magazine.*

Carriero, N., & Gelernter, D. (1986). *The S/Net's Linda kernel.* ACM Transactions on Computer Systems.

Casati, F. (1998). *Models, semantics, and formal methods for the design of workflows and their exceptions.* Unpublished doctoral dissertation, Politecnico di Milano.

Chakrabarti, S. (2000). Data mining for hypertext: A tutorial survey. *Proceedings of the ACM SIGKDD Explorations: Newsletter of the Special Interest Group (SIG) on Knowledge Discovery & Data Mining, 1*(2), 1-11.

Chang, Y. (2010). Consolidation of EPC and heterogeneous home network. *TamKang Journal of Science and Engineering, 13*(1), 21–28.

Chaudhary, K., & Dai, X. (2010). Architecture of CORBA based P2P-Netpay micro-payment system in peer-to-peer networks. *International Journal of Computer Science & Emerging Technologies, 1*(4), 208–217.

Chen, S., Wen, L., Hu, J., & Li, S. (2008). Fuzzy Synthetic Evaluation on Form Mapping in Deep Web Integration. In *Proceedings of the International Conference on Computer Science and Software Engineering.* IEEE.

Chen, Y., Peng, Y., Finin, T., Labrou, Y., & Cost, S. (1999). Negotiating agents for supply chain management. *Proceedings of the AAAI Workshop on Artificial Intelligence for Electronic Commerce.*

Chen, Y., Sion, R., & Carbunar, B. (2009). XPay: Practical anonymous payments for tor routing and other networked service. *WPES '09: Proceedings of the 8th ACM Workshop on Privacy in the Electronic Society*, (pp. 41-50). ACM Press.

Cheng, B., & Atlee, J. (2007). Research Directions in Requirements Engineering. In *Proceedings of the IEEE ICSE Future of Software Engineering*, Minneapolis, MN (pp. 285-303).

Cheng, T., Yan, X., & Chang, K. C. C. (2007). Entity Rank: searching entities directly and holistically. In *Proceedings of the VLDB.*

Chen, H., Houston, A. L., Sewell, R. R., & Schatz, B. R. (1998). Internet browsing and searching: User evaluations of category map and concept space techniques. *Journal of the American Society for Information Science American Society for Information Science, 49*(7), 582–603.

Chun, D. M., & Plass, J. L. (1996). Effects of multimedia annotations on vocabulary acquisition. *Modern Language Journal, 80*(2), 183–198. doi:10.2307/328635

Chung, L., & Cooper, K. (2004). Matching, Ranking, and Selecting Components: A COTS-Aware Requirements Engineering and Software Architecture Approach. In *Proceedings of the First International Workshop on Models and Processes for the Evaluation of Off-the-Shelf Components*, Edinburgh, Scotland (pp. 41-44).

Chung, L., Nixon, B., Yu, E., & Mylopoulos, J. (2000). *Non-Functional Requirements in Software Engineering.* Dordrecht, The Netherlands: Kluwer Academic Publishing.

Cohen, U. (2004). *Inside GigaSpaces XAP - Technical overview and value proposition.* White Paper. New York, NY: GigaSpace Technologies Ltd.

Collanos. (2011). *Products overview: Team enabling professionals.* Retrieved September 8, 2011, from http://www.collanos.com/en/products

Collins, G. E. (1960). A method for overlapping and erasure of lists. *Communications of the ACM, 3*(12), 655–657. doi:10.1145/367487.367501

Collins, J., & Gini, M. (2008). Scheduling tasks using combinatorial auctions: The MAGNET approach In Adomavicius, G., & Gupta, A. (Eds.), *Handbooks in Information Systems series: Business computing.* Elsevier.

Cope, J., Craswell, N., & Hawking, D. (2003). Automated Discovery of Search Interfaces on the web. In *Proceedings of the Fourteenth Australasian Database Conference (ADC2003)*, Adelaide, Australia.

Corso, M., & Mainetti, S. (2008). *Enterprise 2.0: La rivoluzione che viene dal web. Technical Report.* Milano, Italy: Politecnico di Milano, School of Management.

Creeger, M. (2009). CTO roundtable: Cloud computing. *Communications of the ACM, 52*(8), 50–65. doi:10.1145/1536616.1536633

Crescenzi, V., Mecca, G., & Merialdo, P. (2001). RoadRunner: towards automatic data extraction from large Web sites. In *Proceedings of the 27th International Conference on Very Large Data Bases*, Rome, Italy (pp. 109-118).

Cui, Z., Zhao, P., Fang, W., & Lin, C. (2008). *From Wrapping to Knowledge: Domain Ontology Learning from Deep Web.* In *Proceedings of the International Symposiums on Information Processing.* IEEE.

Cultured Code. (2011). *Things Mac*. Retrieved September 8, 2011, from http://culturedcode.com/things

Dai, H., & Mobasher, B. (2004). Integrating Semantic Knowledge with Web Usage Mining for Personalization. *The AAAI 2004 Workshop on Semantic Web Personalization (SWP '04)*, (pp. 276-306). San Jose, California, USA.

Dai, X., & Grundy, J. (2005). Off-line micro-payment system for content sharing in P2P networks. *2ⁿᵈ International Conference on Distributed Computing & Internet Technology (ICDCIT 2005) December 22-24, 2005, Lecture Notes in Computer Science, vol. 3816*, (pp. 297 –307).

Dai, X., Chaudhary, K., & Grundy, J. (2007). Comparing and contrasting micro-payment models for content sharing in P2P networks. *Third, International IEEE Conference on Signal-Image technologies and Internet-Based System (SITIS '07) China, 16 - 19 December 2007*, (pp. 347-354). IEEE Computer Society.

Dai, X., & Grundy, J. (2007). NetPay: An off-line, decentralized micro-payment system for thin-client applications. [Elsevier.]. *Electronic Commerce Research and Applications, 6*(1), 91–101. doi:10.1016/j.elerap.2005.10.009

Daras, P., Palaka, D., Giagourta, V., Bechtsis, D., Petridis, K., & Strintzis, G. M. (2003). *A novel peer-to-peer payment protocol*. IEEE EUROCON 2003, International Conference on Computer as a Tool (Invited paper), Ljubljana, Slovenia, September 2003.

Davis, E. T., Scott, K., Pair, J., Hodges, L. F., & Oliverio, J. (1999). *Can audio enhance visual perception and performance in a virtual environment?* Paper presented at the Proceedings of The Human Factors and Ergonomics Society 43rd Annual Meeting, Houston, TX.

Davis, S. (2006). *Google Map API, V2: Adding Where to You Applications*. The Pragmatic Programmers LLC.

de Araújo Formiga, A., & Lins, R. D. (2007). A new architecture for concurrent lazy cyclic reference counting on multi-processor systems. *Journal of Universal Computer Science, 13*(6), 817–829.

De Floriani, L., Marzano, P., & Puppo, E. (1996). Spatial Queries and Data Models. In A. Frank & I. Campari (Eds.), *Spatial Information Theory: A Theoretical Basis for GIS* (LNCS 716, pp. 113-138).

De Jong, T., Specht, M., & Koper, R. (2008). A reference model for mobile social software for learning. *International Journal of Continuing Engineering Education and Lifelong Learning, 18*(1), 118–138. doi:10.1504/IJCEELL.2008.016079

Deitel, A., Faron, C., & Dieng, R. (2001). *Learning ontologies from rdf annotations*. Paper presented at the IJCAI Workshop in Ontology Learning.

Deng, X. B., Ye, Y. M., Li, H. B., & Huang, J. Z. (2008). An Improved Random Forest Approach For Detection Of Hidden Web Search Interfaces. In *Proceedings of the Seventh International Conference on Machine Learning and Cybernetics*, Kunming, China. IEEE.

Dijkstra, E. W., Lamport, A. J. L., Scholten, C. S., & Steffens, E. F. M. (1978). *On-the-fly garbage collection: An exercise in cooperation*. Communications of ACM.

Dijkstra, E. W. (1959). A note on two problems in connexion with graphs. *Numerische Mathematik, 1*, 269–271. doi:10.1007/BF01386390

Di-Jorio, L., Bringay, S., Fiot, C., Laurent, A., & Teisseire, M. (2008). Sequential Patterns for Maintaining Ontologies over Time. *OTM '08: Proceedings of the OTM 2008 Confederated International Conferences, CoopIS, DOA, GADA, IS, and ODBASE 2008. Part II on On the Move to Meaningful Internet Systems* (pp. 1385-1403). Berlin, Heidelberg: Springer-Verlag.

Dikaiakos, M. D., Pallis, G., Katsaros, D., Mehra, P., & Vakali, A. (2009). Cloud computing. Distributed internet computing for IT and scientific research. *IEEE Internet Computing, 13*(5), 10–13. doi:10.1109/MIC.2009.103

Ding, L., Kolari, P., Ding, Z., & Avancha, S. (2007). A Survey. In Sharman, R., Kishore, R., & Ramesh, R. (Eds.), *Ontologies A Handbook of Principles, Concepts and Applications in Information Systems* (pp. 79–113). Springer, US: Using Ontologies in the Semantic Web.

Do, H. H., & Rahm, E. (2002, August). COMA-a System for Flexible Combination of Schema Matching Approaches. In *Proceedings of the 28th Intl. Conference on Very Large Databases (VLDB)*, Hong Kong.

Doan, A. H., Domingos, P., & Levy, A. (2000). Learning source descriptions for data integration. In *Proceedings of the WebDB Workshop* (pp. 81-92).

Doan, A., Domingos, P., & Halevy, A. (2001). Reconciling schemas of disparate data sources: A machine-learning approach. In *Proceedings of the International Conference on Management of Data (SIGMOD)*, Santa Barbara, CA. New York: ACM Press.

Doan, A., Lu, Y., Lee, Y., & Han, J. (2003). Object matching for information integration: A profiler-based approach. *II Web*, 53-58.

Dodds, A. G. (1982). The Mental Maps of the Blind: The Role of Previous Visual Experience. *Journal of Visual Impairment & Blindness, 76*(1), 5–12.

DoIt.im. (2011). *Smart way to manage tasks*. Retrieved September 8, 2011, from http://www.doit.im

Dourish, P., & Bellotti, V. (1992). Awareness and coordination in shared workspaces. In M. Mantel & R. Baecker (Eds.), *1992 ACM Conference on Computer-Supported Cooperative Work* (pp. 107–114). New York, NY: ACM Press.

Dourish, P. (2001). *Where the action is: The foundations of embodied interaction*. Cambridge, MA: MIT Press.

Downs, R. M., & Stea, D. (1977). *Maps In minds: Reflections on cognitive mapping*. New York: Harper & Row.

Duan, X. (2007). *Method for establishing diameter session for packet flow based charging*. Retrieved from http://www.freshpatents.com/Method-for-establishing-diameter-session-for-packet-flow-based-charging-dt20070816ptan 20070189297.php

Dumas, S. J., & Redish, J. C. (1993). *A practical guide to usability testing*. Norwood, MA: Ablex Publishing Corporation.

Dynamic Markets. (2008). *Corporate social networking in Europe. Independent market research report. Commissioned by AT&T (Technical Report)*. Abergavenny, UK: Dynamic Markets.

Eggen, R., & Eggen, M. (2001). *Efficiency of distributed parallel processing using Java RMI, Sockets, and CORBA*. Retrieved from www.imamu.edu.sa/dcontent/IT_Topics/java/paper3.pdf

Eirinaki, M., Lampos, H., Vazirgiannis, M., & Varlamis, I. (2003). Sewep: Using site semantics and a taxonomy to enhance the web personalization process. *The Ninth ACM SIGKDD International Conference on Knowledge Discovery and Data Mining* (pp. 99-108). Washington, DC, USA: ACM Special Interest Group on Knowledge Discovery in Data.

Elish, M., & Rine, D. (2003). Investigation of Metrics for Object-Oriented Design Logical Stability. In *Proceedings of the 7th European Conference On Software Maintenance And Reengineering*. Washington, DC: IEEE Computer Society.

Elish, M., & Rine, D. (2005). Indicators of Structural Stability of Object Oriented Designs: A Case Study. In *Proceedings of the 29th Annual NASA/IEEE Software Engineering Workshop* (pp. 183-192).

Elkotob, M. (2008). *Autonomic resource management in IEEE 802.11 open access networks*. Dissertation, Lules University of Technology, Sweden. Retrieved from http://epubl.ltu.se/1402-1757/2008/38/LTU-LIC-0838-SE.pdf

Esfahani, A., & Analoui, M. (2008). *Widest K-Shortest Paths Q-Routing: A New QoS Routing Algorithm in Telecommunication Networks*.

Euzenat, J., & Shvaiko, P. (2007). *Ontology matching*. Berlin, Germany: Springer-Verlag.

Faatz, A., & Steinmetz, R. (2002). Ontology enrichment with texts from the WWW. *Proceedings of Semantic Web Mining 2nd Workshop at ECML/PKDD-2002*.

Fähndrich, M., Aiken, M., Hawblitzel, C., Hodson, O., Hunt, G., Larus, J. R., & Levi, S. (2006). Language support for fast and reliable message-based communication in singularity OS. In *Proceedings of the EuroSys 2006 Conference* (pp. 177-190). New York: ACM.

Faiola, A. (2006). Designing humane technologies: A potential framework for human-computer interaction design. *The International Journal of the Humanities, 2*(3), 1877–1886.

Faiola, A., & MacDorman, K. (2008). Exploring the influence of web designer cognitive style on information design: A cross-cultural comparison of a holistic and analytical perspective. *Information Communication and Society, 11*(3), 348–374. doi:10.1080/13691180802025418

Faiola, A., & Matei, S. (2010). Enhancing human–computer interaction design education: teaching affordance design for emerging mobile devices. *International Journal of Technology and Design Education, 20,* 239–254. doi:10.1007/s10798-008-9082-4

FakeSpace. (2005). *FLEX™ and reFLEX™. Innovative designs allow for fast reconfiguration for fully detached module capabilities.* Retrieved September 28, 2005, from http://www.fakespacesystems.com/pdfs/solutions/Fakespace-FLEX-reFLEX.pdf

Feng Office. (2011). *Project management - Easy, powerful collaborative project management.* Retrieved September 8, 2011, from http://www.fengoffice.com

Feras, A. H. H. (2006). *Integrity constraints maintenance for parallel databases.* Unpublished doctoral dissertation, Universiti Putra Malaysia, Malaysia.

Foo, P., Warren, W. H., Duchon, A., & Tarr, M. (2005). Do humans integrate routes into a cognitive map? Map versus landmark-based navigation of novel shortcuts. *Journal of Experimental Psychology, 31*(2), 195–215.

Forlizzi, J. (2008, April 24). The Product Ecology: Understanding Social Product Use and Supporting Design Culture. *International Journal of Design, 2*(1). Retrieved from http://www.ijdesign.org/ojs/index.php/IJDesign/article/view/220/143.

Freeman, E., Hupfer, S., & Arnold, K. (1999). *JavaSpaces: Principles, Patterns, and Practice. The Jini Technology Series.* Reading, MA: Addison-Wesley.

Friedman, D. P., & Wise, D. S. (1979). Reference counting can manage the circular environments of mutual recursion. *Information Processing Letters, 8*(1), 41–45. doi:10.1016/0020-0190(79)90091-7

Gagne, E. D. (1978). Long-term retention of information following learning from prose. *Review of Educational Research, 48*(4), 629–665.

Galler, J., Chun, S. A., & An, Y. J. (2008, September). Toward the Semantic Deep Web. In *Proceedings of the IEEE Computer* (pp. 95-97).

Garcia, F. D., & Hoepman, J. H. (2005). *Off-line karma: A decentralized currency for static peer-to-peer and grid networks.* In 5th International Network Conference (INC2005), July 5-7 2005.

Gelernter, D. (1989). *Multiple tuple spaces in LINDA* (LNCS 366, pp. 20-27). Berlin: Springer Verlag.

Gelernter, D. (1985). Generative communication in LINDA. *ACM Transactions on Programming Languages and Systems, 7*(1), 80–112. doi:10.1145/2363.2433

Gellevij, M., Meij, H. V. D., Jong, T. D., & Pieters, J. (2002). Multimodal versus unimodal instruction in a complex learning context. *Journal of Experimental Education, 70*(3), 215–239. doi:10.1080/00220970209599507

Gerber, S., & Shuell, T. J. (1998). Using the internet to learn mathematics. *Journal of Computers in Mathematics and Science Teaching, 17*(2), 113–132.

Gey, F., Larson, R., Sanderso, M., Joho, H., & Clough, P. (2005). GeoCLEF: the CLEF 2005 cross-language geographic information retrieval track. In *Proceedings of CLEF 2005: Working Notes for the CLEF 2005 Workshop* (pp. 908-919).

Gibson, J. J. (1977). *The theory of affordances.* Hillsdale, NJ: Erlbaum Associates.

Glassman, S., Manasse, M., Abadi, M., Gauthier, P., & Sobalvarro, P. (1995). The millicent protocol for inexpensive electronic commerce. In *4th WWW Conference Proceedings,* (pp. 603-618). New York, NY: O'Reilly.

Gochev, H., Poulkov, V., & Iliev, G. (2010). Uplink power control for LTE improving cell edge throughput. *International Conference on Telecommunications and Signal Processing, TSP'2010,* (pp. 465-467).

Goldstein, M., Nyberg, M., & Anneroth, M. (2003). Providing proper affordances when transferring source metaphors from information appliances to a 3G mobile multipurpose handset. *Personal and Ubiquitous Computing, 7*(6), 372–380. doi:10.1007/s00779-003-0252-9

Golledge, R. G., & Stimson, R. J. (1997). *Spatial behavior: A geographic perspective.* New York: Guilford Press.

Gomez, E. J., Pozo, F. D., & Hernando, M. (1996). Telemedicine for diabetes care: The DIABTel approach towards diabetes telecare. *Informatics for Health & Social Care, 21*(4), 283–295. doi:10.3109/14639239608999290

Good, R., & Ventura, N. (2009). Application-driven policy-based resource management for IP multimedia subsystems. *International Conference on Testbeds and Research Infrastructures for the Development of Networks & Communities, TridentCom'2009*, (pp. 1-9).

Good, R., Gouveia, F., Ventura, N., & Magedanz, T. (2010). Session-based end-to-end policy control in 3GPP evolved packet system. *International Journal of Communication Systems* [NGNs]. *Special Issue: Part 1: Next Generation Networks, 23*(6-7), 861–883.

Google. (2011a). *Google Documents: Create documents, spreadsheets and presentations online.* Retrieved September 8, 2011, from http://www.google.com/google-d-s/tour1.html.

Google. (2011b). *Google Tasks: Keep track of what you need to do.* Retrieved September 8, 2011, from http://mail.google.com/mail/help/tasks

Google. (2011c). *Pubsubhubbub.* Retrieved September 8, 2011, from http://code.google.com/p/pubsubhubbub

Google. (2011d). *Google+.* Retrieved August 18, 2011, from http://www.google.com/+/learnmore/

Gould, P. (1975). Acquiring spatial information. *Economic Geography, 51*(2), 87–99. doi:10.2307/143066

Gouveia, F., Wahle, S., Blum, N., & Magedanz, T. (2009). *Cloud computing and EPC / IMS integration: New value-added services on demand.* 5th International ICST Mobile Multimedia Communications Conference'2009.

Grau, G., Carvallo, J., Franch, X., & Quer, C. (2004). DesCOTS: A Software System for Selecting COTS Components. In *Proceedings of the 30th EUROMICRO Conference*, Rennes, France (pp. 118-126).

Grefen, P. W. P. J. (1993). Combining theory and practice in integrity control: A Declarative approach to the specification of a transaction modification subsystem. In *Proceedings of the 19ᵗʰ International Conference on Very Large Data Bases*, Dublin (pp. 581-591). ISBN 1-55860-152-X

Grigori, D., Casati, F., Dayal, U., & Shan, M. C. (2001). Improving business process quality through exception understanding, prediction, and prevention. In *Proceedings of the 27th International Conference on Very Large Data Bases* (pp. 159-168). San Francisco, CA, USA.

Grimes, A., & Brush, A. (2008). Life scheduling to support multiple social roles. In M. Burnett, M. F. Costabile, T. Catarci, B. de Ruyter, D. Tan, M. Czerwinski, & A. Lund (Eds.), *26th Annual CHI Conference on Human Factors in Computing Systems* (pp. 821–824), New York, NY: ACM Press.

Grosser, D., Sahraoui, H., & Valtchev, P. (2002). Predicting Software Stability Using Case-Based Reasoning. In *Proceedings of the 17ᵗʰ IEEE International Conference on Automated Software Engineering* (pp. 295-298).

Grossnickle, J., Board, T., Pickens, B., & Bellmont, M. (2005, October). *RSS—Crossing into the Mainstream.* Ipsos Insight, Yahoo.

Guo, Z., Qiao, J., Lin, S., & Cai, X. (2009). *A distributed parallel QoS routing algorithm with multi-path probing.* Washington, DC: IEEE Press.

Gupta, A. (1994). *Partial information based integrity constraint checking.* Unpublished doctoral dissertation, Stanford University, Stanford, CA.

Gutwin, C., Greenberg, S., & Roseman, M. (1996). Workspace awareness in real-time distributed groupware: Framework, widgets, and evaluation. In Sasse, M. A., Cunningham, R. J., & Winder, R. L. (Eds.), *HCI 96 People and Computers XI* (pp. 281–298). Berlin, Germany: Springer-Verlag.

Haake, J., Hussein, T., Joop, B., Lukosch, S., Veiel, D., & Ziegler, J. (2010). Modeling and exploiting context for adaptive collaboration. *International Journal of Cooperative Information Systems, 19*(1-2), 71–120. doi:10.1142/S0218843010002115

Habib, S. (2008). Dynamic Evaluation of Server Placement within Network Design Tool by Embedded Monte Carlo Simulator. *International Journal of Business Data Communications and Networking, 4*(2), 38–57.

Haddadou, K., Ghamri-Doudane, S., Ghamri-Doudane, Y., & Agoulmine, N. (2006). Designing scalable on-demand policy-based resource allocation in IP networks. *IEEE Communications Magazine, 44*(3), 142–149. doi:10.1109/MCOM.2006.1607878

Hartson, H. R. (2003). Cognitive, physical, sensory, and functional affordances in interaction design. *Behaviour & Information Technology*, *22*(5), 315–338. doi:10.1080/01449290310001592587

Hauser, R., Steiner, M., & Waidner, M. (1996). Micropayments based on IKP. *Proceedings of 14th Worldwide Congress on Computer and Communications Security Protection*, Paris-La Defense, France, 1996, C.N.I.T, (pp. 67- 82).

Hawking, D., Upstill, T., & Craswell, N. (2004). Toward Better Weighting of Anchors. In *Proceedings of the 27ᵗʰ Annual International ACM SIGIR Conference on Research and Development in Information Retrieval* (pp. 512-53).

He, B., & Chang, K. C. C. (2003). *Statistical schema matching across web query interfaces*. Paper presented at the SIGMOD Conference.

He, B., & Chang, K. C. C. (2006, March). Automatic complex schema matching across web query interfaces: A correlation mining approach. In *Proceedings of the ACM Transaction on Database Systems* (Vol. 31, pp. 1-45).

He, B., Zhang, Z., & Chang, K. C. C. (2005, June). *Meta Querier: Querying Structured Web Sources On the-fly*. Paper presented at SIGMOD, System Demonstration, Baltimore, MD.

He, H., Meng, W., Yu, C. T., & Wu, Z. (2003). WISE-Integrator: An Automatic Integrator of Web search interfaces for e-commerce. In *Proceedings of the 29th International Conference on Very Large Data Bases (VLDB'03)* (pp. 357-368).

He, H., Meng, W., Yu, C., & Wu, Z. (2005). Constructing interface schemas for search interfaces of Web databases. In *Proceedings of the 6th International Conference on Web Information Systems Engineering (WISE'05)* (pp. 29-42).

Health Ministry. (2006). *Health year book 2005-2006*. Health Ministry, Government of Pakistan, 2007. Retrieved January 28, 2011, from http://www.health.gov.pk/

Heo, J., Ham, D., Park, S., Song, C., & Yoon, W. C. (2009). A framework for evaluating the usability of mobile phones based on multi-level, hierarchical model of usability factors. *Interacting with Computers*, *21*(4), 263–275. doi:10.1016/j.intcom.2009.05.006

Hernandez, M., & Stolfo, S. (1995). *The merge/purge problem for large databases*. In *Proceedings of the SIGMOD Conference* (pp. 127-138).

Hislop, D., & Axtell, C. (2007). The neglect of spatial mobility in contemporary studies of work: The case of telework. *New Technology, Work and Employment*, *22*(1), 34–51. doi:10.1111/j.1468-005X.2007.00182.x

Houston, M. S., & Myers, J. D. (1999). Clinical consultations using store-and-forward telemedicine technology. *Mayo Clinic Proceedings*, *74*(8), 764–769. doi:10.4065/74.8.764

Hsu, A., & Imielinski, T. (1985). Integrity checking for multiple updates. In *Proceedings of the 1985 ACM SIGMOD International Conference on the Management of Data*, Austin, TX (pp. 152-168).

Hsu, H. J., & Wang, F. J. (2011). Detecting artifact anomalies in temporal structured workflow as reusable assets. In *Proceedings of the 35th IEEE Annual Computer Software and Applications Conference Workshops* (pp. 362-367). Munich, Germany.

Huan, J., Wang, W., Prins, J., & Yang, J. (2004). SPIN: Mining Maximal Frequent Subgraphs from Graph Databases. *10th ACM SIGKDD International Conference on Knowledge Discovery and Data Mining*, (pp. 581-586).

Huang, P., & Sycara, K. (2002). A computational model for online agent negotiation. *Proceedings of the 35th Hawaii International Conference on System Sciences* (HICSS'02).

Huang, Q., & Zhao, Y. (2009). A secure and lightweight micro-payment scheme in P2P networks. *Proceedings of IEEE International Conference on Industrial and Information Systems(IIS '09)*, (pp. 134 - 137).

Ibrahim, A., Fahmi, S. A., Hashmi, S. I., & Choi, H. J. (2008). Addressing Effective Hidden Web Search Using Iterative Deepening Search and Graph Theory. In *Proceedings of the 8th International Conference on Computer and Information Technology Workshops*. IEEE.

Ibrahim, H. (1998). *Semantic integrity constraints enforcement for a distributed database*. Unpublished doctoral dissertation, University of Wales College of Cardiff, Cardiff, UK.

Ibrahim, H. (2002). A Strategy for semantic integrity checking in distributed databases. In *Proceedings of the Ninth International Conference on Parallel and Distributed Systems* (pp. 139-146). Washington, DC: IEEE.

Ibrahim, H. (2006). Checking integrity constraints – How it differs in centralized, distributed and parallel databases. In *Proceedings of the 17th International Conference on Database and Expert System Application - 2nd International Workshop on Logical Aspects and Applications of Integrity Constraints*, Krakow, Poland (pp. 563-568).

Ibrahim, H., Gray, W. A., & Fiddian, N. J. (2001). Optimizing fragment constraints – A Performance evaluation. *International Journal of Intelligent Systems – Verification and Validation Issues in Databases. Knowledge-Based Systems, and Ontologies, 16*(3), 285–306.

InBox Foundry. (2010). *GTDInbox*. Retrieved March 22, 2010, from http://gtdinbox.com/better_inbox.htm

Inokuchi, A., Washio, T., & Motoda, H. (2000). An Apriori-Based Algorithm for Mining Frequent Substructures from Graph Data. *PKDD '00: Proceedings of the 4th European Conference on Principles of Data Mining and Knowledge Discovery* (pp. 13-23). Lyon, France: Springer.

Intersense. (2005). *InterSense IS-900 Precision Motion Tracker*. Retrieved September 28, 2005, from http://www.isense.com/products/prec/is900/

Iqbal, U., Javed, Y., Rehman, S., & Khanum, A. (2010). SIP-based QoS management framework for IMS multimedia services. *International Journal of Computer Science and Network Security, 10*(5), 181–188.

ISO/IEC. (1998). *9241-11 Ergonomic requirements for office work with visual display terminals (VDT)--Part 11 Guidance on usability*. ISO/IEC 9241-11.

Jacob, J. L., & Wood, A. (2000). *A principled semantics for inp* (LNCS 1906, pp. 51-66). Berlin: Springer Verlag.

Jain, M., & Prokopi, M. (2008). The IMS 2.0 service architecture. *Second International Conference on Next Generation Mobile Applications, Services and Technologies, NGMAST '08,* (pp. 3-9).

James, W. (1890). *The principles of psychology*. New York: Holt.

Jazayeri, M. (2002). On architectural stability and evolution. In *Proceedings of the 7th Ada-Europe International Conference on Reliable Software Technologies* (pp. 13-23). London: Springer Verlag.

Jeffrey, E. F. (2006). *Mastering Regular Expressions* (3rd ed.). Sebastopol, CA: O'Reilly Media Inc.

Jensen, P. A., & Bard, J. F. (2003). *Operations Research: Models and Methods*. New York: Wiley.

Jiang, F., Jia, L., Meng, W., Meng, X., & MrCoM. (2008). A Cost Model for Range Query Translation in Deep Web Data Integration. In *Proceedings of the Fourth International Conference on Semantics, Knowledge and Grid*. IEEE.

Jiehui, J., & Jing, Z. (2007). Remote patient monitoring system for China. *IEEE Potentials, 26*(3), 26–29. doi:10.1109/MP.2007.361641

Jin, X., & Mobasher, B. (2003). Using semantic similarity to enhance item-based collaborative filtering. *The 2nd IASTED International Conference on Information and Knowledge Sharing*. Scottsdale, AZ, US.

Jin, B. S., & Ji, Y. G. (2010). Usability risk level evaluation for physical user interface of mobile phone. *Computers in Industry, 61*(4), 350–363. doi:10.1016/j.compind.2009.12.006

Johansson, B. G. (2001). *Design and implementation of a clinical decision support system based on open standards.*

Jones, C. B., Alani, H., & Tudhope, D. (2001). Geographical Information Retrieval with Ontologies of Place. In *Proceedings of the International Conference on Spatial Information Theory: Foundations of Geographic Information Science* (LNCS 2205, pp. 322-325).

Jones, C. B., Purves, R., Ruas, A., Sanderson, M., Sester, M., van Kreveld, M., & Weibel, R. (2002). Spatial information retrieval and geographical ontologies: An overview of the SPIRIT project. In *Proceedings of SIGIR-02, the 25th Conference on Research and Development in Information Retrieval* (pp. 387-388).

Kaljuvee, O., Buyukkokten, O., Molina, H. G., & Paepcke, A. (2001). Efficient Web form entry on PDAs. In *Proceedings of the 10th International Conference on World Wide Web (WWW'01)* (pp. 663- 672).

Kallitsis, M., Michailidis, G., & Devetsikiotis, M. (2009). Measurement-based optimal resource allocation for network services with pricing differentiation. *Performance Evaluation*, *66*(9-10), 505–523. doi:10.1016/j.peva.2009.03.003

Kammer, P. J., Bolcer, G. A., Taylor, R. N., Hitomi, A. S., & Bergman, M. (2000). Techniques for supporting dynamic and adaptive workflow. *Computer Supported Cooperative Work*, *9*(3-4), 269–292. doi:10.1023/A:1008747109146

Kangas, E., & Kinnunen, T. (2005). Applying User-Centered Design to Mobile Application Development. *Communications of the ACM*, *48*(7), 55–59. doi:10.1145/1070838.1070866

Kareem, S., & Bajwa, I. S. (2011a). *A virtual telehealth framework: Applications and technical considerations*. In IEEE International Conference on Emerging Technologies 2011 (ICET 2011)

Kareem, S., & Bajwa, I. S. (2011b). Clinical decision support system based virtual telemedicine. *3rd International Conference on Intelligent Human-Machine Systems and Cybernetics* (IHMSC 2011) (pp. 78-83). Zhejiang University, China Hangzhou.

Kawamoto, K., Houlihan, C. A., Balas, E. A., & Lobach, D. H. (2005). Improving clinical practice using clinical decision support systems: A systematic review of trials to identify features critical to success. *British Medical Journal*, *330*(7494), 765. doi:10.1136/bmj.38398.500764.8F

Kepner, C. H., & Tregoe, B. B. (1981). *The New Rational Manager*.

Khalid, M. Z., Akbar, A., Kumar, A., Tariq, A., & Farooq, M. (2008). Using telemedicine as an enabler or antenatal care in Pakistan. *2nd International Conference: E-Medical System*, Tunisia, (pp. 1-8).

Khattab, M. A., Fouad, Y., & Rawash, O. A. (2009). Proposed Protocol to Solve Discovering Hidden Web Hosts Problem. *International Journal of Computer Science and Network Security*, *9*(8).

Kim, G., Han, M., Park, J., Park, H., Park, S., Kim, L., & Ha, S. (2009). An OWL-Based Knowledge Model for Combined-Process-and-Location Aware Service. In *Proceedings of the Symposium on Human Interface 2009 on Human Interface and the Management of Information. Information and Interaction. Part II: Held as part of HCI International 2009* (p. 167).

King, A. (1992). Comparison of self-questioning, summarizing, and note-taking review as strategies for learning from lectures. *American Educational Research Journal*, *29*(2), 303–323.

Kirkpatrick, S., Gelatt, S., & Vecchi, M. (1983). Optimization by Simulated Annealing. *Science*, *220*(4598), 671–680. doi:10.1126/science.220.4598.671

Kitchenham, B., Linkman, S., Pasquini, A., & Nanni, V. (1997). The SQUID approach to defining a quality model. *Software Quality Control*, *6*(3), 211–233. doi:10.1023/A:1018516103435

Klein, M. (2001). Combining and relating ontologies: An analysis of problems and solutions. *Proceedings of the Workshop on Ontologies and Information Sharing at the 17th International Joint Conference on Artificial Intelligence*, (pp. 53-62).

Klein, M., & Dellarocas, C. (2000). A knowledge-based approach to handling exceptions in workflow systems. *Journal of Computer Supported Collaborative Work*, *9*(3-4), 399–412. doi:10.1023/A:1008759413689

Klonoff, D. C., & True, M. W. (2009). The missing element of telemedicine for diabetes: Decision support software. *Journal of Diabetes Science and Technology*, 3.

Klopfer, E., Squire, K., & Jenkins, H. (2004). Environmental detectives: PDAs as a window into a virtual world. In Kerres, M., Kalz, M., Stratmann, J., & De Witt, C. (Eds.), *Didactik der notebook-universitat*. Munster, Germany: Waxmann Verlag.

Knublauch, H. (n.d.). Retrieved September 12, 2008, from http://protege.cim3.net/: http://protege.cim3.net/file/pub/ontologies/travel/travel.owl

Koo, S., Rosenberg, C., Chan, H. H., & Lee, Y. C. (2003, March). *Location-based e-campus web services: from design to deployment.* Paper presented at the IEEE International Conference on Pervasive Computing and Communications (PerCom), Dallas, TX.

Kornai, A., & Sundheim, B. (2003). In *Proceedings of the NAACL-HLT Workshop on the Analysis of Geographic References.*

Kou, Y., Shen, D., Yu, G., & Nie, T. (2008). *LG-ERM: An Entity-level Ranking Mechanism for Deep Web Query.* Washington, DC: IEEE.

Koutsopoulou, M., Kaloxylos, A., Alonistioti, A., & Merakos, L. (2007). A platform for charging, billing, and accounting in future mobile networks. *Computer Communications*, *30*, 516–526. doi:10.1016/j.comcom.2005.11.022

Krystkowiak, M., Bucciarelli, B., & Dubois, E. (2003). COTS Selection for SMEs: a report on a case study and on a supporting tool. In *Proceedings of the 1st International Workshop on COTS and Product Software: Why Requirements are so Important (RECOTS)*, Monterey, CA.

Kulhavy, R. W., Caterino, L. C., & Melchiori, F. (1989). Spatially cued retrieval of sentences. *The Journal of General Psychology*, *116*(3), 297–304.

Kulhavy, R. W., Lee, J. B., & Caterino, L. C. (1985). Conjoint retention of maps and related discourse. *Contemporary Educational Psychology*, *10*, 28–37. doi:10.1016/0361-476X(85)90003-7

Kung, D., Gao, J., Hsia, P., Wen, F., Toyoshima, Y., & Chen, C. (1994). Change Identification in Object Oriented Software Maintenance. In *Proceedings of the International Conference on Software Maintenance* (pp. 201-211).

Lage, P. B. G. J. P., Silva, D., & Laender, A. H. F. (2004). Automatic generation of agents for collecting hidden web pages for data extraction. *Data & Knowledge Engineering*, *49*, 177–196. doi:10.1016/j.datak.2003.10.003

Lai, A. M., Nieh, J., & Starren, J. B. (2007). REPETE2: A next generation home telemedicine architecture. *AMIA 2007 Symposium Proceedings*, (pp. 1020-1022).

Lai, S. (2000). Increasing associative learning of abstract concepts through audiovisual redundancy. *Journal of Educational Computing Research*, *23*(3), 275–289. doi:10.2190/XKLM-3A96-2LAV-CB3L

Larson, R. R. (1995). Geographic Information Retrieval and Spatial Browsing. In L. C. Smith & M. Gluck (Eds.), *Geographic Information Systems Patrons Maps and Spatial Information* (pp. 81-123). Champaign-Urbana, IL: University of Illinois.

Laurini, R., & Thompson, D. (1992). *Fundamentals of Spatial Information Systems.* New York: Academic Press.

Lee, L., Lee, Y., Lin, C., & Huang, K. (2007). Automatic journey geocoding. In *Proceedings of the Taiwan Geographic Information Conference.*

Lee, S., Das, S., Pau, G., & Gerla, M. (2009). *A Hierarchical Multipath Approach to QoS Routing.*

Lee, T. B. (2001). *W3C Semantic Web Activity.* Retrieved 2010, from http://www.w3.org/2001/sw: http://www.w3.org/2001/sw

Lee, T., Hendler, J., & Lassila, O. (2001, May). *The semantic web. Scientific American* Retrieved 02 2010, from http://www.sciam.com/article.cfm?id=the-semantic-web

Lee, M., Offutt, J., & Alexander, R. (2000). Algorithmic Analysis of the Impact of Changes to Object-Oriented Software. In. *Proceedings of TOOLS*, *2000*, 61–70.

Lee, S., Leaney, J., O'Neill, T., & Hunter, M. (2005). Open service access for QoS control in next generation networks – Improving the OSA/Parlay connectivity manager. *Journal of Operations and Management in IP-Based Networks*, *3751*, 29–38. doi:10.1007/11567486_4

Lenat, D. B., & Guha, R. V. (1990). *Building large knowledge-based systems: Representation and inference in Cyc Project.* Boston, MA: Addison-Wesley.

Leng, H., Liang, M., Song, J., Xie, Z., & Zhang, J. (2009). *Routing on Shortest Pair of Disjoint Paths with Bandwidth Guaranteed.* Washington, DC: IEEE.

Li, Y., Li, J., Zhang, D., & Tang, J. (2006). *Result of ontology alignment with RiMOM at OAEI'06.* International Workshop on Ontology Matching collocated with the 5th International Semantic Web Conference.

Liang, H., Zuo, W., Ren, F., & Sun, C. (2008). Accessing Deep Web Using Automatic Query Translation Technique. In *Proceedings of the Fifth International Conference on Fuzzy Systems and Knowledge Discovery*. IEEE.

Liao, S.-H., Chen, J.-L., & Hsu, T.-Y. (2009). Ontology-based data mining approach implemented for sport marketing. *Expert Systems with Applications: An International Journal*, 11045-11056.

Liebau, N., Heckmann, O., Kovacevic, A., Mauthe, A., & Steinmetz, R. (2006). Charging in peer-to-peer systems based on a token accounting system. *5th International Workshop on Internet Charging and QoS Technologies, LNCS 4033*, (pp. 49–60).

Li, H., & Yang, Y. (2005). Dynamic checking of temporal constraints for concurrent workflows. *Electronic Commerce Research and Applications*, *4*(2), 124–142. doi:10.1016/j.elerap.2004.09.003

Lin, K. H.-Y., Hou, W.-J., & Chen, H.-H. (2005) Retrieval of Biomedical Documents by Prioritizing Key Phrases. In *Proceedings of the Fourteenth Text REtrieval Conference (TREC 2005)*, Gaithersburg, MD.

Ling, Y.-Y., Liu, W., Wang, Z.-Y., Ai, J., & Meng, X.-F. (2006). Entity identification for deep web data integration. *Journal of Computer Research and Development*, 46-53.

Lins, R. D. (2003). *An efficient multi-processor architecture for parallel cyclic reference counting* (LNCS 2565, pp. 111-139). Berlin: Springer.

Lins, R. D. (2006). *New algorithms and applications of cyclic reference counting* (LNCS 4178, pp. 15-29). Berlin: Springer.

Lins, R. D. (2002). An efficient algorithm for cyclic reference counting. *Information Processing Letters*, *83*(3), 145–150. doi:10.1016/S0020-0190(01)00328-3

Lins, R. D. (2008). Cyclic reference counting. *Information Processing Letters*, *109*(1), 71–78. doi:10.1016/j.ipl.2008.09.009

Liu, X., Chen, J., Wu, Z., Ni, Z., Yuan, D., & Yang, Y. (2010b). Handling recoverable temporal violations in scientific workflow systems: A workflow rescheduling based strategy. In *Proceedings of the 10th IEEE/ACM International Conference on Cluster, Cloud, and Grid Computing* (pp. 534-537). Melbourne, Australia.

Liu, X., Yang, Y., Jiang, Y., & Chen, J. (2010a). (Accepted for publication). Preventing temporal violations in scientific workflows: Where and how. *IEEE Transactions on Software Engineering*.

Lloyd, R. (1994). Learning spatial prototypes. *Annals of the Association of American Geographers. Association of American Geographers*, *84*(3), 418–440. doi:10.1111/j.1467-8306.1994.tb01868.x

Lloyd, R., & Heivly, C. (1987). Systematic distortions in urban cognitive maps. *Annals of the Association of American Geographers. Association of American Geographers*, *77*(2), 191–207. doi:10.1111/j.1467-8306.1987.tb00153.x

Lu, S., Bernstein, A., & Lewis, P. (2006). Automatic workflow verification and generation. *Theoretical Computer Science*, *353*(1-3), 71–92. doi:10.1016/j.tcs.2005.10.035

Lynch, K. (1960). *Image of the City*. Cambridge, MA: The MIT Press.

MacEachren, A. M. (1992). Application of environmental learning theory to spatial knowledge acquisition from maps. *Annals of the Association of American Geographers. Association of American Geographers*, *82*(2), 245–274. doi:10.1111/j.1467-8306.1992.tb01907.x

Madhavan, J., Bernstein, P. A., & Rahm, E. (2001). *Generic Schema Matching with Cupid*. Paper presented at the 27th VLBB Conference, Rome.

Madiraju, P., Sunderraman, R., & Haibin, W. (2006). A Framework for global constraint checking involving aggregates in multidatabases using granular computing. In *Proceedings of IEEE International Conference on Granular Computing*, Atlanta (pp. 506-509).

Mansour, N., & Salem, H. (2006). Ripple Effect in Object Oriented Programs. *Journal of Computational Methods in Sciences and Engineering*, *6*(5-6), 23–32.

Marti, S., & Schmandt, C. (2005). Physical embodiments for mobile communication agents. In *Proceedings of the 18th annual ACM symposium on User interface software and technology*, Seattle, WA (pp. 231-240). New York: ACM.

Martinenghi, D. (2005). *Advanced techniques for efficient data integrity checking*. Unpublished doctoral dissertation, Roskilde University, Roskilde, Denmark.

Martins, B., & Silva, M. J. (2005). A graph-ranking algorithm for geo-referencing documents. In *Proceedings of the ICDM-05, the 5th IEEE International Conference on Data Mining* (pp. 741-744).

Matei, S. A., Madsen, L., Arns, L., Bertoline, G., & Davidson, D. (2005, October 5-9). *Socio-spatial cognition and community identification in the context of the next generation of location-aware information systems.* Paper presented at The Association of Internet Researchers Annual Conference, Chicago.

Matei, S. A., Miller, C. C., Arns, L., Rauh, N., Hartman, C., & Bruno, R. (2007). *Visible Past: Learning and discovering in real and virtual space and time (Vol. 12).* First Monday.

Mattsson, M., & Bosch, J. (2000). Stability assessment of evolving industrial object oriented Frameworks. *Journal of Software Maintenance: Research and Practice, 12,* 79–102. doi:10.1002/(SICI)1096-908X(200003/04)12:2<79::AID-SMR204>3.0.CO;2-A

May, W.-Y., & Chang, K.-J. (2003). Introduction to CKIP Chinese Word Segmentation System for the First International Chinese Word Segmentation Bakeoff. In *Proceedings of the ACL, Second SIGHAN Workshop on Chinese Language Processing* (pp. 168-171).

Mayer, R. E., & Anderson, R. B. (1991). Animations need narrations: An experimental test of a dual-coding hypothesis. *Journal of Educational Psychology, 83*(4), 484–490. doi:10.1037/0022-0663.83.4.484

Mazumdar, S. (1993). Optimizing distributed integrity constraints. In *Proceedings of the 3rd International Symposium on Database Systems for Advanced Applications,* Taejon, Korea (Vol. 4, pp. 327-334).

McBeth, J. H. (1963). On the reference counter method. *Communications of the ACM, 6*(9), 575. doi:10.1145/367593.367649

McCarroll, N. F. (1995). *Semantic integrity enforcement in parallel database machines.* Unpublished doctoral dissertation, University of Sheffield, Sheffield, UK.

McCune, W. W., & Henschen, L. J. (1989). Maintaining state constraints in relational databases. A Proof theoretic basis. *Journal of the Association for Computing Machinery, 36*(1), 46–68.

McCurley, K. S. (2001). Geospatial mapping and navigation of the Web. In *Proceedings of the 10th International World Wide Web (WWW10)* (pp. 221-229).

Medina, A., Lakhina, A., Matta, I., & Byers, J. (2001). BRITE: An Approach to Universal Topology Generation. In *Proceedings of MASCOTS* (Vol. 1).

Melnik, S., Garcia-Molina, H., & Rahm, E. (2002) Similarity Flooding: A Versatile Graph Matching Algorithm. In *Proceedings of the 18th International Conference on Data Engineering (ICDE),* San Jose, CA.

Melville, P., Mooney, R. J., & Nagarajan, R. (2002). Content-boosted collaborative ltering for improved recommendations. *Eighteenth National Conference on Artificial Intelligence(AAAI-2002)* (pp. 187-192). Edmonton, Alberta, Canada: ACM Special Interest Group on Artificial Intelligence.

Menezes, R. (2000). *Resource Management in Open Tuple Space Systems.* Unpublished doctoral dissertation, University of York, York, UK.

Menezes, R., & Wood, A. (1997). Garbage collection in open distributed tuple space systems. In *Proceedings of the 15th Brazilian Computer Networks Symposium (SBRC'97)* (pp. 525-543).

Menezes, R., & Wood, A. (1998). Using tuple monitoring and process registration on the implementation of garbage collection in open LINDA-like systems. In *Proceedings of the 10th IASTED International Conference Parallel and Distributed Computing and Systems,* Las Vegas, NV (pp. 490-495).

Meng, X., Yin, S., & Xiao, Z. (2006). A Framework of Web Data Integrated LBS Middleware. *Wuhan University Journal of Natural Sciences, 11*(5), 1187–1191. doi:10.1007/BF02829234

Merrick, I. (2001). *Scope-Based Coordination for Open Systems.* Unpublished doctoral dissertation, University of York, York, UK.

Metropolis, N., & Ulam, S. (1949). The Monte Carlo Method. *Journal of the American Statistical Association, 44*(247), 335–341. doi:10.2307/2280232

Mielnik, J., Lang, B., Laurière, S., Schlosser, J., & Bouthors, V. (2003). eCots Platform: An Inter-industrial Initiative for COTS-Related Information Sharing. In *Proceedings of the 2ⁿᵈ International Conference on COTS-Based Software System (ICCBSS)*, Ottawa, Canada (pp. 157-167).

Mineau, G. W., Moulin, B., & Sowa, J. F. (1993). Conceptual graphs for knowledge representation. *Lecture Notes in Artificial Intelligence 699*.

Mitra, S., & Acharya, T. (2003). *Data Mining: Multimedia, Soft Computing and Bioinformatics*. New York: John Wiley & Sons.

Mobasher, B., Jin, X., & Zhou, Y. (2004). Web Semantically enhanced collaborative filtering on the web. *Web Mining: From Web to Semantic*, 57-76.

Momtazpour, M., & Khadivi, P. (2009). New Routing Strategies for RSP Problems with Concave Cost. *Advances in Computer Science and Engineering*, 412-418.

Monk, A., Fellas, E., & Ley, E. (2004). Hearing only one side of normal and mobile phone conversations. *Behaviour & Information Technology*, 23(5), 301–305. doi:10.1080/01449290410001712744

Mostafa, J. (2005). Seeking better web searches. *Scientific American*, 292(2), 66–73. doi:10.1038/scientificamerican0205-66

Mounin, G. (1980). The semiology of orientation in urban space. *Current Anthropology*, 21(4), 491–501. doi:10.1086/202498

Mozilla Labs. (2010). *RainDrop*. Retrieved March 22, 2010, from http://mozillalabs.com/raindrop

Musthaq, S., Salem, O., Lohr, C., & Gravey, A. (2008). Policy-based QoS management for multimedia communication. Retrieved from https://portail.telecom-bretagne.eu/publi/public/download.jsp?id...542.6

Nair, S. K., Zentveld, E., Crispo, B., & Tanenbaum, A. S. (2008), Floodgate: A micropayment incentivized P2P content delivery network. *Proceedings of 17th IEEE International Conference on Computer Communications and Networks, 2008,* (ICCCN '08), (pp. 1 – 7).

Nam, Q. H. (1998). Maintaining global integrity constraints in distributed databases. *Constraints: An International Journal*, 2(3-4), 377–399.

Ncube, C., & Maiden, N. (1999). PORE: Procurement-Oriented Requirements Engineering Method for the Component-Based Systems Engineering Development Paradigm. In *Proceedings of the International Workshop on Component-Based Software Engineering*, Los Angeles, CA (pp. 130-140).

Nicolas, J. M. (1982). Logic for improving integrity checking in relational data bases. *Acta Informatica*, 18(3), 227–253.. doi:10.1007/BF00263192

Nie, Z., Ma, Y., Shi, S., Wen, J., & Ma, W. (2007). Web object retrieval. In *Proceedings of the WWW Conference*.

Nielsen, J. (2005). Severity ratings for usability problems. Retrieved February 25, 2009, from http://www.useit.com/papers/heuristic/severityrating.html

Nielsen, J. (2009). *Social networking on intranets*. Retrieved March 22, 2010, from http://www.useit.com/alertbox/social-intranet-features.html

Nielsen, J. (1993). *Usability engineering*. Boston, MA: AP Professional.

Nielsen, J. (1994). Heuristic evaluation. In Nielsen, J., & Mack, R. L. (Eds.), *Usability inspection methods*. New York, NY: John Wiley & Sons.

Nielsen, J., & Mack, R. L. (1994). *Usability Inspection Methods* (1st ed.). New York: John Wiley & Sons, Inc.

Niepert, M., Buckner, C., & Allen, C. (2007). A Dynamic Ontology for a Dynamic Reference Work. In *Proceedings of the (JCDL'07)*, Vancouver, British Columbia, Canada.

Niles, I., & Pease, A. (2001). Towards a standard upper ontology. *Proceedings of the International Conference on Formal Ontology in Information Systems*.

Norman, D. (1998). *The invisible computer: why good products can fail, the personal computer is so complex, and information appliances are the solution*. Cambridge, MA: MIT Press.

Norman, D. A. (2002). *The design of everyday things*. New York: Basic Books.

O'Reilly, T. (2007). What is Web 2.0: Design patterns and business models for the next generation of software. *Communications & Strategies, 1*(65), 17–37.

Olfati-Saber, R., Fax, J. A., & Murray, R. M. (2007). Consensus and cooperation in networked multiagent systems. *Proceedings of the IEEE, 95*(1), 215–233. doi:10.1109/JPROC.2006.887293

OrbiTeam Software GmbH & Co. KG. (2011). *Be smart - cooperative, worldwide*. Retrieved August 18, 2011, from http://public.bscw.de/en/about.html

Ouellette, S., Marchand, L., & Pierre, S. (2011). A potential evolution of the policy and charging control/QoS architecture for the 3GPP IETF-based evolved packet core. *IEEE Communications Magazine, 49*(5), 231–239. doi:10.1109/MCOM.2011.5762822

Oulasvirta, A., & Saariluoma, P. (2004). Long-term memory and interrupting messages in human-computer interaction. *Behaviour & Information Technology, 23*(1), 53–64. doi:10.1080/01449290310001644859

Packer, S. H., Gibbins, N., & Jennings, N. R. (2009). *Ontology evolution through agent collaboration*. Workshop on Matching and Meaning: Automated Development, Evolution and Interpretation of Ontologies.

Paivio, A. (1990). *Mental representations: a dual coding approach*. New York: Oxford University Press.

Panangaden, P. (2009). *Notes on labelled transition systems and bisimulation*. Retrieved from http://www.cs.mcgill.ca/~prakash/Courses/comp330/Notes/lts09.pdf.

Papadopoulos, G., & Arbab, F. (1998). Coordination models and languages. In Zelkowitz, M. (Ed.), *Advances in computers* (pp. 329–400). San Diego, CA: Academic Press.

Parker, R. D. (1997). The architectonics of memory: On built form and built thought. *Leonardo, 30*(2), 147–152. doi:10.2307/1576426

Pedersen, T. (1996). Electronic payments of small amounts. In M. Lomas (Ed.), *Proceedings of 1996 International Workshop on Security Protocols, Lecture Notes in Computer Science, vol. 1189*, (pp. 59 - 68). Berlin, Germany: Springer Verlag.

Perednia, D. A., & Allen, A. (1995). Telemedicine technology and clinical applications. *Journal of the American Medical Association, 273*(6), 483–488. doi:10.1001/jama.273.6.483

Pérez-Iglesias, J. (2008). *Integrating BM25 & BM25F into Lucene*. Retrieved from http://nlp.uned.es/~jperezi/Lucene-BM25/

Periakaruppan, R., & Nemeth, E. (1999). GTrace – A Graphical Traceroute Tool. In *Proceedings of the 13th Systems Administration Conference (LISA '99)* (pp. 69-78).

Peterson, R. A., & Merino, M. C. (2003). Consumer information search and the internet. *Psychology and Marketing, 20*(2), 99–121. doi:10.1002/mar.10062

Piaget, J., & Inhelder, B. (1956). *The child's conception of space*. London: Routledge & K. Paul.

Pica, T. (1987). Second-language acquisition, social interaction and the classroom. *Applied Linguistics, 8*(1), 3–21. doi:10.1093/applin/8.1.3

Pierrakos, D., Paliouras, G., Papatheodorou, C., & Spyropoulos, C. (2003). Web usage mining as a tool for personalization: A survey. *User Modeling and User-Adapted Interaction, Kluwer Academic Publishers*, 311-372.

Prinz, W., Löh, H., Pallot, M., Schaffers, H., Skarmeta, A., & Decker, S. (2006). ECOSPACE - Towards an integrated collaboration space for eProfessionals. In *2nd International Conference on Collaborative Computing: Networking, Applications and Worksharing* (pp. 39–45). New York, NY: IEEE Press.

Purves, R. R., Sanderson, A., Sester, M. M., Kreveld, M. V., & Weibel, R. (2002). Spatial information retrieval and geographical ontologies an overview of the SPIRIT project. In *Proceedings of the 25th Annual International ACM SIGIR Conference on Research and Development in Information Retrieval* (pp. 387-388).

Puskin, D. S. (2006). *Telemedicine, telehealth, and health Information Technology*. An ATA Issue Paper, The American Telemedicine Association, May 2006.

Puskin, D. S. (1995). Opportunities and challenges to telemedicine in rural America. *Journal of Medical Systems, 19*(1), 59–67. doi:10.1007/BF02257191

Qian, X. (1989). Distribution design of integrity constraints. In *Proceedings of the 2nd International Conference on Expert Database Systems*, Vienna, VA (pp. 205-226).

Qiang, B., Xi, J., Qiang, B., & Zhang, L. (2008). *An Effective Schema Extraction Algorithm on the Deep Web.* Washington, DC: IEEE.

Radvansky, G. A., & Copeland, D. E. (2000). Functionality and spatial relations in memory and language. *Memory & Cognition, 28*(6), 987–992.

Raghavan, S., & Garcia-Molina, H. (2001). Crawling the hidden Web. In *Proceedings of 27th International Conference on Very Large Data Bases (VLDB '01)* (pp. 129-138).

Rajapaksha, S., & Kodagoda, N. (2008). Internal Structure and Semantic Web Link Structure Based Ontology Ranking. *ICIAFS 2008. 4th International Conference on Information and Automation for Sustainability*, (pp. 86-90). Colombo.

Rajlich, V. (2001). *Propagation of changes in Object Oriented Programs (Tech. Rep.).* Detroit, MI: Wane State University.

Rashid, E., Ishtiaq, O., & Gilani, S. (2003). Comparison of store and forward method of teledermatology with face-to-face consultation. *Journal of Ayub Medical College, Abbottabad, 15*(2), 34–36.

Raskin, J. (2000). *The humane interface: New directions for designing interactive systems.* Reading, MA: Addison-Wesley.

Reid, J. A. (2003). geoXwalk – A Gazetteer Server and Service for UK Academia. In Koch & Sølvberg (Eds.), *Proceedigns of the Research and Advanced Technology for Digital Libraries: 7th European Conference (ECDL 2003)* (pp. 387-392).

Rijsbergen, V. C. J. (1979). *Information Retrieval* (2nd ed.). London: Butterworth.

Robertson, S. E., Walker, S., Jones, S., Hancock-Beaulieu, M., & Gatford, M. (1994). Okapi at TREC-3. In *Proceedings of the Third Text REtrieval Conference (TREC 1994)* (pp. 109-126).

Ryder, B., & Tip, F. (2001). Change Impact Analysis for Object- Oriented programs. In *Proceedings of the ACM Workshop on Program Analysis for Software Tools and Engineering* (Vol. 10, pp. 46-53).

Saaty, T. L. (1980). *The Analytical Hierarchy Process.*

Salton, G. (1968). *Automatic Information Organization and Retrieval.* New York: McGraw-Hill Inc.

Salton, G., & Buckley, C. (1988). Term-Weighting Approaches in Automatic Text Retrieval. *Information Processing & Management, 24*(5), 513–523. doi:10.1016/0306-4573(88)90021-0

Sandor, C., Kitahara, I., Reitmayr, G., Feiner, S., & Ohta, Y. (2009). Let's go out: Research in outdoor mixed and augmented reality. In *Proceedings of the 2009 8th IEEE International Symposium on Mixed and Augmented Reality* (p. 229). Washington, DC: IEEE Computer Society.

Santoni, D., & Katchabaw, M. (2007). Resource matching in a peer-to-peer computational framework. *International Conference on Internet Computing 2007*, (pp. 89-95).

Sauro, J., & Kindlund, E. (2005). A method to standardize usability metrics into a single score. In *Proceedings of the Computer Human Interaction*, Portland, OR (pp. 401-409).

Scharl, A. (2007). Towards the Geospatial Web: Media Platforms for Managing Geotagged Knowledge Repositories. In A. Scharl & K. Tochtermann (Eds.), *The Geospatial Web – How Geo-Browsers, Social Software and the Web 2.0 Shaping the Network Society* (pp. 3-14). London: Springer.

Scharl, A. (2008). Annotating and Visualization Location Data in Geospatial Web Application. In *Proceedings of the First International Workshop on Location and The Web.*

Selvakumar, S., Xavier, S., & Balamurugan, V. (2009). Policy based service provisioning system for WiMAX network: An approach. *ICSCN International Conference on Signal Processing, Communications and Networking*, (vol. 4-6, pp 177-181).

Shepperd, M. (1993). Software Engineering Metrics: *Vol. I. Measures and Validations.* London: McGraw Hill.

Shneidman, J., & Parkes, D. (2003). Rationality and self-interest in peer-to-peer networks. In *Proceedings of 2nd International Workshop on Peer-to-Peer Systems (IPTPS '03)*, Berkeley, CA, USA, February 2003.

Shvaiko, P. (2004). A classification of schema-based matching approaches. *Proceedings of the Meaning Coordination and Negotiation workshop at International Semantic Web Conference* (ISWC).

Shvaiko, P., & Euzenat, J. (2005). A survey of schema-based matching approaches. *Journal on Data Semantics*, *4*, 146–171.

Simon, E., & Valduriez, P. (1989). Integrity control in distributed database systems. In *Proceedings of the 19th Hawaii International Conference on System Sciences*, HI (pp. 622-632).

Sitemaps. (2009). *Sitemaps Protocol*. Retrieved from http://www.sitemaps.org

Snyder, C. (2003). *Paper prototyping*. San Francisco: Elsevier Science.

Song, H., Giri, S., & Ma, F. (2004). Data Extraction and Annotation for Dynamic Web Pages. In *Proceedings of EEE* (pp. 499-502).

Son, J. H., Kim, J. S., & Kim, M. H. (2005). Extracting the workflow critical path from the extended well-formed workflow schema. *Journal of Computer and System Sciences*, *70*(1), 86–106. doi:10.1016/j.jcss.2004.07.001

Son, J. H., & Kim, M. H. (2001). Improving the performance of time-constrained workflow processing. *Journal of Systems and Software*, *58*(3), 211–219. doi:10.1016/S0164-1212(01)00039-5

Soumya, B., Madiraju, P., & Ibrahim, H. (2008). Constraint optimization for a system of relation databases. In *Proceedings of the IEEE 8th International Conference on Computer and Information Technology*, Sydney, Australia (pp. 155-160).

Souza, J. F., Siqueira, S. W., & Melo, R. N. (2009). Adding meaning negotiation skills in multiagent systems. *Proceedings of the IEEE International Conference on Intelligent Computing and Intelligent Systems*, (pp. 663-667).

Souza, L., Davis, C. J., Borges, K., Delboni, T., & Laender, A. (2005). The role of gazetteers in geographic knowledge discovery on the web. In *Proceedings of the LA-Web-05, the 3rd Latin American Web Congress* (p. 157).

Souza, J. F., Melo, R. N., & Siqueira, S. W. (2010). Improving software agent communication with structural ontology alignment methods. *International Journal of Information Technology and Web Engineering*, *5*, 49–64. doi:10.4018/jitwe.2010070103

Souza, J. F., Paula, M., Oliveira, J., & Souza, J. M. (2006). Meaning negotiation: Applying negotiation models to reach semantic consensus in multidisciplinary teams. *Group Decision and Negotiation*, 297–300.

Souza, J. M. (1986). *Software tools for conceptual schema integration*. University of East Anglia.

Stojanovic, L., Stojanovic, N., & Volz, R. (2002). Migrating data intensive web sites into the semantic web. In *Proceedings of the 17th ACM Symposium on Applied Computing* (pp. 1100-1107).

Stojanovic, M., Rakas, S., & Acimovic-Raspopovic, V. (2010). End-to-end quality of service specification and mapping: The third party approach. *Computer Communications*, *1*, 1354–1368. doi:10.1016/j.comcom.2010.03.024

Stylite. (2011). *eGroupware*. Retrieved September 8, 2011, from http://www.egroupware.org

Suchaman, L., Randall, T., & Blomberg, J. (2002). Working artefacts: Ethnomethods of the prototype. *The British Journal of Sociology*, *53*(2), 163–179. doi:10.1080/00071310220133287

Szigeti, T., & Hattingh, C. (2004). *End-to-End QoS Network Design*. San Jose, CA: Cisco Press.

Tanenbaum, A. S. (2002). *Computer Networks*. Upper Saddle River, NJ: Prentice Hall.

Tatnall, A., & Burgess, S. (2007). Experiences in building and using decision- support systems in postgraduate university courses. *Interdisciplinary Journal of Information, Knowledge, and Management, 2*.

Tejada, S., Knoblock, C. A., & Minton, S. (2002). Learning domain-independent string transformation weights for high accuracy object identification. In *Proceedings of the World Wide Web conference (WWW)* (pp. 350-359).

Tezuka, T., & Tanaka, K. (2005). Landmark Extraction: A Web Mining Approach. In *Proceedings Of the COSIT'2005* (pp. 379-396).

Tezuka, T., Kurashima, T., & Tanaka, K. (2006). Toward tighter integration of web search with a geographic information system. In *Proceedings of the 15th international conference on World Wide Web* (pp. 277-286).

The Open Archives Initiative Protocol for Metadata Harvesting. (2003). *Protocol Version 2.0.* Retrieved from http://www.openarchives.org/OAI/2.0/openarchivesprotocol.htm

Thomas, S. N. (2007). Meeting the health care needs of California's children: The role of telemedicine. *Digital Opportunity for Youth Issue Brief, Number 3,* September 2007.

Thomas, H., Cormen, C. E. L., Rivest, R. L., & Stein, C. (2001). *Introduction to Algorithms.* Cambridge, MA: MIT Press.

Tijerino, Y. A., Al-Muhammed, M., & Embley, D. W. (2004). Toward a flexible human-agent collaboration framework with mediating domain ontologies for the Semantic Web. *Proceedings of the ISWC'04 Workshop on Meaning Coordination and Negotiation,* (pp. 131-142).

Tijerino, Y. A., Embley, D. W., Lonsdale, D. W., Ding, Y., & Nagy, G. (2005). Towards ontology generation from tables. *World Wide Web Journal,* 261-285.

Toutanova, K., & Manning, C. D. (2000) Enriching the knowledge sources used in a maximum entropy part-of-speech tagger. In the *Joint SIGDAT Conference on Empirical Methods in Natural Language Processing and Very Large Corpora* (pp. 63-70).

Truszkowski, W. F. (2006). What is an agent? And what is an agent community? In Rouff, C. A., Hinchey, M., Rash, J., Truszkowski, W., & Gordon-Spears, D. (Eds.), *Agent technology from a formal perspective* (pp. 3–24). London, UK: Springer-Verlag. doi:10.1007/1-84628-271-3_1

Tsumaki, T., & Tamai, T. (2005). A Framework for Matching Requirements Engineering Techniques to Project Characteristics and Situation Changes. In *Proceedings of the International Workshop on Situational Requirements Engineering Processes - Methods, Techniques and Tools to Support Situation-Specific Requirements Engineering Processes,* Paris (pp. 45-55).

Tsyrklevich, E., & Tsyrklevich, V. (2007). *Single sign-on for the Internet: A security story.* White Paper. Seattle, WA: Black Hat.

Tuan, Y.-F. (1975). Images and mental maps. *Annals of the Association of American Geographers. Association of American Geographers,* 65(2), 205–213. doi:10.1111/j.1467-8306.1975.tb01031.x

Turner, M., Budgen, D., & Brereton, P. (2003). Turning software into a service. *Communications & Strategies,* 36(10), 38–44.

Udzir, N. I. (2006). *Capability-Based Coordination for Open Distributed Systems.* Unpublished doctoral dissertation, University of York, York, UK.

Udzir, N. I., Muda, Z., Sulaiman, M. N., Zulzalil, H., & Abdullah, R. (2008). Refined garbage collection for open distributed systems with multicapabilities. In *Proceedings of the 3rd International Symposium on Information Technology 2008 (ITSim'08).*

Udzir, N. I., Wood, A. M., & Jacob, J. L. (2005). *Coordination with multicapabilities* (LNCS 3454, pp. 79-93). Berlin: Springer Verlag.

Udzir, N. I., Wood, A. M., & Jacob, J. L. (2007). Co-ordination with multicapabilities. *Science of Computer Programming: Special Issue on Coordination Models and Languages,* 64(2), 205–222.

Ullah, N., Khan, P., Sultana, N., & Kwak, K. S. (2009). A telemedicine network model for health applications in Pakistan: Current status and future prospects. *International Journal of Digital Content Technology and its Applications,* 3(3).

van Braak, J. P. (2004). Domains and determinants of university students' self-perceived computer competence. *Computers & Education,* 43(3), 299–312. doi:10.1016/j.compedu.2003.09.006

Van der Aalst, W. M. P., & Ter Hofstede, A. H. M. (2005). YAWL: Yet another workflow language. *Information Systems, 30*(4), 245–275. doi:10.1016/j.is.2004.02.002

Van der Aalst, W. M. P., Ter Hofstede, A. H. M., Kiepuszewski, B., & Barros, A. P. (2003). Workflow patterns. *Distributed and Parallel Databases, 14*(1), 5–51. doi:10.1023/A:1022883727209

Vekiri, I. (2002). What is the value of graphical displays in learning? *Educational Psychology Review, 14*(3), 261–312. doi:10.1023/A:1016064429161

Verdi, M. P., & Kulhavy, R. W. (2002). Learning with maps and text: An overview. *Educational Psychology Review, 14*(1), 27–46. doi:10.1023/A:1013128426099

Vestavik, Ø. (2008). *Geographic Information Retrieval: An Overview.* Retrieved August 20, 2008 from http://www.idi.ntnu.no/~oyvindve/article.pdf

Vishnumurthy, V., Chandrakumar, S., & Sirer, E. G. (2003). KARMA: A secure economic framework for P2P resource sharing. *Proceedings of the First Workshop on Economics of Peer-to-Peer Systems (P2PEcon'03).*

Visiongain. (2009). *Mobile email 2009: Challenging Blackberry and succeeding in the consumer market.* Retrieved December 10, 2009, from http://www.visiongain.com/Report.aspx?rid=377

Vitek, J., Bryce, C., & Oriol, M. (2003). Coordinating processes with secure spaces. *Science of Computer Programming, 46*(1-2), 163–193. doi:10.1016/S0167-6423(02)00090-4

Vivrou, M., & Kabbasi, K. (2002). Reasoning about users' actions in a graphical user interface. *Human-Computer Interaction, 17*(4), 369–398. doi:10.1207/S15327051HCI1704_2

Vogel, D., Bickel, S., Haider, P., Schimpfky, R., Siemen, P., Bridges, S., & Scheffer, T. (2005). Classifying search engine queries using the Web as background knowledge. *SIGKDD Explorations Newsletter, 7*(2), 117–122. doi:10.1145/1117454.1117469

Wallace, B., Ross, A. J., & Davies, J. B. (2003). Applied hermeneutics and qualitative safety data: The CIRAS project. *Human Relations, 56*(5), 587–607. doi:10.1177/0018726703056005004

Wang, Y., Liu, W., & Guo, W. (2010). Architecture of IMS over WIMAX PCC and the QoS mechanism. *IET 3rd International Conference on Wireless, Mobile and Multimedia Networks, ICWMNN'10,* (pp. 159-162).

Wang, B., Zhang, J., Guo, Y., & Zhou, J. (2009). *A Study of Fast Network Self-Healing Mechanism for Distance Vector Routing Protocol.* Washington, DC: IEEE.

Wang, Z., & Crowcroft, J. (1996). Quality-of-service routing for supporting multimedia applications. *Selected Areas in Communications, 14,* 1228–1234. doi:10.1109/49.536364

Washio, T., & Motoda, H. (2003). State of the art of graph-based data mining. *ACM SIGKDD Explorations Newsletter,* 59-68.

Waxman, B. M. (1988). Routing of multipoint connections. *Selected Areas in Communications, 6,* 1617–1622. doi:10.1109/49.12889

Webb, J. M., & Saltz, E. D. (1994). Conjoint influence of maps and auded prose on children's retrieval of instruction. *Journal of Experimental Education, 62*(3), 195–208.

Webb, J. M., Thornton, N. E., Hancock, T. E., & McCarthy, M. T. (1992). Drawing maps from text: A test of conjoint retention. *The Journal of General Psychology, 119*(3), 303–313.

Wegiel, M., & Krintz, C. (2008). XMem: type-safe, transparent, shared memory for cross-runtime communication and coordination. In *Proceedings of the 2008 ACM SIGPLAN conference on Programming language design and implementation.*

Wei, K., Smith, A. J., Chen, Y. R., & Vo, B. (2006). WhoPay: A scalable and anonymous payment system for peer-to-peer environments. In *Proceedings of 26th IEEE Intl. Conf. on Distributed Computing Systems,* (p. 13). Los Alamitos, CA: IEEE Computer Society Press.

Weiler, A. (2005). Information-seeking behavior in generation y students: Motivation, critical thinking, and learning theory. *Journal of Academic Librarianship, 31*(1), 46–53. doi:10.1016/j.acalib.2004.09.009

Weiser, M. (1993). Some Computer Science Issues in Ubiquitous Computing. *Communications of the ACM, 36*(7), 75–84. doi:10.1145/159544.159617

Wheatley, D. (2007). User-Centered Design and Evaluation of a Concurrent Voice Communication and Media Sharing Application. In *Human-Computer Interaction. HCI Intelligent Multimodal Interaction Environments* (LNCS 4552, pp. 990-999). Berlin: Springer.

Wheeler, M. A., Ewers, M., & Buonanno, J. F. (2003). Different rates of forgetting following study versus test trials. *Psychology Press, 11*(6), 571–580.

WHO. (2010). *World health statistics*. World Health Organization (WHO). Retrieved December 27, 2010, from http://www.who.int/whosis/whostat/2007/en/

Winston, W. L. (1994). *Operations Research: Applications and Algorithms*. New York: International Thomson Publishing.

Wireless and Mobile News. (n.d.). *Android Top Users Fastest Growing Smartphone Makers, IPhone Declines, Says iSuppli*. Retrieved October 21, 2010, from http://www.wirelessandmobilenews.com/2010/10/android-top-users-fastest-growing-smartphone-makers-iphone-declines-says-isuppli.html

Woodruff, A. G., & Plaunt, C. (1994). GIPSY: Georeferenced Information Processing System. *Journal of the American Society for Information Science American Society for Information Science, 45*(9), 645–655. doi:10.1002/(SICI)1097-4571(199410)45:9<645::AID-ASI2>3.0.CO;2-8

Wu, W., Yu, C., Doan, A., & Meng, W. (2004). An interactive clustering-based approach to integrating source query interfaces on the Deep Web. In *Proceedings of the ACM SIGMOD International Conference on Management of Data (SIGMOD'04)* (pp. 95-106).

Wyckoff, P., McLaughry, S., Lehman, T., & Ford, D. (1998). TSpaces. *IBM Systems Journal, 37*(3), 454–474. doi:10.1147/sj.373.0454

Xie, T., Yu, Y., & Kuang, G. (2009). A time exception handling algorithm of temporal workflow. In *Proceedings of the 2009 IEEE International Symposium on Parallel and Distributed Processing with Applications* (pp. 641-646). Chengdu, China.

Xueshun, W., Shao-Hua, Y., & Ting, L. (2009). *A Multiple Constraint Quality of Service Routing Algorithm Base on Dominating Tree*.

Yahoo. (2011). *Pipes*. Retrieved August 18, 2011, from http://pipes.yahoo.com/pipes/

Yang, B., & Garcia-Molina, H. (2003). PPay: Micropayments for peer-to-peer systems. In *Proceedings of the 10th ACM Conference on Computer and Communication Security*, (pp. 300-310). ACM Press.

Yang, J., & Park, H. (2008). A design of open service access gateway for converged Web service. *10th International Conference on Advanced Communication Technology 2008*, (pp.1807- 1810).

Yang, J., Shi, G., Zheng, Y., & Wang, Q. (2007). Data Extraction from Deep Web Pages. In *Proceedings of the International Conference on Computational Intelligence and Security*. IEEE.

Yang, S. J., Okamoto, T., & Tseng, S. S. (2008). Context-aware and ubiquitous learning. *Educational Technology & Society, 11*(2), 1–2.

Yau, S., Collofello, J., & McGregor, T. (1978). Ripple effect analysis of software maintenance. In. *Proceedings of COMPSAC, 78*, 60–65.

Yuan, H. T., Ding, B., & Sun, Z. X. (2008). Workflow exception forecasting method based on SVM theory. In *Proceedings of the 2008 International Symposium on computational Intelligence and Design* (pp. 81-86). Hefei, China.

Zghaibeh, M., & Harmantzis, F. C. (2006). Lottery-based pricing scheme for peer-to-peer networks. *ICC APOS; 06. IEEE International Conference on Communications, 2006*, (vol. 2, pp. 903–908).

Zhang, Z., He, B., & Chang, K. (2004). Understanding Web query interfaces: Best-effort parsing with hidden syntax. In *Proceedings of the ACM SIGMOD International Conference on Management of Data (SIGMOD'04)* (pp. 107-118).

Zhang, Z., He, B., & Chang, K. C. C. (2005). Light-weight Domain-based Form Assistant: Querying Web Databases On the Fly. In *Proceedings of the VLDB Conference*, Trondheim, Norway (pp. 97-108).

Zhao, F., Jiang, L., & He, C. (2008). Policy-based radio resource allocation for wireless mobile networks. *Proceedings on International Conference on Neural Networks and Signal Processing*, (pp. 476-481).

Zhao, P., Lin, C., Fang, W., & Cui, Z. (2007). A Hybrid Object Matching Method for Deep Web Information Integration. In *Proceedings of the International Conference on Convergence Information Technology*. IEEE.

Zhao, P., & Cui, Z. (2007). Vision-based deep web query interfaces automatic extraction. *Journal of Computer Information Systems*, 1441–1448.

Zhong, X., Fu, Y., Liu, Q., Lin, X., & Cui, Z. (2007). A Holistic Approach on Deep Web Schema Matching. In *Proceedings of the International Conference on Convergence Information Technology*. IEEE.

Zhou, C., & Huang, L. (2010). *Study on the Improvement of Analytic Hierarchy Process under College Course Evaluation System*. Washington, DC: IEEE.

Zou, E. J., Si, T., Huang, L., & Dai, Y. (2005). A new micro-payment protocol based on P2P networks. *Proceedings of the 2005 IEEE International Conference on e-Business Engineering (ICEBE'05)*, (pp. 449 – 455). IEEE Computer Society Press.

About the Contributors

Ghazi Alkhatib is an assistant professor of software engineering at the College of Computer Science and Information Technology, Applied Science University (Amman, Jordan). In 1984, he obtained his Doctor of Business Administration from Mississippi State University in information systems with minors in computer science and accounting. Since then, he has been engaged in teaching, consulting, training and research in the area of computer information systems in the US and gulf countries. In addition to his research interests in databases and systems analysis and design, he has published several articles and presented many papers in regional and international conferences on software processes, knowledge management, e-business, Web services and agent software, workflow and portal/grid computing integration with Web services.

* * *

Mehdi Adda, professor of computer science at the University of Quebec in Rimouski, Rimouski, Canada. His principal research interests lie in the fields of Software and Web engineering, data mining and knowledge discovery, Aspect Oriented Programming and Distributed Computing, Web Personalization and Recommendation.

Asmaa A. Alsumait received BS in Computer Engineering from Kuwait University. She earned her MS and PhD degree from Concordia University, Canada in 2001 and 2004, respectively. Currently, she is an assistant professor at Kuwait University, Computer Engineering department. Her research interest includes user interface design, usability evaluation methods and measures, and designing and evaluating e-learning programs for children.

Ali A. Alwan is currently a Ph.D. candidate at the Faculty of Computer Science and Information Technology, Universiti Putra Malaysia. He obtained his Master in Computer Science from the Universiti Putra Malaysia, Malaysia in 2008. His current research interests include context aware database, multi-objective query processing for database, distributed and mobile databases.

Liliana Ardissono is an Associate Professor at the Computer Science Department of the Università di Torino, where she obtained her University Degree and her Ph.D in Computer Science. She teaches Computer Science courses at the Università di Torino, and she has been the advisor of several Ph.D students during the last 15 years. Her research interests include user modeling, adaptive hypermedia, service oriented computing and cloud computing, as well as intelligent agents. She is Secretary of the

Board of Directors of User Modeling Inc., and she is a member of the Editorial Board of *User Modeling and User-Adapted Interaction - The Journal of Personalization Research*. Moreover, she has been a senior member of the Scientific Committee of many international research conferences related to her research topics, among which the UMAP series of conferences on adaptation and user modeling, and the IJCAI conferences on Artificial Intelligence. She has also cooperated to several national and European projects focused on user modeling and context adaptation, as well as autonomic and autonomous systems.

Ivaylo Atanasov received his M.S. degree in Electronics from Technical University of Sofia, Sofia, Bulgaria, and PhD degree in Communication Networks. His current position is Associate Professor at Faculty of Telecommunications, Technical University of Sofia. His main research focus is set on open service platforms for next generation networks.

Imran Sarwar Bajwa is Assistant Professor in the Islamia University of Bahawalpur. He has been teaching and doing research in various universities in Pakistan since 2005. He has also been part of faculty and active member of research community in the University of Coimbra, Portugal in 2006-07. His major areas of research are natural language processing, information retrieval (text & image), and Information Systems. He is member of different professional societies i.e. IEEE, AAAI, IASCIT, ACA, EATCS, SCTA, et cetera.

Robert P. Biuk-Aghai is an Assistant Professor at the University of Macau. His research interests include wikis and other collaboration systems, and information visualization. He is one of the founders of the Data Analytics and Collaborative Computing Group at the University of Macau.

Gianni Bosio received his Bachelor's degree in Communication Science in 2004 (thesis: Analysis of the Free Software movement) and his Master's degree "cum laude" in Communication for the Information Society in 2007 (thesis: Usability study on iPod and iTunes). He is currently a PhD student in Computer Science at the Computer Science Department of the Università di Torino. During his PhD studies, he attended postgraduate courses in Advanced Data structures, Data mining, Intelligent agents, Markovian and stochastic models, and Distributed Algorithms. His research activity comprises the following areas: human computer interaction, usability studies, collaborative technologies, user centric technologies, cloud architectures for computer supported collaborative work, notification systems, and user interfaces.

Robert Bruno is a communication doctoral candidate at Purdue University, with a focus on advanced technology's effects on cognition and learning. He previously worked for several years in Silicon Valley, most recently at Sun Microsystems.

Kaylash Chaudhary received his BSc degree in Computing Science from University of the South Pacific in Fiji in 2004. He has completed his MSc degree from the School of Computing, Information & Mathematical Sciences, University of the South Pacific in September 2009. Currently, he is an Assistant Lecturer at the University of the South Pacific. His research interests include software engineering, distributed system design and implementation, software architecture, and electronic micro-payment systems for file-sharing in peer-to-peer networks.

Xiaoling Dai is currently a part-time Lecturer of Computer Science at Charles Sturt University, Australia. She holds the BSc (Hons) in Mathematics, from Hebei University in China, and the PhD degree in Computer Science, from the University of Auckland in New Zealand. Her current research areas include component-based software engineering, distributed system design and implementation, software architecture, electronic micro-payment systems for e-commerce, file-sharing, or m-commerce in client-server, peer-to-peer, and mobile networks, Web service security, and service-oriented software engineering. She has been a PC member for the ACM International Conference on Information Integration and Web-based Applications & Services.

Sileshi Demesie graduated from Addis Ababa University with his Bachelor of Computer Science in 2005. He received his Master of Computer Science degree in 2009 from Universiti Putra Malaysia (UPM), majoring in Distributed Computing. He is currently working in the Faculty of Engineering, Bahir Dar University, Ethiopia, and currently a member of the System Design and Development case team there.

Anthony Faiola is the Executive Associate Dean for the School of Informatics at IUPUI, Director of the Human-Computer Interaction program, and an Associate Professor of Informatics. As EAD, Dr. Faiola has day-to-day responsibility for the administration of the school on the IUPUI campus; and as Director of HCI, his administrative responsibilities include the oversight of all HCI programs: Ph.D., MS, and Certificate in Graduate Studies. Dr. Faiola brings to his current post, 19 years of experience (1979-1998) in the industrial sector and an array of academic experience in teaching and research, including his many published papers and books in the field of HCI and CMC; and as a three-time Fulbright Scholar to Russia in Communication Technology, Usability, and New Media.

Jairo Francisco de Souza is an Assistant Professor at the Department of Computer Science, Federal University of the Juiz de Fora (UFJF), Brazil, where he teaches courses in Software Engineering and Semantic Web. He holds a M.Sc. (2007) from Federal University of Rio de Janeiro (UFRJ) and he is a Ph.D. candidate in the Pontifical Catholic University of Rio de Janeiro (PUC-Rio), Brazil, both in Computer Science area. His research interests include knowledge representation, Semantic Web, ontologies, information integration, and semantic models.

Anna Goy is a Researcher at the Computer Science Department of the Università di Torino where she works in the area of Web-based systems. She got the *Diplôme d'Etudes Supérieures* in Philosophy of the Linguistics in 1994 at the University of Genève, and she obtained the Ph.D in Cognitive Science at the Università di Torino in 1998 with studies in the area of Lexical Semantics. She worked in a three years Telecom Italia project about adaptive websites at the Computer Science Department of the same university, followed by a post-doc grant. She is a Computer Science Researcher since 2002; she carries on her research and development activity about intelligent Web-based systems providing personalized services, distributed Web-based applications, "open" architectures in the Web 2.0, and context-aware systems at the Computer Science Department. Moreover, she teaches Web programming and cloud technologies at the University of Torino. She has been member of Program Committees of many international events, reviewer for international journals and conferences, and she is associate editor of the *International Journal of Cloud Applications and Computing*.

John Grundy is Professor of Software Engineering at the Swinburne University of Technology, Australia. He has published over 200 refereed papers on software engineering tools and methods, automated software engineering, visual languages and environments, collaborative work systems and tools, aspect-oriented software development, user interfaces, software process technology, and distributed systems. He has made numerous contributions to the field of collaborative software engineering including developing novel process modeling and enactment tools; collaborative editing tools, both synchronous and asynchronous; thin-client project management tools; software visualization tools for exploratory review; sketching-based UML and software design tools providing hand-drawn diagram support for collaborative review and annotation; and numerous event-based collaborative software architectures for building collaborative software engineering tools. He has been Program Chair of the IEEE/ACM Automated Software Engineering conference, the IEEE Visual Languages and Human-Centric Computing Conference, and has been a PC member for the International Conference on Software Engineering.

Sami J. Habib received BS in Computer Engineering from Iowa State University. Then, he spent a year working as a Lab Engineer in the Department of Electrical and Computer Engineering at Kuwait University. Then, he pursued a graduate study at the University of Southern California, where he earned MS and PhD in Computer Engineering. Currently he is a vice-dean for academic affairs at the College of Graduate Studies and an associate professor at the Department of Computer Engineering at Kuwait University. His researches are focused on developing computer-aided design methodologies and performance analysis techniques for designing/redesigning computer and distributed systems.

Hamidah Ibrahim is currently an Associate Professor at the Faculty of Computer Science and Information Technology, Universiti Putra Malaysia. She obtained her Ph.D. in Computer Science from the University of Wales Cardiff, UK in 1998. Her current research interests include distributed databases, transaction processing, and knowledge based systems.

Shazia Karim is a final year student of M.Phil (Computer Science) in Department of Computer Science & IT, The Islamia University of Bahawalpur, Pakistan. She is currently writing her M.Phil thesis. The topic of her thesis is virtual tele-health. Before MS, she graduated in Computer Science from the same department.

Iok Fai-Leong holds a MSc degree in E-Commerce Technology from the Department of Computer and Information Science, Faculty of Science and Technology, University of Macau. His research interests are in the area of workflow management systems and artificial intelligence.

Lance Madsen received his master's degree in communication at Purdue University. He currently is a Radiology Information Systems Administrator at InnerVision Advanced Medical Imaging in Lafayette, Indiana.

Nashat Mansour is a professor of computer science and Assistant Dean of the School of Arts & Sciences in Beirut. Dr. Mansour's research interests include software engineering, applications of metaheuristics and data mining, and biomedical informatics. He is the director of the Software Institute at LAU and President of the Arab Computer Society.

Sorin A. Matei, an Associate Professor in the Communication Department at Purdue University, has been involved in research and development of user experiences and has developed a number of research methodologies for understanding spatial orientation and learning in location aware situations. He teaches multimedia design classes and has designed and implemented Web applications that utilize location aware (http://ubimark.com or http://visiblepast.net) or 3D components (http://www.mentalmaps.info). His work has been published in *Journal of Communication, Journal of Computer Mediated Communication, Communication Research,* and *Communication Monographs.*

Rubens Nascimento Melo is a Senior Professor and Researcher in the field of Databases at the Computer Science Department of the Pontifical Catholic University of Rio de Janeiro, PUC-Rio. He holds a B.Sc. in Electronic Engineering (1968), a M.Sc. (1971), and a Ph.D. (1976) degree in Computer Science from the Air Force Institute of Technology (ITA) in Sao Paulo. Currently, he is Associate Professor at PUC-Rio where he leads the Database Research Lab (TecBD). One of his current interests is the application of Database Technology to distance learning. In this field he has also served as Director of the Centre of Distance Education under the Vice-Rectory for Academic Affairs of PUC-Rio.

Evelina Pencheva received her M.S. degree in Mathematics from University of Sofia "St. Kliment Ohridski," Sofia, Bulgaria and PhD degree in communication networks from Technical University of Sofia. Her current position is Professor at Faculty of Telecommunications, Technical University of Sofia. Her scientific interests include next generation mobile applications and middleware platforms.

Giovanna Petrone is a researcher of Computer Science at the Università di Torino. Her recent research interests concern two main areas: multi-agent systems, with specific interest for distributed software architectures, Web Services, Web Services Choreography, and Cloud Computing; she is also interested in Intelligent User Interfaces, with specific attention to personalization in Web-based services and, most recently, Virtual desktop for Web-based Collaboration environments within a personal Cloud. Previously, she has worked for several years as a software engineer and architect in US and Italian computer companies. She had first-line responsibility in the development of a CASE product for a startup, as well as managing large product offerings, such as their X11-based Windowing System, for Sun Microsystem and Digital Equipment. She cooperated with the X Consortium and the MIT on the design and development of the new features of the X Server R11R6. Major fields of activity have been: object-oriented programming and design, multimedia, document publishing tools, CASE tools, programming environments and compilers, user interfaces, and expert systems shells.

Marino Segnan is a Researcher at the Computer Science Department, Università di Torino, working with the Advanced Service Architectures group. His recent research activities deal with open collaboration environments including context awareness, context-dependent notifications, activity visualization, collaborative activity and workflow execution, personalized models for cloud computing, interaction models and protocol mediation for Web Services both for orchestrated and choreographed execution models, and Web Service monitoring. His previous activity focused on the development of a qualitative simulation tool for model-based diagnosis. He has developed his research also by taking part to both national and UE funded projects. Previously, he worked with several companies, main projects being: user interface design and development for an integrated multi language CASE tool, compilers for object-oriented languages, a 2PC transaction manager, a graphic constraint-based workflow design tool, an expert system for decision support, and an Information System for the local health service. He is currently teaching Information Systems at the Università di Torino.

A. K. Sharma received his MTech (Computer Sci. & Tech) with Hons. From University of Roorkee (Presently I.I.T. Roorkee) in the year 1990 and PhD (Fuzzy Expert Systems) from JMI, New Delhi in the year 2000. From July 1992 to April 2002, he served as Assistant Professor and became Professor in Computer Engg. at YMCA Institute of Engineering, Faridabad, Haryana in April 2002. He obtained his second PhD in Information Technology form Indian Institute of Information Technology & Management, Gwalior in 2004. Presently he is working as Dean, Faculty of Engineering and Technology at YMCA University of Science and Technology, Faridabad. His research interest includes Fuzzy Systems, Object Oriented Programming, Knowledge Representation and Internet Technologies. He has guided 8 Ph.D. thesis and 8 more are in progress with about 175 research publications in International and National journals and conferences. The author of 7 books, is actively engaged in research related to Fuzzy logic, Knowledge based systems, MANets, Design of crawlers. Besides being member of many BOS and Academic councils, he has been Visiting Professor at JMI, IIIT&M, and I.I.T. Roorkee. He is also project coordinator of Technical Education Quality Improvement Program under World Bank scheme.

Dilip Kumar Sharma is BSc BE(Honors)(CSE), MTech (IT), MTech (CSE) and pursuing PhD in Computer Engineering. He is life member of CSI Mumbai, IETE New Delhi, ISTE New Delhi, ISCA Kolkata, SSI and member of CSTA, USA. He has 18 short term courses/workshops/seminars organized by various esteemed originations such as IIT Kanpur, IIT Roorkee, IIT Delhi, IIT Kharagpur, Microsoft, WIPRO, IBM, TCS etc. He has published 18 research papers in International Journals /Conferences of repute and participated in 18 International/National conferences. Presently he is working as Reader in Department of Computer Science & Engineering (NBA Accredited) at GLA Institute of Technology & Management, Mathura (U.P.), (AICTE Approved & Affiliated to U.P. Technical University, Lucknow) since March 2003. His research interests are deep web information retrieval, e-commerce, Digital Watermarking and Software Engineering. He has guided various projects and seminars undertaken by the students of undergraduate/postgraduate.

Sean Wolfgand Matsui Siqueira is an Assistant Professor at the Department of Applied Informatics, Federal University of the State of Rio de Janeiro (UNIRIO), Brazil, where he teaches courses in Databases, Information Systems, and Semantic Web. He holds a M.Sc. (1999) and a Ph.D. (2005) in Computer Science, both from the Pontifical Catholic University of Rio de Janeiro (PUC-Rio), Brazil. His research interests include knowledge representation, Semantic Web, ontologies, information integration, semantic models, user models, e-learning, social Web,and music Information Systems. He has experience in the Computer Science area, with focus on Information Systems and Technology Enhanced Learning. He has participated in some international research projects and has published more than 70 papers for conferences, journals, and books.

Ming-Cheng Tsou received the MS degree in computer science and PhD degree in spatial information science from Oklahoma City University and National Taiwan University, Taiwan, in 1992 and 2004, respectively. Before 2009, he served as an assistant professor in the Department of Information

Management, Shieh-Chien University, Kaohsiung, Taiwan, working on spatial data mining, e-business applications and database systems. He is currently an assistant professor in the Department of Shipping Technology, National Kaohsiung Marine University, Taiwan. His currently research interests include Location-Based service (LBS), GIS, Intelligent Transportation System (ITS) and spatio-temporal data mining.

Nur Izura Udzir is a Senior Lecturer at the Faculty of Computer Science and Information Technology, Universiti Putra Malaysia (UPM). She received her Bachelor of Computer Science (1996) and Master of Science (1998) from UPM, and her PhD in Computer Science from the University of York, UK (2006). She is a member of IEEE Computer Society. Her areas of specialization are access control, secure operating systems, coordination models and languages, and distributed systems. She is a member of the Information Security Group at the faculty.

Yain Whar-Si is an Assistant Professor at the University of Macau. His research interests are in the areas of business process management and decision support systems. He is one of the founders of the Data Analytics and Collaborative Computing Group at the University of Macau.

Index